T0400084

Oxygen Transport to Tissue XXX

Advances in Experimental Medicine and Biology

Recent Volumes in this Series

A Continuation Order Plan is available for this series. A continuation order will bring delivery of each new volume immediately upon publication. Volumes are billed only upon actual shipment. For further information please contact the publisher.

Per Liss · Peter Hansell · Duane F. Bruley
David K. Harrison

Editors

Oxygen Transport to Tissue XXX

 Springer

Editors

Per Liss
Uppsala University
Uppsala, Sweden
per.liss@akademiska.se

Duane F. Bruley
University of Maryland
Baltimore County
Baltimore, MD, USA
bruley@umbc.edu

Peter Hansell
Uppsala University
Uppsala, Sweden
peter.hansell@mcb.uu.se

David K. Harrison
University Hospital of
North Durham
Durham, UK
d.k.harrison@ncl.ac.uk

ISBN 978-0-387-85997-2   e-ISBN 978-0-387-85998-9
DOI 10.1007/978-0-387-85998-9

Library of Congress Control Number: 2008941245

Printed on acid-free paper

springer.com

INTERNATIONAL SOCIETY ON OXYGEN
TRANSPORT TO TISSUE 2007-2008

The International Society on Oxygen Transport to Tissue is an interdisciplinary society comprising about 250 members worldwide. Its purpose is to further the understanding of all aspects of the processes involved in the transport of oxygen from the air to its ultimate consumption in the cells of the various organs of the body.

The annual meeting brings together scientists, engineers, clinicians and mathematicians in a unique international forum for the exchange of information and knowledge, the updating of participants on the latest developments and techniques, and the discussion of controversial issues within the field of oxygen transport to tissue.

Examples of areas in which members have made highly significant contributions include electrode techniques, spectrophotometric methods, mathematical modeling of oxygen transport, the understanding of local regulation of oxygen supply to tissue and fluorocarbons/blood substitutes.

Founded in 1973, the society has been the leading platform for the presentation of many of the technological and conceptual developments within the field both at the meetings themselves and in the proceedings of the society. These are currently published by Springer in the *Advances in Experimental Medicine and Biology* series which is listed with an impact factor in the Science Citation Index.

Officers

Per Liss (Sweden)	President
Kyung A. Kang (USA)	Past President
Eiji Takahashi (Japan)	President Elect
Oliver Thews (Germany)	Secretary
Peter E. Keipert (USA)	Treasurer
Duane F. Bruley (USA)	Chairperson, Knisely Award Committee

Executive Committee

Chris Cooper (UK)
Jerry D. Glickson (USA)
Fahmeed Hyder (USA)
Paul Okunieff (USA)
Martin Wolf (Switzerland)
Akitoshi Seiyama (Japan)
Joseph LaManna (USA)
Christopher B. Wolff (UK)

Local Committee

Fredrik Palm
Peter Hansell
Angelica Fasching
Jenny Edlund
Malou Friederich
Lina Nordquist
Anna Dufflin

New Members of Executive Committee
Elected ISOTT 2007

William Welch (USA)
Artur Fournell (Germany)

Awardees

Melvin H. Knisely Award:	Ilias Tachtsidis (UK)
Dietrich W. Lübbers Award:	Helga Blockx (Belgium)
Britton Chance Award:	Eric Mellon (USA)
Duane F. Bruley Awards:	Dominique De Smet (Belgium) Thomas Ingram (UK) Nicola Lai (USA) Andrew Pinder (UK) Joke Vanderhaegen (Belgium)

PANEL OF REVIEWERS

Donald Buerk, University of Pennsylvania, USA
Chris Cooper, University of Essex, UK
Clare Elwell, University College London, UK
Jerry Glickson, University of Pennsylvania, USA
David Harrison, University Hospital of North Durham, UK
Göran Hedenstierna, University of Uppsala, Sweden
Louis Hoofd, University of Nijmegen, Netherlands
Georg Iliakis, University of Essen, Germany
Philip James, University of Cardiff, UK
Peter Keipert, Sangart Inc., USA
Joe LaManna, Case Western Reserve University, USA
Avraham Mayevsky, Bar Ilan University, Israel
Masaomi Nangaku, University of Tokyo Medical School, Japan
Edwin Nemoto, University of Pittsburch, USA
Paul Okunieff, University of Rochester, USA
Fredrik Palm, University of Uppsala, Sweden
Richard Porter, Trinity College Dublin, Ireland
Akitoshi Seiyama, Osaka University, Japan
John Severinghaus, University of California San Francisco, USA
Jens Sörensen, University of Uppsala, Sweden
Hrold Swartz, Dartmouth Medical School, USA
Ilias Tachtsidis, University College London, UK
Eiji Takahashi, Yamagata University, Japan
Oliver Thews, University of Mainz, Germany
Peter Vaupel, University of Mainz, Germany
Martin Wolf, University Hospital Zurich, Switzerland

PREFACE

In 1772 in Uppsala the Swedish chemist Karl Wilhelm Scheele discovered the element Oxygen. Two hundred and one years later, in 1973, the International Society on Oxygen Transport to Tissue (ISOTT) was founded. Since then there has been an annual ISOTT meeting. After 24 years of international ISOTT meetings it was decided, at the 2005 summit in Bary, Italy, that the 2007 meeting was to be held in Uppsala, Sweden.

Thus, after the Louisville meeting we, in the Uppsala group, withdrew to the Edgewater Resort at Taylorsville Lake outside Louisville and prepared the Uppsala ISOTT meeting by tasting Kentucky Bourbons, smoking cigars while bathing in a jacuzzi in the hot dark Kentucky night full of fire flies and a sky full of stars.

The ISOTT program should include different aspects of oxygen - however, it is accepted that each meeting has its own local "touch". We decided to focus the Uppsala ISOTT meeting on the theme of "Imaging and measuring oxygen changes". With this in mind we invited scientists within and outside the ISOTT society. We then also received lots of good abstracts from ISOTT members that were included in the program.

Lars-Olof Sundelöf introduction speech "AIR AND FIRE" concerned how oxygen was discovered in Uppsala in 1772 by Karl Wilhelm Scheele. After the introduction speech a get together event took place in the magnificent and spacious foyer of Uppsala University main building. The vice chancellor Ulf Pettersson welcomed all delegates to Sweden and Uppsala.

The conference started on a Monday at the Atrium Conference facility, centrally situated in Uppsala. On the evening, after a full day of science, we embarked the ship m/s Kung Carl Gustaf, built in 1892, for a dinner cruise and went the same route along Fyris River as the Vikings used in the 9th century on their way to conquer Europe.

On Tuesday evening we met at the garden of Carolus Linnaeus for dinner and entertainment. Carolus Linnaeus is perhaps the most well known resident of Uppsala of all times and his garden was the first botanical garden in Sweden founded in 1655 by Olof Rudbeck the elder. This evening we actually met Carolus Linnaeus himself at the garden.

All too soon, the final lecture was delivered, the banquet, held at Östgöta Nation building which completed in 1885, was upon us and the prizes awarded. Time arrived to say our farewells, each delegate carrying home a Dala horse, originating from the presidents home county, to remind us to attend the next meeting of ISOTT, and many more after that.

The editors thank all participants in this conference as well as those who worked behind the scenes to ensure that things ran smoothly. We especially want to acknowledge the assistance of our technical editors Larraine Visser and Eileen Harrison in the preparation of this volume.

TABLE OF CONTENTS

AIR AND FIRE

Carl Wilhelm Scheele, Torbern Bergman, The Royal Society of Sciences and the Discovery of Oxygen in Uppsala in the year 1772

Lars-Olof Sundelöf [*]

It was an early morning a couple of weeks ago in the deep forest not so far from here. The aged fur trees stood closely, their twisted roots hidden under the wavy fur of green, almost fluorescent moss. Soft after the night rain. Gleaming in the few beams of light from the raising sun that tight branches let pass. As I stood there suddenly a light wind swept between the grey trunks and embraced my face with a refreshing touch of air, cool and fragrant in memory of the night. Essence of a fading Nordic summer, an almost holy moment. And all of a sudden I recalled a similar sensation, fifty years ago outside the Biochemistry Department in Berkeley. I had come for a meeting in the laboratory and was early. I took a walk among the eucalyptus trees, their violet stems towering in the sky sending whirls of exotic smell to the ground. I stopped. And the morning breeze came in from the bay carrying the sea to the shadows of the park. A deep breath filled my lungs with this magnificent moment. No closer we may come – I believe – to the sense of oxygen.

In 1710, almost three hundred years ago, the plague came to Stockholm. Times were heavy for the Swedish people. The great Nordic war had raged for a long time. The army under king Charles XII beaten by Russia the year before and now the king and a handful of his soldiers imprisoned in Turkey. Hunger and sickness was the lot in the homeland. Enemies around all borders. The plague moved slowly north through the land and threatened Uppsala. The university was closed and the students sent home. This was the moment of Eric Benzelius junior, university librarian, professor of theology but with broad interests and competence in all of science.

---

[*] Opening lecture at the ISOTT 2007 conference in Uppsala August 26, 2007, by
Lars-Olof Sundelöf, Professor of Physical Chemistry Emeritus, former Deputy Vice Chancellor of Uppsala University, Department of Medicinal Chemistry, BMC, Box 574, SE-751 23 Uppsala, Sweden.

P. Liss et al. (eds.), *Oxygen Transport to Tissue XXX*, DOI 10.1007/978-0-387-85998-9_1,
© Springer Science+Business Media, LLC 2009

Towards the end of the 17th century, Eric Benzelius had travelled a few years in Europe and made good connections with several well known learned people. In particular he had been intrigued by Leibnitz and his Preussische Societät der Wissenschaften in Leipzig, one of the earlier learned societies founded to promote knowledge and acting to make it useful to the public in general. The first such societies were The Royal Society in London and Académie des Sciences in Paris. Their coming about was largely governed by on one hand the new era in the sciences during the early 17th century when direct observations and planned experiments together with a careful analysis of cause and reaction replaced the more contemplated attitude of the Middle Ages. On the other hand the universities of that time were much more of educational establishments than places for development of knowledge in the modern sense. Thus the learned societies brought an entirely new attitude to natural philosophy and essentially they laid the ground for the marvellous technical development during the 18th century.

The idea to create something similar to Leibnitz academy in Uppsala haunted Eric Benzelius when he returned home but times were against him and nothing came about. Then in 1710 came the plague and in the dark month of November that year he collected some colleagues at the University Library – which would have been Gustavianum, the building at the other end of the park from here – and the first learned society in Sweden was founded under the name Collegium curiosorum. It lived through some crises during the next two decades but evolved in 1728 under royal protection as the Royal Society of Sciences at Uppsala. The dream of Eric Benzelius was fulfilled and the Society was in the good hands of a young Secretary, the physicist Anders Celsius.

The learned societies had many missions. An important one was to correspond with colleagues and practitioners of technology as well as inventors trying to find possibilities to explore new knowledge for the benefit of people but also to collect curiosity items. – There was still a mixture of experimental truth and speculativeimagination. – Some important scientific friends from abroad were made corresponding members. The early lists of members of the learned societies show a considerable overlap and a network was developed – oral or in the form of letters written on paper – that although slow nevertheless was an early form of internet.

Another important mission was to try to support research financially. It may have been building heavy equipment of that day in the form of astronomic observatories with their brass instruments or it could be provisions for travelling. Since this is the year when we celebrate the 300th anniversary of Carl von Linné a few words may be spent on him since he is a portal figure for the 18th century science at Uppsala and since it was the Royal Society of Sciences that first saw his capacity and helped him to develop it.

Carl von Linné arrived in Uppsala from Lund in 1728 and from the very beginning he was recognised by members of the Society, in particular by the Dean Olof Celsius, who put his well-equipped library at the disposal of the young student. Already in 1729 the Society gave a financial contribution to Linné to explore the flora of Gräsön, an island in the Baltic, not so far from Uppsala. One year later the Society expressed its warmest recommendations concerning the publication of a new study by Linné. In 1732 a considerable sum of money was allocated by the Society to support Linné on his famous journey to Lapland and this support was even increased upon his return. He then went abroad and when Linné returned from Holland in 1738 he was elected a member of the Society. This token of his scientific achievements Linné considered as the most important one from his colleagues at home and he expressed his deep gratitude to the members of the Society for their unanimous recognition. After the death of Anders Celsius in 1744 Linné became Secretary of the Society.

One remarkable early corresponding member of the Royal Society of Sciences at Uppsala was Emmanuel Swedenborg, in many circles more known as founder of a spiritual religion than for his scientific skills, although he probably is one of the most outstanding scientists Sweden has produced. He realised the need for a more formalised means of communication and preservation of facts and compiled himself the first scientific journal in Sweden, Dædalus Hyperboræus, issued in the period 1715-1718. However, the financial means did not permit a continuation and it was not until the middle of the 18th century that the economy allowed for some issues of a new journal. From 1773 the journal of the Society Nova Acta Regiæ Societatis Scientiarium Upsaliensis exists. Until the 1960's it served as a place for publication of dissertations in the natural sciences including those of the Swedish Nobel laureates. Since then the way of disseminating scientific knowledge has changed dramatically and journals like "Nova Acta" no longer play a role as we all know.

Chemistry has old roots at the university and already around 1650 some chemical laboratory activities are indicated at the University Dispensary. However, when in 1750 a chair in Chemistry, Pharmacy and Metallurgy was created at the Faculty of Arts and Sciences at Uppsala University, a glorious period was started.

The first professor in chemistry was Johan Gotschalk Wallerius. He built the very first laboratory building for chemistry – still in existence at Västra Ågatan 24 – now used for other purposes. In a chemical laboratory of those days an oven with open fire was necessary and often this also served as the kitchen stove. Wallerius made excellent contributions to agricultural chemistry and was for a long period an international authority in this field. Due to illness he resigned already in 1766. Maybe a contributing reason was that fire had ravaged his laboratory. Fire, a necessity for chemical operations but a process not yet understood, was a disastrous thing when it went out of hand. In 1702 the whole city of Uppsala was more or less demolished by fire. Now the chemical laboratory of Wallerius was hurt.

Torbern Bergman succeeded Wallerius as professor of chemistry. He was a scientist of unusually broad capabilities. His earlier papers dealt with noxious insects. His doctoral thesis concerned astronomy, he became teacher of physics and at the time he was appointed professor of chemistry he was associate professor in mathematics. When it came to chemistry he had only a small paper on alum manufacture. But due to the direct intervention of the Chancellor, Crown-Prince Gustaf – later King Gustav III – Torbern Bergman got the professorship mainly for his acknowledged intellectual capacity. Bergman's first task was to restore the fire-ravaged laboratory. He also enlarged the building and gave it essentially the shape it has today. As professor, Bergman devoted his undivided attention to chemistry and very soon became an international authority in the field. He should probably be regarded as one of the greatest names of the University of Uppsala through all time.

About this time Christian Lokk from Pommern got the priviledges to run a dispensary in Uppsala and in 1770 his house built for this purpose stood ready at the Main Square of the city. The dispensary carried the name Upplands Vapen and during the years 1770 to 1775 Carl Wilhelm Scheele lived and worked in this building.

Even Carl Wilhelm Scheele came from the old Swedish possessions in Northern Germany, namely the city of Stralsund. At the age of fifteen he moved as a trainee to a dispensary in Gothenburg where he stayed for six years. Entirely devoted to the craftsmanship of the pharmaceutical trade, which then was basically chemical laboratory work, these years formed the bases for his coming scientific achievements. After Gothenburg he spent a few years in Lund and Stockholm and finally arrived in Uppsala in 1770. Soon he made acquaintance with Torbern Bergman and they became close friends. This partnership turned out to be extremely important for Bergman. Scheele was the self-taught man very much at home in the laboratory. Bergman mastered wide areas of science and had the ability to see complex interrelations and to understand the wide perspectives of new results. Even another famous Uppsala chemist, Johan Gottlieb Gahn, worked in close collaboration with Scheele.

The friendship with Torbern Bergman was probably just as important for Scheele as it was for Bergman. One can easily imagine the discussions they had and how the theoretical interpretations by Bergman, based on the laboratory results of Scheele, induced new work in the laboratory. And so it went on. It may be anticipated that the dispensary at the Main Square played a more important role for the development of chemistry than the chemical laboratory at Västra Ågatan.

Although Scheele wrote some important publications in Swedish as well as in other languages they did not reach an international audience until long after the discoveries described were made. After his death they were collected, translated and published. Torbern Bergman, however, who mastered latin and was a member of the Royal Society of Sciences, published in Nova Acta, the journal of the Society, many of the important results of Scheele soon after their discovery. Thus they were made available to a broader audience at an early date. It is with special feelings one opens these volumes from the 1770's with their coarse, thick, wavy paper and warm-black Gothic printing. The sound of the turning pages. No figures. Just text. And the language sounds beautiful even for someone like me who does not reed latin fluently. It is in one of these volumes that Torbern Bergman describes how Scheele discovered oxygen and this passage is essential in showing that Scheele was the first one to describe this important element. It appears in Vol. II of Nova Acta Reg. Soc. Ups. (1775) at page 235. Scheele himself had finished the manuscript of his famous dissertation "Chemische Abhandlung von det Luft und dem Feuer" in 1775 just before he left Uppsala for the little country town Köping and its dispensary but the printing in Leipzig was delayed until 1777 whereby he missed many priorities to important discoveries. On the other hand this also indicates that the activities in chemistry in Uppsala were on the scientific frontier.

The priority issue of the discovery of oxygen has been much debated and for some time Joseph Priestly was held to be the first one. A key object in this discussion was a letter that Scheele claimed to have written to Antoine Lavoisier in Paris in 1774 describing how oxygen was evolved from manganese oxide upon heating. This letter, however, was lost for a long time but finally recovered among the once belongings to Mme Lavoisier and donated to Académie des Sciences in Paris in 1993. Now it is an accepted fact that Scheele was the very first one and that the discovery was made in 1772. This thrilling story has been told in a beautiful lecture by one of the participants of this meeting, John Severinghaus, who generously made it available to me.

It is remarkable to see what Scheele managed to do during those few years in Uppsala. In 1772 he discovered oxygen – "fire-air" as he called it. Oxygen was the name introduced by Lavoisier who was the first to see the far reaching consequences of the discovery. The same year Scheele also discovered nitrogen – at the same time as Daniel Rutherford, but independently. In 1774 he discovered barium and chlorine. The discovery of manganese by Johan Gottlieb Gahn in 1774 was made after important previous work by Scheele. His basic work also paved the way for the discovery of molybdenum and tungsten in the 1780's. And this list only mentions the discovery of new elements. His general work in chemistry was just as important.

However important the contributions by Scheele Torbern Bergman also carried out his own lines of research. In the volume on Chemistry published in 1977 to the centennial celebration of the founding of Uppsala University this is very well described, and I quote:

"Bergman made great contributions in analytical chemistry where his work on quantitative analysis in particular was of fundamental importance. He clarified the composition of many compounds, performed important investigations concerning carbonic acid and was first to work on the preparation of mineral waters. He compiled extensive affinity tables in order to systematise and predict the course of reactions and also proposed a chemical notation in which the symbols of the alchemists from the middle ages were modified and complemented to meet the current needs. Bergman also devoted much time to teaching, both theoretical and practical, and assembled a cabinet of models so that one could study the apparatus and mechanical arrangements used in industry and mining."

With the death of Torbern Bergman in 1784 this golden age of chemistry at Uppsala was finished. It was not until the appointment in 1853 of Lars Fredrik Svanberg, a pupil of Berzelus, that new ideas were introduced that gradually allowed for a new expansive period in chemistry at Uppsala.

How could it come that the latter half of the 18th century saw this remarkable scientific period in Uppsala? We have only touched upon chemistry but biology and physics had similar qualities. One reason is certainly that there happened to be at the university a number of very talented persons with an eye for large perspectives urging to tackle difficult but productive problems. But just as important, I believe, was the sense of living in an innovative and creative atmosphere formed by the freedom of mind that the new natural philosophy had shown. Something of this can be read from the proceedings of the Royal Society of Sciences during this time. More important than the actual items recorded is the overall impression one gets of curious minds that ask questions, propose explanations, report curiosities, discuss reports. Read in this way the minutes of the meetings, though sparse in words, tell us about people who lived for their science – in this distant country town way up north – creating an intellectual atmosphere contagious to students and much more valuable than the victories during the days of the Great Power of Sweden in the previous century.

Yet the achievements were made by very simple means. The chemical laboratory of those days rested on simple vessels for volume measuring, maybe a balance to record mass. And fire to increase the rate of reactions, to melt, to make boil. But it was the ingenuity by which the means were utilised and combined that made the thing. The thinking was instrumental. Strangely enough quite a bit of this situation still was there in the early 1950's when I began my studies. Still the analytical balance was in a sense the main instrument. It took a few more years for spectroscopy in all its forms to conquer the chemistry labs.

Oxygen is a complex molecular structure with unpaired electrons and having a specific set of energy levels. Only in a very thin layer close to the surface of the earth does it occur in a concentration sufficient to support human life and after a thunderstorm we may feel the scent of its triplet, ozone. In the view of a modern chemist oxygen is a much more intriguing entity than the fire-supporting substance Carl Wilhelm Scheele anticipated. This will be further substantiated during this conference.

One late November evening in the mid fifties – more than fifty years ago – I left my studies and went downtown to relax and entered the café then occupying the old premises of the dispensary Upplands Vapen where Carl Wilhelm Scheele had worked almost two hundred years earlier and where he among other things made his discovery of oxygen. I went to the upper floor with the nicely sized windows facing the Main Square. Flowers in the windows. Simple wooden furniture with candles. I had a cup of coffee. Not many people were there. I saw the traffic come and go in the darkness of the square where the magnificent iron lantern only left barely noticeable light marks on the faces that moved around. There was no music, the only sound came from the scrambling vehicles down at the square. Time stopped. I was here for the first – and only – time. I knew the history of the place from the sign posted on the outside wall. I also knew that the building was doomed to be torn down to give room for more efficient constructions. Time went back. Over the entrance of the chemistry building where I just had had my elementary training were three busts: those of Torbern Bergman, Carl Wilhelm Scheele and Jöns Jakob Berzelius. For my inner imagination Scheele was green – as I had seen him on the stamps during my boyhood. Now I was sitting where he had worked. In the dusk of the room the fittings changed to those of a laboratory and in the background those soft warm eyes from the stamp looked at me for a moment when the quiet man shifted some things on the table in front of him. It had all gadgets of an ancient laboratory: sooty retorts, jars with chemicals, spoons. Behind the table was a small fireplace against which the tiny silhouette moved back and forth... At the sudden screaming of tyres from a car the scene evaporated and the fittings of a café were restored. I closed my eyes realising that a bit of history was also soon to go. The beautiful proportions of this 18th century house had to be preserved in memory.

It was raining when I left for home. Streams of water in the gutter reflected the faint incandescent light from the lantern. The wind swept the streets and carried the clicks of my steps over the river. Dark trees without leaves stretched their branches towards moving clouds. For a short moment the network of that golden age of chemistry had returned as a token that over crushed walls and smashed windows prevails something not destroyable: The achievements of an intellect.

REFERENCE

Fredga A, Chemistry at Uppsala until the Beginning of the Twentieth Century *in* Acta Universitatis Upsaliensis Uppsala University 500 Years 9. Faculty of Science at Uppsala University, Chemistry, ed. Lars-Olof Sundelöf

MATHEMATICAL MODELING OF THE INTERACTION BETWEEN OXYGEN, NITRIC OXIDE AND SUPEROXIDE

Donald G. Buerk[*]

Abstract: Computer simulations were performed based on a multiple chemical species convection-diffusion model with coupled biochemical reactions for oxygen (O_2), nitric oxide (NO), superoxide ($O_2^{\bullet-}$), peroxynitrite ($ONOO^-$), nitrite (NO_2^-) and nitrate (NO_3^-) in cylindrical geometry with blood flow through a 30 μm diameter arteriole. Steady state concentration gradients of all chemical species were predicted for different $O_2^{\bullet-}$ production rates, superoxide dismutase (SOD) concentrations, and blood flow rates. Effects of additional $O_2^{\bullet-}$ production from dysfunctional endothelial nitric oxide synthase (eNOS) were also simulated. The model predicts that convection is essential for characterizing O_2 partial pressure gradients (PO_2) in the bloodstream and surrounding tissue, but has little direct effect on NO gradients in blood and tissue.

1. INTRODUCTION

Mathematical mass transport models can provide insight into extremely rapid, diffusion-limited interactions between reactive O_2 and nitrogen species in blood and tissue. Buerk et al.[1] developed a steady-state mathematical model with coupled reactions for O_2, NO, $O_2^{\bullet-}$, $ONOO^-$, hydrogen peroxide (H_2O_2), NO_2^- and NO_3^- for an arteriole. Gradients due to diffusion through five concentric regions (blood, plasma layer, endothelium, vascular wall, perivascular tissue) for NO, PO_2, $O_2^{\bullet-}$, and $ONOO^-$, as well as production rates for H_2O_2, NO_2^-, and NO_3^- were predicted for varying $O_2^{\bullet-}$ production rates, SOD concentrations, carbon dioxide (CO_2) levels, hydrogen ion (pH), and other conditions. The model included reversible inhibition of mitochondrial O_2 consumption by NO, and predicted spatial variations in $ONOO^-$ due to competition between scavenging of $O_2^{\bullet-}$ by NO and inactivation of $O_2^{\bullet-}$ by SOD. A sensitivity analysis for model parameters was done. However, convection by blood flow through the vessel was not included in this model. Production rates for H_2O_2, NO_2^-, and NO_3^- were followed, but their concentration gradients could not be predicted since convective transport was not included for those species. Our most recent NO and O_2 transport models include convective transport in the

[*]Donald G. Buerk, Departments of Physiology and Bioengineering, University of Pennsylvania, Philadelphia, PA 19104.

P. Liss et al. (eds.), *Oxygen Transport to Tissue XXX*, DOI 10.1007/978-0-387-85998-9_2,

bloodstream.[2,3] These models were modified to develop the multiple species convection-diffusion model described in this paper.

2. MODEL DEVELOPMENT

The arteriole for this simulation has a 30 μm internal diameter. The previous model by Buerk et al.[1] was modified by adding convective terms for blood flow and an additional layer (thin glycocalyx coating on the endothelial surface). The model now consists of 6 concentric radial layers, including the blood core (layer 1; 0<r<13 μm), red blood cell (rbc) free plasma layer (layer 2; 13<r<14.5 μm), glycocalyx (layer 3; 14.5<r<15 μm), endothelium (layer 4; 15<r<16 μm), vascular wall (layer 5; 16<r<19 μm), and surrounding avascular tissue (layer 6; 19<r<120 μm). Blood flow through the lumen was assumed to have a parabolic shape, with the maximum velocity at the centerline (1000 μm s^{-1} at r =0). All simulations were for a 300 μm axial length, and comparisons for different conditions were made for radial concentration gradients computed at the center of the segment (z =150 μm).

Convection-diffusion-reaction mass transport partial differential equations for concentrations C_i of multiple species were defined for each layer:

$$D_i \nabla^2 C_i - v \cdot \nabla C_i \pm \sum_m R_{i,j} = 0$$

where i = O_2, $O_2^{\bullet-}$, NO, NO_2^-, NO_3^-, H_2O_2, and $ONOO^-$ with diffusion coefficient D_i for each species, convective velocity field v, and the sum of all possible chemical reactions $R_{i,j}$ involving species of interest (i) and other reacting chemical species (j). Additional chemical species in the model, CO_2, SOD, and pH, were set at constant values. The resulting set of 7 coupled partial differential equations was written in cylindrical coordinates and solved numerically by finite element computational methods using commercial software (Flex-PDE, PDE Solutions, Inc., Antioch, CA, USA).

2.1. Rate Constants for Coupled Reactions and Other Parameters

Reactions and rate constants are identical to those used previously by Buerk et al.[1] Rate constants determined by Huie and Padmaja[4] for the reaction between NO and $O_2^{\bullet-}$ (6.7×10^9 M^{-1} s^{-1}), by Pfeiffer et al.[5] for the reaction between NO and $ONOO^-$ (9.1×10^4 M^{-1} s^{-1}), and from Fridovich[6] for scavenging of $O_2^{\bullet-}$ by SOD (1.6×10^9 M^{-1} s^{-1}) were used. Reactions described by Chen and Deen[7] were used for decomposition of $ONOO^-$ with CO_2. The second order rate constant from Lewis and Dean[8] for autooxidation of NO (2.9×10^6 M^{-2} s^{-1}) was used. These reactions occurred in all 6 layers of the model. The reaction between NO and guanylate cyclase occurred only in the vascular wall and tissue, and was held constant (0.01 s^{-1}) in these layers.

D used for O_2, $O_2^{\bullet-}$, NO, NO_2^-, NO_3^-, H_2O_2, and $ONOO^-$ were 2800 μm^2 s^{-1}, 2800 μm^2 s^{-1}, 3300 μm^2 s^{-1}, 3000 μm^2 s^{-1}, 3000 μm^2 s^{-1}, 3000 μm^2 s^{-1}, and 2600 μm^2 s^{-1}, respectively. pH was 7.4 in blood, plasma, and glycocalyx layers, and 7.0 in all other layers. CO_2 was 1100 μM in all regions. The input blood PO_2 at the centerline (r=0) was 60 Torr, with zero concentrations for all other species in blood entering the arteriole. Zero fluxes were assumed at outer boundaries, assuming that there were no interactions with tissue perfused by capillaries beyond the outer radius (r=120 μm) of the modeled region.

2.2. Reactions with Hemoglobin (Hb)

Scavenging of NO by Hb (382.5 s^{-1}) is based on measurements of NO uptake by human rbcs by Carlsen and Comroe,[9] as detailed in our previous models.[1,10-12] NO scavenged by Hb was assumed to be rapidly and completely converted to NO$_3^-$. Scavenging of ONOO$^-$ by Hb (30 s^{-1}) is based on the rate used in another model by Savill et al.[13] The most important function of Hb is the transport of O$_2$. The oxyhemoglobin equilibrium curve was modeled using the algorithm of Buerk and Bridges.[14]

2.3. Reversible Inhibition of O$_2$ Consumption by NO

Evidence that NO regulates microvascular O$_2$ delivery and metabolism is reviewed by Buerk.[15] Our previous models[1-3,10-12] use the relationship proposed by Buerk[16] for reversible inhibition of mitochondrial O$_2$ consumption by NO

$$RO_2 = RO_{2max} \times PO_2 / (PO_2 + appK_m) ; appK_m = 1 \text{ Torr} \times (1 + NO/27 \text{ nM})$$

where the apparent K_m varies linearly with NO. In the absence of NO, $appK_m = 1$ Torr. RO_{2max} was 1 µM s^{-1} and 20 µM s^{-1} for the vascular wall and tissue, respectively.

2.4. O$_2$-Dependent Production of NO by Endothelium

As pointed out in an earlier review by Buerk,[16] O$_2$ is required for the production of NO catalyzed by eNOS. Michaelis-Menten kinetics were used for this simulation

$$RNO = RNO_{max} \times PO_2 / (PO_2 + K_m)$$

assuming $RNO_{max} = 100$ µM s^{-1} with $K_m = 4.7$ for eNOS[17], produced uniformly across the endothelium. Since production of each NO molecule requires 2 O$_2$ molecules, RO_{2max} for the endothelium was 200 µM s^{-1} for this value of RNO_{max}.

2.5. Simulation of Endothelial Dysfunction

O$_2^{\bullet-}$ produced by dysfunctional endothelium was represented as a fraction of RNO, with simulations ranging from 0% (normal endothelium) up to 5% endothelial dysfunction, corresponding to maximum O$_2^{\bullet-}$ production rates ranging from 0–5 µM s^{-1}. RNO_{max} was proportionally reduced with increasing endothelial dysfunction, down to 95 µM s^{-1} for case with 5% endothelial dysfunction, taking decreases in O$_2$ consumed by The model conservatively assumed that one half of the O$_2$ that would have been used by the endothelium for producing NO was instead converted to O$_2^{\bullet-}$. RO_{2max} was held constant. In the vascular wall and tissue layers O$_2^{\bullet-}$ production was assumed to be 2% of RO_2 in each layer (maximum rates of 0.02 µM s^{-1} and 0.4 µM s^{-1}, respectively) following Michaelis-Menten kinetics with $appK_m$ depending on NO as described above. Simulations were performed for SOD levels ranging from 1–30 µM, assumed to be uniformly distributed in all regions.

3. RESULTS

Effects of endothelial dysfunction on radial concentration gradients due to increasing O$_2^{\bullet-}$ are shown in **Figure 1** for simulations with SOD = 1 µM in blood and tissue.

Increased $O_2^{\bullet-}$ in the endothelium is shown for a detailed section of the model (A). The combination of smaller RNO_{max} and greater amounts of NO reacting with $O_2^{\bullet-}$ has a relatively small effect on NO profiles (B), with the peak NO decreasing (inset). $ONOO^-$ increases progressively, with the peak shifting from tissue into the vascular wall and endothelium (C). There is a net decrease in NO_3^- with increasing endothelial dysfunction (D), primarily due to decreased NO scavenging by Hb, even though there is a small increase in NO_3^- formed by decomposition of $ONOO^-$ with CO_2. Peak NO_3^- levels are found in blood next to the rbc-free plasma layer, due to the steep decrease in NO in the bloodstream. NO_2^- is 3 orders of magnitude smaller, decreasing with greater endothelial dysfunction, with higher levels in tissue than in blood (E). There is a minor effect on the PO_2 gradient with the outermost PO_2 (at r=120 µm) 0.075 Torr lower with 5% endothelial dysfunction, due to slightly lower NO and less inhibition of RO_2 (F). H_2O_2 gradients in the 3.3–3.8 µM range were predicted, with higher levels in tissue than in blood (not shown); however, these values are overestimated since catalase and glutathione peroxidase activity was not modeled. Preliminary simulations were performed to investigate effects of catalase, which showed that H_2O_2 could be reduced by 50% or more using values for the product of the rate constant x [catalase] in the 0.05–0.2 s^{-1} range (not shown). Scavenging H_2O_2 generates O_2, but the effects on PO_2 gradients were found to be negligible (< 0.05 Torr increase).

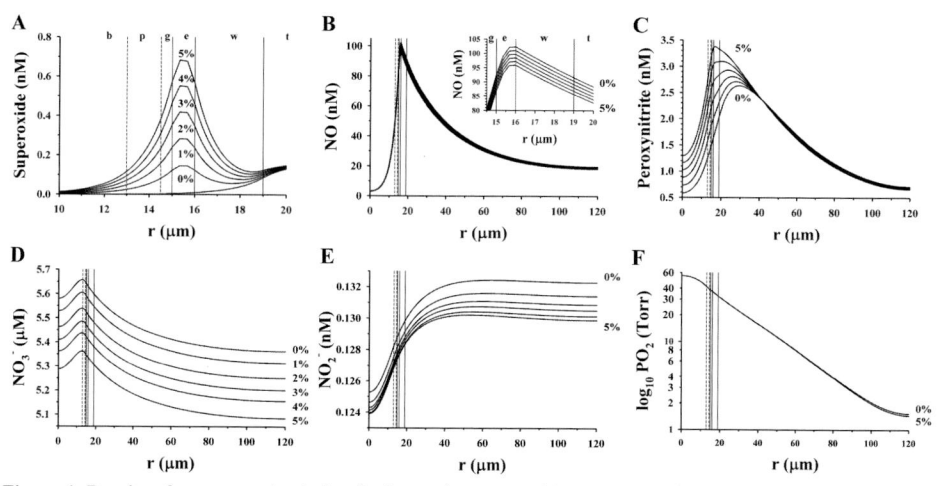

Figure 1. Results of computer simulation for increasing superoxide generation from dysfunctional endothelium in a 300 µm long cylindrical model of an arteriole with SOD = 1 µM in blood and tissue. Radial gradients at the midpoint (z=150 µm) are shown for (A) superoxide, (B) NO with inset for detail in endothelium and vascular wall, (C) peroxynitrite, (D) nitrate, (E) nitrite, and (F) PO_2 for increasing endothelial dysfunction. Radial layers in the model are b=blood, p=plasma layer, g=glycocalyx, e=endothelium, w=vascular wall, and t=tissue.

Simulations were also done for higher (2x) and lower (0.5x) blood flow rates for the 5% endothelial dysfunction case with 1 µM SOD in blood and tissue (not shown). The centerline blood PO_2 exiting the arteriole dropped from the 60 Torr entrance value to 55.6 Torr, 51.9 Torr, and 45.3 Torr for high, baseline, and low blood flow rates, respectively, at the end of the arteriole. Small blood flow-related changes in NO, $O_2^{\bullet-}$, and $ONOO^-$ gradients were observed, which can be attributed to the effect of endothelial PO_2 changes on NO production by eNOS, with negligible direct effect of convective transport. Gradients for NO_2^-, NO_3^-, and H_2O_2 were affected by changes in blood flow as expected.

Effects of increasing SOD on radial concentration gradients with 5% endothelial dysfunction are shown in **Figure 2**. Decreasing $O_2^{\bullet-}$ in the endothelium with increasing SOD is shown in detail (A). An increase in NO is predicted as SOD increases, with values > 90 nM in the vascular wall and tissue for SOD > 15 µM (B). ONOO$^-$ decreased with increasing SOD, from a peak of 4.7 nM to 0.2 nM in the endothelium (C). The increase in NO and greater inhibition of mitochondrial RO_2 with increasing SOD increases PO_2 in the vascular wall and tissue (D). Although there was an increase in NO entering the bloodstream with increasing SOD, there was a small net decrease in NO_3^- since contributions from decomposition of ONOO$^-$ were reduced (not shown).

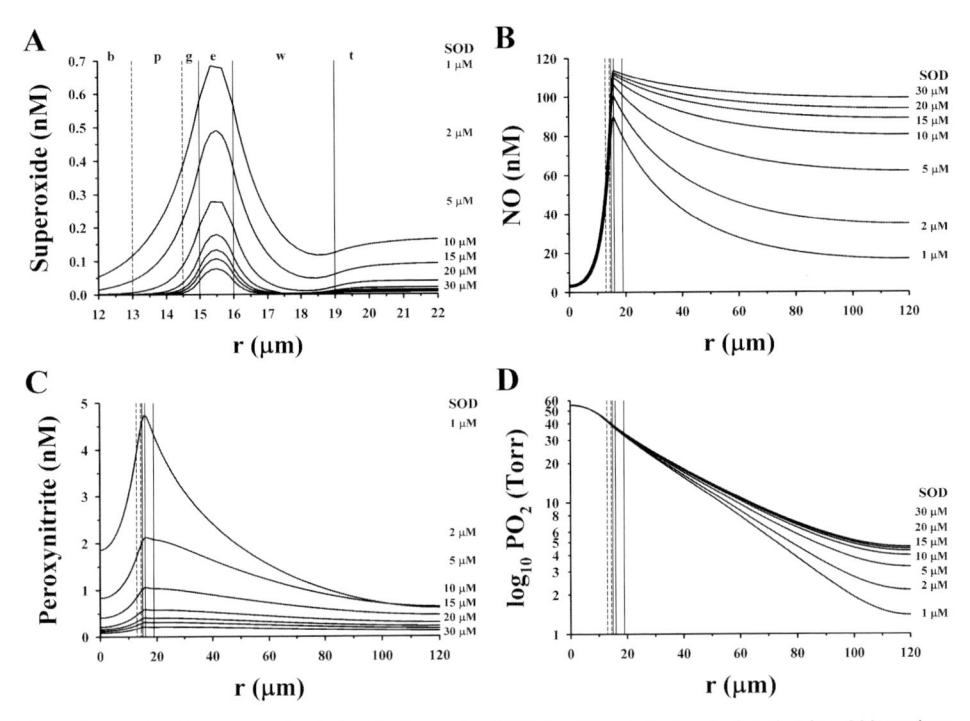

Figure 2. Results of computer simulation for increasing SOD for 5% endothelial dysfunction in a 300 µm long cylindrical model of an arteriole. Effects of SOD on radial gradients at the midpoint of the segment are shown for (A) superoxide, (B) NO, (C) peroxynitrite, and (D) PO_2.

4. DISCUSSION

These simulations support the concept, described by Thomas et al.,[18] that interactions between NO and $O_2^{\bullet-}$ co-regulate their concentrations and thereby control their signal transduction pathways. However, the current model is highly simplified, focusing only on major reactions. Furthermore, sources of NO from other NOS isoforms, or other sources of $O_2^{\bullet-}$ production (e.g., xanthine oxidoreductase, NADPH oxidase), and the active or selective transport of charged species across membranes was not considered. More complex biochemical pathways will need to be explored and added to the model as new information becomes available. As an example, Quijano et al.[19] describe a compartmental modeling approach with a total of 22 reactions for predicting $ONOO^-$ mediated tyrosine nitration as a function of the ratio of $O_2^{\bullet-}/NO$ fluxes. Their model suggests that there can be a decrease in nitration when the flux of one radical is in excess over the other. They show that these relationships were altered when the reaction between $O_2^{\bullet-}$ and SOD was added, along with loss of NO by diffusion out of the compartment.

The present model has not considered other hypothesized pathways such as those described in the review by Robinson and Lancaster.[20] There is very limited information about reaction rates for alternative hypotheses to allow testing in the present model.

5. ACKNOWLEDGMENTS

Supported in part by grants HL 068164 from NIH and BES 0301446 from NSF.

6. REFERENCES

1. D.G. Buerk, K. Lamkin-Kennard, and D. Jaron, Modeling the influence of superoxide dismutase on superoxide and nitric oxide interactions, including reversible inhibition of oxygen consumption, *Free Radic Biol Med* **34**(11),1488-1503 (2003).
2. X. Chen, D.G. Buerk, K.A. Barbee, and D. Jaron, A model of NO/O_2 transport in capillary-perfused tissue containing an arteriole and venule pair, *Ann Biomed Eng* **35**(11),517-529 (2007).
3. X. Chen, D. Jaron, K.A. Barbee, and D.G. Buerk, The influence of radial RBC distribution, blood velocity profiles, and glycocalyx on coupled NO/O_2 transport, *J Appl Physiol* **100**(4),482-492 (2006).
4. R.E. Huie, and S. Padmaja, The reaction of NO with superoxide, *Free Radic Res Commun* **18**(4),195-199 (1993).
5. S. Pfeiffer et al., Metabolic fate of peroxynitrite in aqueous solution. Reaction with nitric oxide and pH-dependent decomposition to nitrite and oxygen in a 2:1 stoichiometry, *J Biol Chem* **272**(6),3465-3470 (1997).
6. I. Fridovich, Superoxide radical and superoxide dismutases, *Annu Rev Biochem* **64**(6), 97-112 (1995).
7. B. Chen, and W.M. Deen, Analysis of the effects of cell spacing and liquid depth on nitric oxide and its oxidation products in cell cultures, *Chem Res Toxicol* **14**(1),135-147 (2001).
8. R.S. Lewis, and W.M. Deen. Kinetics of the reaction of nitric oxide with oxygen in aqueous solutions. *Chem Res Toxicol* **7**(4),568-574 (1994).
9. E. Carlsen, and J.H. Comroe, Jr., The rate of uptake of carbon monoxide and of nitric oxide by normal human erythrocytes and experimentally produced spherocytes, *J Gen Physiol* **42**(1),83-107 (1958).
10. K. Lamkin-Kennard, D. Jaron, and D.G. Buerk, Modeling the regulation of oxygen consumption by nitric oxide, *Adv Exp Med Biol* **510**,145-149 (2003).
11. K.A. Lamkin-Kennard, D.G. Buerk, and D, Jaron. Interactions between NO and O_2 in the microcirculation: a mathematical analysis, *Microvasc Res* **68**(1),38-50 (2004).
12. K.A. Lamkin-Kennard, D. Jaron, and D.G. Buerk, Impact of the Fåhraeus effect on NO and O_2 biotransport: a computer model, *Microcirculation* **11**(11),337-349 (2004).
13. N.J. Savill, R. Weller, and J.A. Sherratt, Mathematical modelling of nitric oxide regulation of rete peg formation in psoriasis, *J Theor Biol* **214**(1),1-16 (2002).
14. D.G. Buerk, and E.W. Bridges, A simplified algorithm for computing the variation in oxyhemoglobin saturation with pH, PCO_2, T and DPG, *Chem Eng Commun* **47**,113-124 (1986).
15. D.G. Buerk, Nitric oxide regulation of microvascular oxygen, *Antioxid Redox Signal* **9**(7),829-843 (2007).
16. D.G. Buerk, Can we model nitric oxide biotransport? A survey of mathematical models for a simple diatomic molecule with surprisingly complex biological activities, *Annu Rev Biomed Eng* **3**,109-143 (2001).
17. A. Rengasamy, and R.A. Johns, Determination of K_m for oxygen of nitric oxide synthase isoforms, *J Pharmacol Exp Ther* **276**(1),30-33 (1996).
18. D.D. Thomas, et al., Superoxide fluxes limit nitric oxide-induced signaling, *J Biol Chem* **281**(36),25984-25993 (2006).
19. C. Quijano, N. Romero, and R. Radi, Tyrosine nitration by superoxide and nitric oxide fluxes in biological systems: modeling the impact of superoxide dismutase and nitric oxide diffusion, *Free Radic Biol Med* **39**(6),728-741 (2005).
20. J.M. Robinson, and J.R. Lancaster, Jr., Hemoglobin-mediated, hypoxia-induced vasodilation via nitric oxide: mechanism(s) and physiologic versus pathophysiologic relevance, *Am J Respir Cell Mol Biol* **32**(4), 257-261 (2005).

HAEMOGLOBIN SATURATION CONTROLS THE RED BLOOD CELL MEDIATED HYPOXIC VASORELAXATION

Andrew G. Pinder[1][†], Stephen C. Rogers[1]*, Keith Morris [2] and Philip E. James[1]

Abstract: The vasorelaxant properties of red blood cells (RBCs) have been implicated in both the control of normal vascular tone and the protection of tissues from ischemic events. The identity of the vasorelaxant released from RBCs has yet to be elucidated, however growing evidence suggests that nitric oxide bound to the ß93 cysteine residue of haemoglobin (SNO-Hb) may be responsible. The vasorelaxant moiety is released during the transition of haemoglobin from its R (oxygenated) to T (deoxygenated) state. We subsequently chose to assess the significance of haemoglobin saturation on the capacity of RBCs to mediate hypoxic vasorelaxation.

Human RBC samples suspended in saline were manipulated in a thin film rotating tonometer, designed to rapidly change haemoglobin saturation within the time frame of circulatory transit. Various cycles of oxygenation and deoxygenation were performed. The vasorelaxant properties of the RBCs were analysed using an aortic ring bioactivity assay, wherein changes in isometric tension were recorded to study vessel relaxation. The rabbit aortic rings were preconstricted with phenylephrine under hypoxic conditions (~1% O_2) prior to RBC addition.

Highly saturated RBCs (98.22% ± 0.45 HbO_2) elicited significantly (P<0.001) more relaxation of hypoxic blood vessels compared to those partially saturated (20.40% ± 5.28 HbO_2). Upon re-oxygenation, previously de-oxygenated RBCs were also capable of eliciting vessel relaxation, which was not significantly different from that observed with the original oxygenated RBC relaxation response. Interestingly, the relaxant capability

---

[1]Department of Cardiology, Wales Heart Research Institute, School of Medicine, Cardiff University, Cardiff, CF14 4XN, United Kingdom.
[2] School of Applied Sciences, University of Wales Institute Cardiff, Cardiff, UK.
† Corresponding author: Email. PinderAG@CF.AC.UK, Tel. +44 (0)29 20 742912.
*Present address, The Department of Paediatrics and Biochemistry & Molecular Biophysics, Washington University in St Louis, Campus Box 8116, 1 Children's Place, St Louis, Missouri, 63110, United States.

P. Liss et al. (eds.), *Oxygen Transport to Tissue XXX*, DOI 10.1007/978-0-387-85998-9_3,
© Springer Science+Business Media, LLC 2009

was not simply returned from extracellular milieu upon re-oxygenation. This data provides further evidence that the conformational switch of haemoglobin from the R-state (oxygenated) to the T-state (deoxygenated) is essential for the release of the vasoactive moiety contained within red blood cells.

1   INTRODUCTION

Nitric oxide (NO) represents one of the most prominent, endogenous regulators of vascular tone and vessel health. A proportion of endothelial derived NO ultimately enters the vessel lumen where it can undergo a number of reactions with so-called 'scavenging' entities, perhaps the most noteworthy of which being haemoglobin. Haemoglobin contained with RBCs has the capacity to scavenge NO in both its deoxy and oxy conformations at a considerable rate, of approximately 10^7 M^{-1} s^{-1}[1, 2]. The products of these reactions are largely NO bound to the oxygen binding site of haemoglobin (HbNO) and the nitrate anion (NO_3^-) with met-haemoglobin respectively. These products alone are of little bioactive consequence, however HbNO is proposed to provide a source of NO for the formation of a second haemoglobin bound NO entity, SNO-Hb. Following the conformational switch in haemoglobin from its T-state (deoxy) to its R-state (oxy) across the lung, NO is belived to be passed from the oxygen binding site (HbNO) to the ß93 cysteine residue of haemoglobin (SNO-Hb)[3].

Evidence exists to suggest that an NO species released from SNO-Hb may represent the RBC vasorelaxant, whose effectiveness is proportionate to haemoglobin oxygen saturation[3, 4]. At present the precise identity of this vasorelaxant remains to be established, however the RBC induced relaxant is ODQ sensitive[4], demonstrating cGMP and thus NO dependence. In addition, the response is also independent of both endothelium and nitric oxide synthase (e.g. L-NMMA insensitive)[4] suggesting that the relaxant can directly stimulate smooth muscle cells. The release of an NO species from SNO-Hb is understood to occur via a mechanism that is essentially opposite to its formation. When oxygenated RBCs (R-state) encounter an oxygen demand, oxygen is liberated causing their haemoglobin to switch from the R-state to the T-state resulting in the simultaneous release of an NO moiety. Here we provide *ex vivo* evidence to support this *in vivo* theory in addition to data that may help to confirm the identity of this elusive species, utilising the well established organ chamber bioassay. This system provides a highly sensitive and reproducible investigation tool, allowing the relaxant capacity of RBCs (following different interventions) to be accurately determined.

2   MATERIALS AND METHODS

2.1   Aortic Ring preparation

Male New Zealand white rabbits (2 to 2.5 Kg) were terminally anaesthetised by intravenous injection of sodium pentobarbitone (0.75ml/Kg). The aorta was harvested and endothelium intact rings of thoracic aorta were prepared for isometric tension recordings. Aortic rings were mounted at 2g resting tension, 37°C in 5ml Krebs buffer (NaCl 109.0 nM, KCl 2.7 nM, KH_2PO_4 1.2nM, $MgSO_4.7H_2O$ 1.2 nM, $NaHCO_3$ 25.0 nM,

$C_6H_{12}O_6$ 11.0 nM, $CaCl_2.2H_2O$ 1.5 nM, all Fisher Scientific). Appropriate gas mixes were used to bubble gas into the bottom of the baths to achieve desired O_2 concentrations as previously described[4]. Tissue was pre-conditioned at 95% O_2 with phenylephrine (PE) 10^{-6}mol/L and Acetylcholine (Ach) 10^{-5}mol/L (both from Sigma UK, Poole). At ~1% O_2 PE 3×10^{-6}mol/L was used to induce constriction before 20μL of RBC pellet was injected into each bath.

2.2   Blood Collection and Preparation

Fresh blood samples were collected by venepucture of the antecubital vein from normal male and female subjects immediately prior to requirement. Samples were centrifuged (1200g, 5mins, 4°C) and the plasma and buffy coat were removed. RBCs were then diluted (1 in 10 to give approx. 3g/dL haemoglobin) in isotonic saline (0.9% w/v, Fresenius Kabi) before loading into the thin film rotating tonometer[5] (Figure 1) maintained at 37°C. The rotating tonometer (which is constantly purged with gas) creates a thin film of RBC in solution, providing optimum conditions for rapid gas exchange. This devise allows the manipulation of oxygen gradients between ~ 70-100% HbO_2 within a physiological circulatory transit time. A RBC sample can be fully deoxygenated (100% to ~ 20% HbO_2) within minutes. Different gas mixes were used (medical grade 95% O_2/5% CO_2 and 95%N_2/5% CO_2) along with different oxygenation-deoxygenation cycles (Figure 2). HbO_2% and haemoglobin content were measured using a blood gas analyser (OSM 3 Hemoximer, Radiometer, Copenhagen).

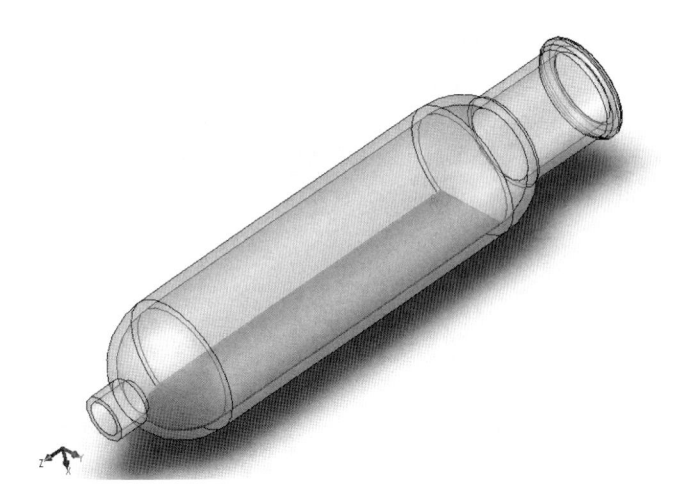

Figure 1. Diagram of the thin film rotating tonometer.

To identify whether the vasorelaxant moiety was simply passed between the RBC and the extracellular milieu upon de-oxygenation and re-oxygenation, RBC samples taken to low saturation had the extracellular milieu removed prior to re-oxygenation by repeated washing with fresh oxygenated bottled saline, prior to having saturation and haemoglobin content retested (Figure 2). Finally all samples were centrifuged (1200g, 3mins 4°C) and the saline removed before addition of RBC pellet to the aortic ring preparation (see section 3.1).

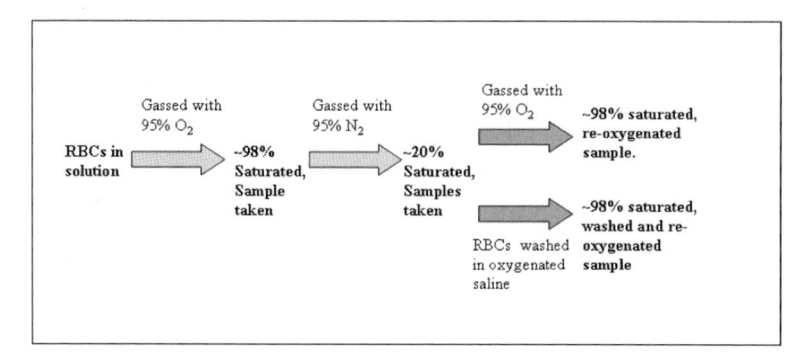

Figure 2. Flow chart of experimental RBC preparation.

2.3   Data Collection and Analysis

Isometric tension was recorded from transducers linked to a Powerlab 8Sp/octal bridge (AD Instruments) connected to a PC running 'Chart for Windows'. For each experiment, eight aortic ring preparations were run in parallel, n = 1 represents data averaged from 2 individual aortic ring preparations given identical treatment. % relaxation is the induced decrease in gram tension as a percentage of the maximum contraction.   Groups were compared with a standard t test or a one-way analysis of variance using a Tukey's multiple comparison *post hoc* test. Assessment of correlations was made using Pearson's correlation coefficient. All data are reported as mean ± standard deviation, in all cases $p < 0.05$ was considered significant.

3   RESULTS

3.1   The Effect of Saturation on RBC Induced Relaxation

Highly saturated RBCs (~ 98% HbO_2) elicited significantly greater relaxation than highly saturated RBC rapidly deoxygenated to ~ 50% and ~ 20% HbO_2 ($p < 0.01$ and $p < 0.001$, respectively; Figure 3). RBC relaxation was found to directly correlate with HbO_2 % (Figure 4), with relaxation increasing with HbO_2 ($r = 0.815$, $P<0.0001$).

3.2   The Effect of Saturation Cycling on RBC Induced Relaxation

Figure 5 demonstrates the effect of saturation cycling on RBC induced relaxation. RBCs immediately taken to a high saturation (98.22% ± 0.45 HbO_2) induced a mean relaxation of 12.00% ± 3.60. Following the de-saturation of these RBCs to ~ 20% and their return to a high saturation (97.70% ± 0.39 HbO$_2$) a mean relaxation of 8.28% ± 2.97 was observed. Although the mean relaxation induced by re-oxygenated RBCs was lower than that of RBCs with a high saturation, the trend was not statistically significant.

3.3 Potential Cross-Membrane Transfer of the Vasorelaxant

RBCs taken from a high saturation to a low saturation level, followed by removal of the extracellular milieu and re-oxygenation (washed in oxygenated saline) induced a mean relaxation of 12.97% ± 1.89 compared to RBCs taken immediately to a high saturation 12.0% ± 3.60 (Figure 6). This demonstrates that the removal of the initial extracellular milieu and replenishment with fresh saline has no significant effect on RBC induced relaxation. In addition, preliminary data suggest that the presence of NEM (a thiol blocker which prevents transnitrosation) reduces RBC induced relaxations when incubated with highly saturated RBCs (data not shown).

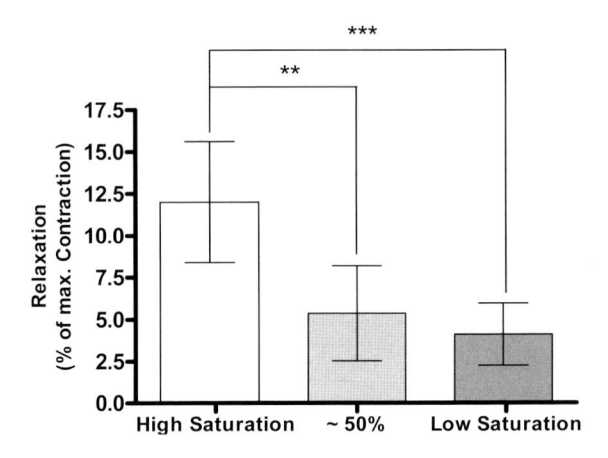

Figure 3. Red blood cell induced relaxations produced by highly saturated red cells (average saturation 98.22% ± 0.45, n=13) compared to red cells approximately 50% saturated (average saturation 51.43% ± 6.16, n=4) and to red cells of low saturation (average saturation 20.40% ± 5.28, n=13). Both the 50% and low saturation RBCs elicited a significantly lower level of relaxation than the highly saturated RBCs **P<0.01, ***P<0.001. Groups were compared by one way ANOVA and Tukey's multiple comparison *post hoc* test, bars represent standard deviation.

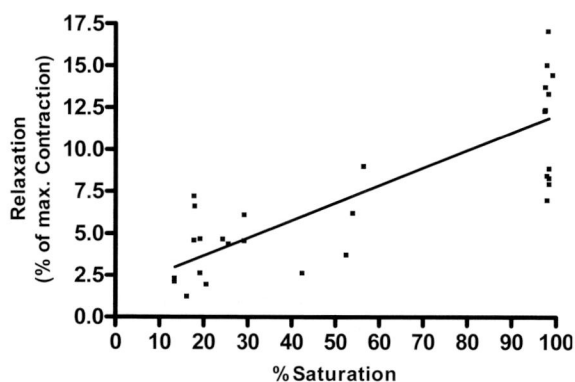

Figure 4.Plot of RBC induced relaxations vs. % RBC saturation A significant positive correlation (R=0.815, p<0.0001, n = 30, between he two variables.

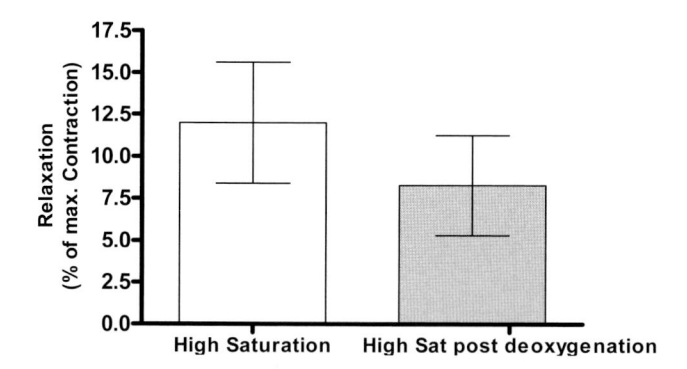

Figure 5. Red blood cell induced relaxations produced by highly saturated red cells (average saturation 98.22% ± 0.45, n=13) compared to highly saturated red cells which were previously taken down to a deoxygenated saturation of ~ 20% (average saturation 97.70% ± 0.39, n=4). No significant difference between groups, bars represent standard deviation.

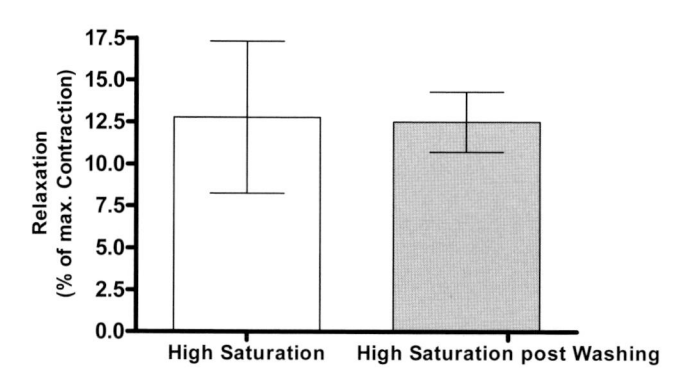

Figure 6. Red blood cell induced relaxations produced by highly saturated red cells (average saturation 98.22% ± 0.45, n=13) compared to Highly saturated red cells which were previously taken down to a deoxygenated saturation of ~ 20% then centrifuged at 1200g and had the extracellular solution removed and were then washed 3 times in oxygenated saline to bring the saturation up (average saturation 98.05% ± 0.65, n=4). No significant difference between groups, bars represent standard deviation.

4   DISCUSSION

The ability of fully oxygenated native human RBCs to induce relaxation in the hypoxic organ chamber bioassay is well known. This *ex vivo* response, demonstrated to be at least in part mediated by NO, has been attributed to several mediators within the RBC, including SNO-Hb[6], ATP[7] and/or nitrite [8, 9]. More recently however increasing evidence has supported SNO-Hb as the RBC vasorelaxant.

SNO-Hb is proposed to be relatively stable under oxygenated conditions. Only upon the switch in haemoglobin from the R-state (oxygenated) to the T-state (deoxygenated) conformation is the NO moiety purportedly made available for release. We therefore examined the effect of RBC haemoglobin saturation on the capacity of RBCs to induce relaxation in hypoxic vessels. We now demonstrate:

o   RBC induced vasorelaxation is dependent on (and directly proportional to) haemoglobin oxygen saturation. The release of oxygen must occur to allow the conformational switch to take place, releasing the vasoactive moiety contained within RBCs. The presence and liberation of oxygen from haemoglobin purely acting as a 'release switch'.

o   RBCs contain a considerable (potentially able to be replenished) store of vasorelaxant which is not released in its entirety during an individual hypoxic exposure/event. RBCs deoxygenated and subsequently re-oxygenated were in fact still able to elicit a notable relaxation. This raises the question as to why only a proportion of vasorelaxant is released upon each hypoxic exposure? We are currently investigating this further.

o   The RBC vasorelaxant is not simply passed back and forth between the RBC supernatant and the RBC following a deoxygenation-re-oxygenation cycle. Replacing the extracellular milieu (saline) had no effect upon the capacity of the RBC to relax the vessels.

o   The relaxant species released from the RBC must be capable of avoiding re-capture by or reaction with haemoglobin in the local region of release. The proposed mechanism of exit in the case of an NO moiety released from SNO-Hb is via transnitrosation reactions (thiol to thiol transfer) across the RBC membrane[10]. In this bound state the NO moiety is effectively protected from haemoglobin re-capture/reaction. We have carried out some preliminary work that supports this hypothesis. RBCs transiently incubated with the thiol blocker NEM did not appear to generate relaxations of the same magnitude as those that were not exposed. It is possible that NEM is blocking potential 'carrier' thiols on proteins located in the RBC membrane, reducing the amount of vasorelaxant reaching the tissue. At the very least this demonstrates that thiols are involved in the release mechanism of the NO moiety.

This study is the first of its kind to investigate the oxygen dependency of ex vivo blood vessel relaxation responses utilising intact RBCs. In conclusion, we provide further evidence that the degree of haemoglobin saturation is a crucial factor in the hypoxic RBC vasorelaxant response, supporting the idea that the vasorelaxant response is associated with a change from the R to the T-state conformation of the haemoglobin molecule.

5   ACKNOWLEDGEMENTS

We would like to thank Ross Pinder for his help creating figure 1 and Dr Doctor of the Department of Paediatrics and Biochemistry & Molecular Biophysics, Washington University in St Louis for the use of the thin film rotating tonometer. We also thank the British Heart Foundation for their continued support.

6   REFERENCES

1. C.E. Cooper: Nitric oxide and iron proteins. *Biochim Biophys Acta* 1999, 1411(2-3):290-309.
2. A.J .Hobbs, M.T. Gladwin, R.P. Patel, D.L. Williams, A.R. Butler: Haemoglobin: NO transporter, NO inactivator or NOne of the above? *Trends Pharmacol Sci* 2002, 23(9):406-411.
3. T.J. McMahon, R.E. Moon, B..P. Luschinger, M.S. Carraway, A.E. Stone, B.W. Stolp, A.J. Gow, J.R.Pawloski, P. Watke, D.J. Singel, *et al*: Nitric oxide in the human respiratory cycle. *Nat Med* 2002, 8(7):711-717.
4. P.E. James, D. Lang , T. Tufnell-Barret, A.B. Milsom, M.P. Frenneaux: Vasorelaxation by red blood cells and impairment in diabetes: reduced nitric oxide and oxygen delivery by glycated hemoglobin. *Circ Res* 2004, 94(7):976-983.
5. A. Doctor, R. Platt, M.L. Sheram, A. Eischeid, T. McMahon, T. Maxey, J. Doherty, M. Axelrod, J. Kline, M. Gurka *et al*: Hemoglobin conformation couples erythrocyte S-nitrosothiol content to O2 gradients. *Proc Natl Acad Sci U S A* 2005, 102(16):5709-5714.
6. J.S. Stamler, L. Jia, J.P. Eu, T.J. McMahon , I.T. Demchenko, J. Bonaventura , K. Gernert, C.A. Piantadosi: Blood flow regulation by S-nitrosohemoglobin in the physiological oxygen gradient. *Science* 1997, 276(5321):2034-2037.
7. M.L. Ellsworth: Red blood cell-derived ATP as a regulator of skeletal muscle perfusion. *Med Sci Sports Exerc* 2004, 36(1):35-41.
8. K. Cosby, K.S. Partovi, J.H. Crawford, R.P.Patel, C.D. Reiter, S. Martyr, B.K. Yang, M.A. Waclawiw, G. Zalos, X. Xu *et al*: Nitrite reduction to nitric oxide by deoxyhemoglobin vasodilates the human circulation. *Nat Med* 2003.
9. J.H. Crawford, T.S. Isbell, Z. Huang, S. Shiva, B.K. Chacko, A.N. Schechter, V.M. Darley-Usmar, J.D. Kerby, J.D. Lang, D. Jr., Kraus *et al*: Hypoxia, red blood cells, and nitrite regulate NO-dependent hypoxic vasodilation. *Blood* 2006, 107(2):566-574.
10. J.R. Pawloski, D.T. Hess, J.S. Stamler: Export by red blood cells of nitric oxide bioactivity. *Nature* 2001, 409(6820):622-626.

BLOOD VESSEL SPECIFIC VASO-ACTIVITY TO NITRITE UNDER NORMOXIC AND HYPOXIC CONDITIONS

Thomas E Ingram[*], Andrew G Pinder, Alexandra B Milsom, Stephen C Rogers, Dewi E Thomas and Philip E James

Abstract: This study uses an organ chamber bioactivity assay to characterise the direct effect of sodium nitrite upon rabbit blood vessels (aorta (Ao), inferior vena cava (IVC) and pulmonary artery (PA)) in a haemoglobin independent/variable oxygen environment.

In 95% oxygen constriction to 8g (Ao), 6g (PA) and 4g (IVC) was achieved using 1μM phenylephrine. The same constriction in 1% oxygen required 3μM. During 95% oxygen constriction was consistent and sustained for all vessels. However under 1% oxygen PA was quick to constrict but rapidly gave up this tension whereas Ao was slower to constrict but exhibited a more sustained response.

Relaxation of each vessel was assessed post constriction using 10μM sodium nitrite. Results were expressed as a percentage loss in tension compared to the maximum achieved and corrected by controls which received no nitrite. At 95% oxygen PA relaxed greater than Ao (10.04% ± 2.28% vs. 5.25% ± 1.51%). IVC response was varied (2.26% ± 9.43%). At 1% oxygen all vessels relaxed more. However the pattern was reversed with both IVC (14.20% ± 3.63%) and PA (16.55% ± 0.93%) relaxing less than Ao (42.20% ± 5.21%).

These results suggest that relatively low concentrations of sodium nitrite can vaso-dilate blood vessels. This effect is independent of haemoglobin and tissue specific.

1. INTRODUCTION

There has been much recent interest in the vasodilator activity of nitrite under different environmental conditions[1]. Nitrite has long been established as a, relatively

[*] Department of Cardiology, Wales Heart Research Institute, University of Cardiff, Heath Park, Cardiff, CF14 4XN, United Kingdom

P. Liss et al. (eds.), *Oxygen Transport to Tissue XXX*, DOI 10.1007/978-0-387-85998-9_4,
© Springer Science+Business Media, LLC 2009

weak, vasodilator when given in a standard biological setting[2, 3]. More recently it has been shown that modulation of the tissue environment into which nitrite is given can increase its vaso-activity potential. In particular when a near-physiological infusion of nitrite is given under hypoxic conditions there is a greater effect upon blood flow than when it is given during normoxia[4]. Several of the models proposed to explain this rely upon a haemoglobin dependent mechanism. It has been suggested that deoxyhaemoglobin acts as an enzymatic catalyst to promote nitric oxide formation from nitrite and therefore the resulting vasodilatation[1]. Furthermore it has been shown that the hypoxic response to nitrite of both arteries and veins in-vivo is not the same and this has been suggested to be due to different local deoxyhaemoglobin concentrations[5].

We sought to test this model by characterising the effect that sodium nitrite has upon different animal blood vessels during both normoxia and hypoxia in an in-vitro, haemoglobin independent environment. Three different types of blood vessel were examined: the abdominal aorta (Ao); inferior vena cava (IVC) and pulmonary artery (PA). A concentration of sodium nitrite was used that is sub-pharmacological but supra-physiological; in order to mimic the effective plasma levels of nitrite achieved in the previous in-vivo studies.

2. METHODS

New Zealand White rabbits (male, 2-2.5 kg, 6-8 weeks old) were terminally anaesthetised with intravenous pentobarbitone (150mg/kg). The abdominal aorta, inferior vena cava and pulmonary artery were dissected and prepared into 2mm wide rings. These were then mounted using hooks facing opposing directions, as shown in figure 1, and attached to a force transducer. 4 rings of each vessel type were hung and categorised into 2 pairs. Rings were placed into 8ml individual organ chamber bioactivity assays maintained at 37°C by surrounding warm water. Each ring was bathed in 5mls of Krebs solution and perfused with 95% O_2. After 30 minutes equilibration time a basal tension of 2g was set. Next three successive constriction / relaxation cycles were performed using a concentration of 1μM phenylephrine (PE) and then 10μM acetylcholine. The bath was emptied and washed out four times over a 30 minute period between each cycle.

After the rings had been prepared as outlined above they were again constricted using the same concentration of PE. Subsequently one pair of rings within each vessel group received 10μM sodium nitrite and the other pair had no additions. The resulting relaxed tensions at 20 minutes were measured. Each bath was then washed four times as described earlier.

In order to create hypoxic conditions the perfusing gas was changed from 95% oxygen to a mixture of 95%N_2/5% CO_2. Given that each organ chamber was exposed at its surface to room air this resulted in an effective local tissue environment of 1% oxygen. After a 10 minute equilibration period constriction was again performed but this time using 3μM PE. The increased dose was necessary in order to achieve the same quantitative level of constriction as that gained during normoxia. 10μM sodium nitrite was added as during normoxia to half of the rings and the subsequent relaxed tension at 20 minutes was recorded.

The relaxation response of each vessel to nitrite was then expressed as a percentage of the maximal tension achieved, averaged between each pair and corrected for the corresponding control rings. Five experiments were performed and analysis of results was done using an unpaired t-test; expressed ± the standard error of the mean.

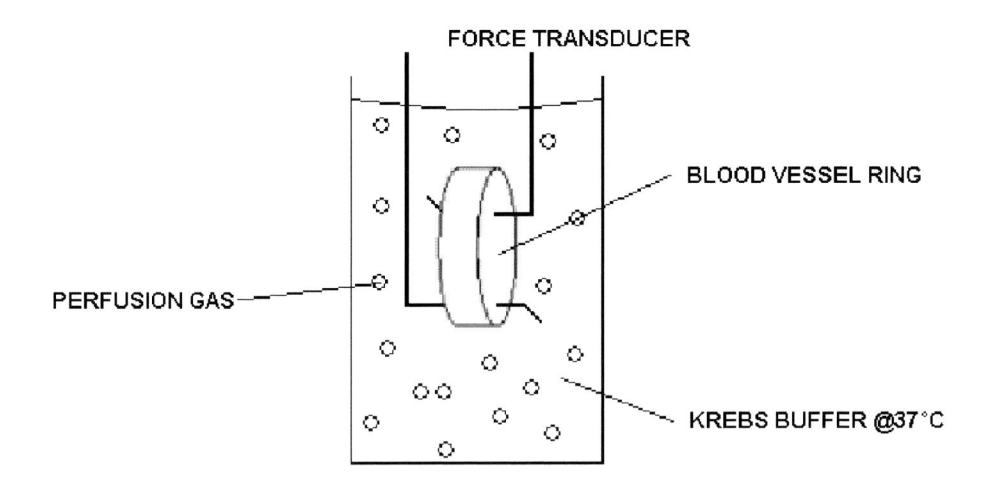

Figure 1. Diagram of the organ chamber bioactivity assay

3. RESULTS

3.1 95% oxygen

In the 95% oxygen perfused buffer constriction to 8g (Ao), 6g (PA) and 4g (IVC) was rapidly achieved using PE. This constriction was sustained once peaked, with little drop off in tension noted in the control vessels during the course of the protocol. Figure 2 displays the results observed. The aorta relaxed by 5.25% ± 1.51% (n=5) with the addition of nitrite. PA relaxation was greater at 10.04% ± 2.28% (n=5) though the difference with aorta was not significant. The IVC response was variable at 2.26% ± 9.43% (n=4). This reflected both technical difficulty with handling of the tissue and the fact that the small absolute tension changes which are inducible in veins result in a greater margin of error in the subsequent responses observed.

3.2 1% oxygen

In a 1% oxygen environment the same absolute values of constriction were achieved using the increased concentration of PE. The time to peak constriction of the PA was faster than that of the Ao however once achieved the PA rapidly lost this tension whereas the Ao exhibited a more sustained constrictor response.

Relaxation corrected for controls receiving no nitrite was 42.20% ± 5.21% (n=5) for the Ao. This was significantly greater than the relaxation seen in both the PA (16.55% ± 0.93%, n=5, p<0.001) and IVC (14.20% ± 3.63%, n=4, p<0.001)

Within vessel groups both the Ao and PA showed a significantly increased potential to dilate in response to nitrite during hypoxia compared to normoxia (p<0.05 for both). However hypoxia caused an eight-fold increase in relaxation of the Ao but only a 60% increased in relaxation of the PA.

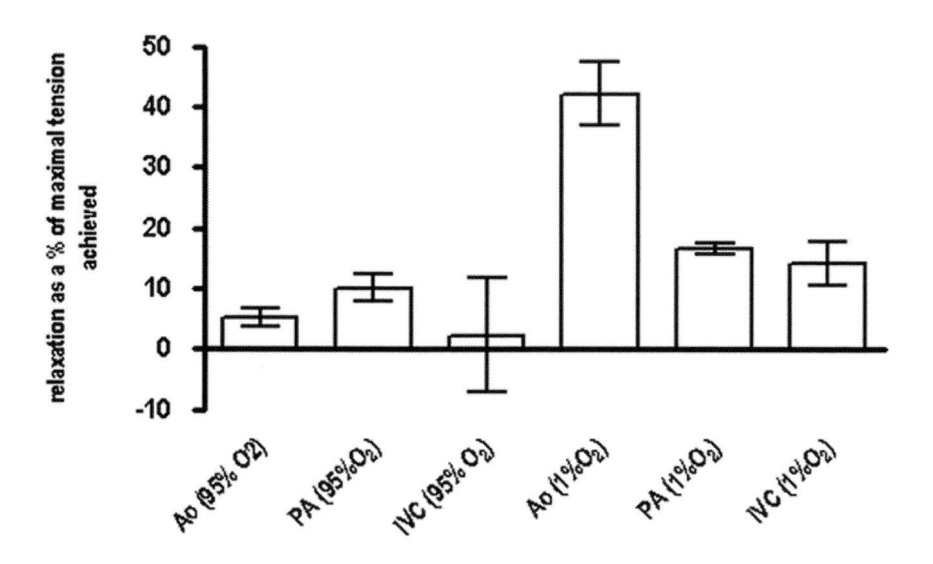

Figure 2. Relaxation of the Ao, PA and IVC in response to 10μM sodium nitrite. Results are expressed as a percentage relaxation of the total maximal constriction initially achieved in response to PE ± SEM for the 5 experiments.

4. DISCUSSION

The basal plasma concentration of nitrite lies in the range of 100-700nmol. The approximately ten-fold increased level of nitrite used in this study was designed to be a sub-pharmacological dose but one which mimics the plasma concentrations of nitrite that have been shown in previous studies to be capable of causing hypoxic vasodilatation.

This experiment suggests that sodium nitrite has a similar action upon blood vessels to those observed in-vivo but in an environment which is independent of haemoglobin. In all vessels hypoxia resulted in an enhanced vasodilatation potential of nitrite. The magnitude of the effect that hypoxia had was much greater in the Ao than the PA. These results suggest that blood vessels become more responsive to the dilator action of sodium nitrite during hypoxia, but that this change in vascular responsiveness is markedly different in the Ao compared to the PA. One potential explanation for this discrepancy is that as nitrite diffuses into each blood vessel it is reduced to nitric oxide by the acidic environment present within the hypoxic musculature. This would explain the difference observed between the Ao and PA/IVC as the former has a greater muscle mass.

In summary any attempts to explain the in-vivo effects of nitrite upon hypoxic blood vessels need to take into account the primary effect observed in this study which hypoxia has upon vascular responsiveness.

5. REFERENCES

1. J. O. Lundberg and E. Weitzberg, NO Generation From Nitrite and Its Role in Vascular Control, *Arterioscler Thromb Vasc Biol.* **25**, 915-922 (2005).
2. E. Reichart and S. Mitchell, On the physiological action of potassium nitrite, *Amer J Med Sci.* **154**, 158-180 (1880).
3. R. F. Furchgott and S. Bhadrakom, Reactions of strips of rabbit aorta to epinephrine, isopropylarterenol, sodium nitrite and other drugs, *J Pharmacol ExpTher.* **108**, 129-143 (1953).
4. K. Cosby and M. T. Gladwin, Nitrite reduction to nitric oxide by deoxyhemoglobin vasodilates the human circulation, *Nat med.* **9**, 1498-1505 (2003).
5. A. R. Maher, P. E. James and M. P. Frenneaux, Hypoxic modulation of exogenous nitrite induced vasodilation in man, *In press* (2007).

NITRITE-INDUCED IMPROVED BLOOD CIRCULATION ASSOCIATED WITH AN INCREASE IN A POOL OF RBC-NO WITH NO BIOACTIVITY

Joseph M. Rifkind[1*], Enika Nagababu[1], Zeling Cao[1], Efrat Barbiro-Michaely[2], and Avraham Mayevsky[2]

Abstract: The reduction of nitrite by RBCs producing NO can play a role in regulating vascular tone. This hypothesis was investigated in rats by measuring the effect of nitrite infusion on mean arterial blood pressure (MAP), cerebral blood flow (CBF) and cerebrovascular resistance (CVR) in conjunction with the accumulation of RBC-NO. The nitrite infusion reversed the increase in MAP and decrease in CBF produced by L-NAME inhibition of e-NOS. At the same time there was a dramatic increase in RBC-NO. Correlations of RBC-NO for individual rats support a role for the regulation of vascular tone by this pool of NO. Furthermore, data obtained prior to treatment with L-NAME or nitrite are consistent with a contribution of RBC reduced nitrite in regulating vascular tone even under normal conditions. The role of the RBC in delivering NO to the vasculature was explained by the accumulation of a pool of bioactive NO in the RBC when nitrite is reduced by deoxygenated hemoglobin chains. A comparison of R and T state hemoglobin demonstrated a potential mechanism for the release of this NO in the T-state present at reduced oxygen pressures when blood enters the microcirculation. Coupled with enhanced hemoglobin binding to the membrane under these conditions the NO can be released to the vasculature.

[1] *Corresponding author, email: rifkindj@mail.nih.gov. [1]Molecular Dynamics Section, National Institute of Aging, 5600 Nathan Shock Drive Baltimore, MD 21224.[2]The Mina and Everard Goodman Faculty of Life Sciences, Bar-Ilan University Ramat Gan, Israel.

P. Liss et al. (eds.), *Oxygen Transport to Tissue XXX*, DOI 10.1007/978-0-387-85998-9_5,
© Springer Science+Business Media, LLC 2009

1. INTRODUCTION

Nitric oxide (NO) synthesized from L-arginine by endothelial nitric oxide synthase plays an important role in regulating vascular tone and maintaining blood flow [1]. NO has a short half-life and needs to be produced/released in close proximity to the site where it is needed. Recent data report that the reduction of nitrite to NO may play an important role in supplying NO for the regulation of blood flow [2,3].

Although mammals do not possess enzymes specifically designed for nitrite reduction, it has been shown that nitrite is reduced to NO in mammals. A physiological role for RBC nitrite reduction in regulating blood flow is supported by the observation that nitrite infusion results in an increase in RBC-NO in conjunction with improved forearm blood flow, inhibition of hypoxic pulmonary vasoconstriction and in the prevention of delayed cerebral vasospasm [3-5]. In this study we have again confirmed a role for the RBC in delivering NO to the vasculature. Cerebral blood flow (CBF) and mean arterial blood pressure (MAP) were measured in a group of rats prior to any treatment, after injection of L-NAME to inhibit nitric oxide synthase and during the subsequent infusion of nitrite. The potential involvement of the RBC in the observed vascular changes [6] was indicated by a dramatic increase in the levels of RBC-NO and their correlation with the observed physiological changes when results from different rats were compared.

The ability to release NO from the RBC, however, requires that the NO formed by nitrite reduction is not scavenged by the well studied rapid reactions with both oxyhemoglobin (oxyHb) and deoxyhemoglobin (deoxyHb) before it can be released to the vasculature. In earlier studies we have shown that during nitrite reduction a metastable delocalized intermediate is formed [2,7] that makes it possible to build up a pool of potentially bioactive NO in the RBC that is not scavenged by deoxyHb or oxyHb. We have now extended theses studies to show how this NO can be released to the vasculature at low oxygen pressures when hemoglobin (Hb) assumes the T-unliganded conformation that has an increased affinity for the RBC membrane.

2. MATERIALS AND METHODS

2.1. Animal Preparation and Experimental Design

All experiments were performed in accordance with the Animal Care Committee of Bar-Ilan University. A total of 8 male Wistar rats of 250-300g (2.5-3 months old) were used. Two of these rats were controls, which were not injected with L-NAME or infused with nitrite. The rats were anesthetized with Equithesin (0.3 ml/100gr body weight, i.p.) (each ml contains: pentobarbital 9.72 mg, chloral hydrate 42.51 mg, magnesium sulfate 21.25 mg, propylene glycol 44.34 % w/v, alcohol 11.5 % and distilled water). During the duration of the experiment steady anesthesia was maintained by 0.1ml Equithesin injection every 30 minutes. Additionally, the rats were placed on a warming tray and body temperature was maintained at $37^{\circ}C$. Polyethylene catheters were introduced into the femoral vein for drug administration and into the femoral artery for systemic MAP monitoring. The jugular vein was also cannulated for obtaining blood samples.

After a period of stabilization, a control sample of 0.6 ml blood was withdrawn and L-NAME (50mg/kg body wt. I.V.) was injected. Approximately 10 min after L-NAME administration (when MAP became stabilized at an elevated value) a second sample of blood was taken followed by a 10 min period of nitrite infusion. The rate of $NaNO_2$

infusion was 1 μmol/kg body wt/min. At the end of the infusion period a third sample of blood was taken. CBF and MAP were continuously monitored through the entire experimental period. For the control animals the injections of saline and the withdrawal of blood had no effect on the CBF or MAP.

2.2. Cerebral Blood Flow

CBF was monitored by laser Doppler flowmetry (LDF), which measures relative changes in microcirculatory blood flow (0-100% range) [8]. To attach the Doppler probe to the brain cortex, the rat was placed on an operation table with a special mouth holder. A 2 mm diameter hole was drilled in the parietal bone, and the bone was removed (the dura matter remained intact). The Doppler probe, which uses light of 632.8 nm, was then placed on the brain cortex using a special micromanipulator and fixated by dental acrylic cement [9,10]. The relative changes in cerebrovascular resistance (CVR) during nitrite infusion was determined from MAP/CBF [11].

2.3. Analysis of Red Cell NO

Blood samples were immediately centrifuged at 3000 rpm for 10 min and the plasma removed. The RBC pellet was stored in liquid nitrogen without further washing. NO was determined by a Nitric Oxide Analyzer (Sievers model 280) on thawed hemolysate as described earlier [2]. Total RBC-NO was determined by lysing the cells in 4 volumes of deoxygenated distilled water in a septum sealed cuvette. Oxygen stable NO was determined by lysing the cells in 4 volumes of distilled water. 100 μl of the sample was injected into the purge vessel containing 5.5 ml glacial acetic acid, 20mM sulfanilamide (to react with any free nitrite present) and 100mM potassium ferricyanide (to release NO tightly bound to Fe(II) Hb). While sulfanilamide reacts with free nitrite, it may not react with nitrite bound to deoxyHb or methemoglobin (metHb). The released NO was flushed through the NO chemiluminescence analyzer and quantitated.

2.4. Preparation of Hemoglobin

Hb was prepared from fresh RBCs as described earlier [2]. Cleavage of the terminal histidine and tyrosine in the β-chains of Hb was accomplished by digesting oxygenated Hb with carboxypeptidase A obtained from Sigma Chemical Co. at pH 7.4 for 30 min [12].

Hb samples in 50mM NaCl and 4mM phosphate buffer, pH 7.4 were deoxygenated in an anaerobic Coy Laboratory glove box. The glove box uses hydrogen and a palladium catalyst to remove any residual oxygen. The oxygen level in the Coy box was < 1ppm. The sample was placed in a septum sealed cuvette in the glove box and then removed for spectral analysis. The total Hb concentrations in the samples were ~100μM. Stock solutions of nitrite in the same buffer were also deoxygenated in the glove box. The reaction was initiated by using a gas tight syringe to add nitrite to the sealed cuvette containing Hb.

The spectra of Hb were recorded from 490 nm to 640 nm on a Perkin Elmer Lambda 35 spectrophotometer for 35 min at a 1:1 molar ratio of nitrite to heme. Parent spectra of deoxyHb, metHb, Hb(II)NO, and nitrite bound metHb were prepared. The series of spectra obtained in any experiment were analyzed using a least squares multi-component fitting program (Perkin Elmer Spectrum Quant C v 4.51) including these 4 spectra.

2.5. Statistical Analysis

Origin 6.1 (Microcal Software, Northampton, MA) was used for analysis of the data. The paired Student's t test was used for comparing samples before treatment with samples after L-NAME injection and after nitrite infusion. Linear regression analysis was used to compare parameters for the different rats studied. Two-tailed values of p<0.05 were considered statistically significant.

3. RESULTS

3.1. The contribution of RBCs to Nitrite induced Vasoactivity in Rats

The effect of nitrite infusion on vasoactivity was obtained by investigating the effect of 10-min nitrite infusion after e-NOS was inhibited by L-NAME [6]. MAP measured in the femoral artery was used as a measure of systemic vasoactivity. The inhibition of NO synthesis by L-NAME produced a very significant (p<0.001) almost 100% increase in MAP that was completely reversed by nitrite infusion. The changes in relative CBF and relative CVR, which reflect the highly regulated cerebral circulation, were less pronounced. L-NAME decreased CBF and increased CVR, but the changes were not significant. However, the increase in CBF and decrease in CVR obtained when nitrite was infused after L-NAME were significant (p<0.05).

After nitrite infusion (Fig. 1) there is a 20 fold increase in oxygen stable RBC-NO (p<0.0001) and a 40 fold increase in total RBC-NO (p<0.0001) indicating that a major fraction of the infused nitrite is taken up by the RBC and reduced to a potentially

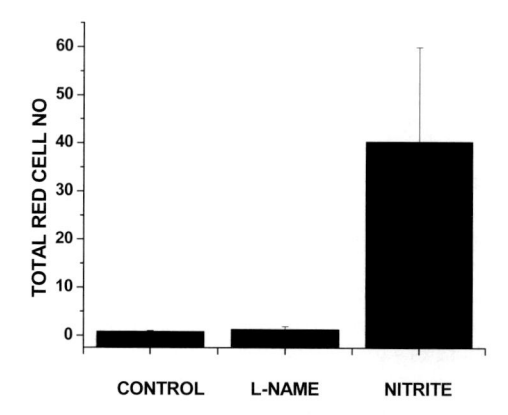

Figure 1. Total RBC-NO before any treatment, after injection of L-NAME and after infusion with nitrite. Error bars are the SE of the mean.

bioactive form of NO. A role for this NO in regulating vascular tone is supported by the correlation between the level of oxygen stable RBC-NO in each rat and both MAP (Fig. 2A; r = -0.83; p<0.05), as a measure of peripheral vasoactivity, and CVR(Fig. 2B; r = -0.843; p<0.05) , as a measure of cerebral vasoactivity.

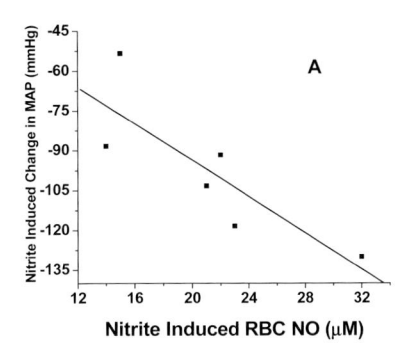 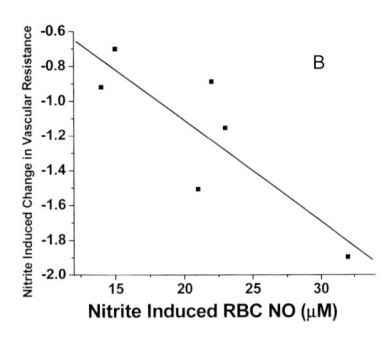

Figure 2. Relationship with oxygen stable RBC-NO after nitrite infusion. (A), mean arterial blood pressure (MAP); (B), cerebrovascular resistance (CVR)

3.2. Release of Bioactive NO from the RBC

In the RBC, NO reacts very rapidly with oxyHb forming nitrate.

$$Hb(II)O_2 + NO \rightarrow Hb(III) + NO_3^- \qquad (1)$$

And with deoxyHb producing Hb(II)NO [13].

$$Hb(II) + NO \rightarrow Hb(II)NO \qquad (2)$$

A role for RBC-NO in the regulation of vascular tone, however, requires that NO bioactivity be released from the RBC without being scavenged by Hb. In earlier studies we have shown that during nitrite reduction intermediates including a metastable delocalized species are formed that makes it possible to build up a pool of potentially bioactive NO in the RBC that is not scavenged by deoxyHb or oxyHb [2,7]. We have recently developed a method to quantitate the intermediates using visible spectroscopy. This procedure involves a comparison of the consumption of deoxyHb with the formation metHb and nitrosylhemoglobin, the final products in the nitrite reduction process.

$$2Hb + NO_2^- + H^+ \rightarrow Intermediates \rightarrow Hb(III) + Hb(II)NO + OH^- \qquad (3)$$

The concentration of intermediates is then given by the equation

$$[intermediates] = [decrease\ in\ deoxyHb] - 2 * [Hb(II)NO] \qquad (4)$$

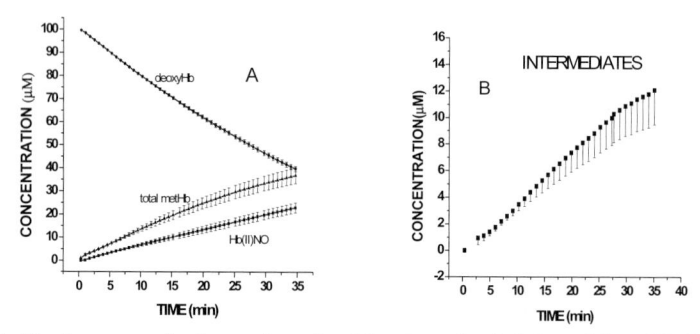

Figure 3. A: The time course for the reaction of a 1:1 molar ratio of nitrite and deoxyHb. (●), deoxyHb; (■), Hb(II)NO; (▲), Total metHb (Hb(III) + Hb(III)NO$_2^-$); B: the formation of intermediates from eq. 4.

From the time course (Fig. 3) for the consumption of deoxyHb and the formation of Hb(II)NO and the concentration of the intermediates the rate constants for the initial formation of intermediates (k_1) and for the release of NO from the intermediate (k_2) can be calculated.

$$-\{d[Hb(II)]/dt\}_0 = k_1[Hb(II)]_0[NO2^-]_0 \qquad (5)$$

$$\{d[Hb(II)NO]/dt\}_{final} = k_2[intermediate]_{final} \qquad (6)$$

The release of NO from the RBC requires that the release of NO from the intermediate is coupled to the transport of the NO out of the RBC. To investigate a possible mechanism for such a process we have investigated the effect of the Hb quaternary conformation that is involved in cooperative uptake and release of oxygen, on the nitrite reaction. Fully deoxygenated Hb is in the low oxygen affinity T-state. By removing the histidine and tyrosine on the amino terminus of the Hb β-chains using carboxypeptidase A Hb remains in the high oxygen affinity R-state even when fully deoxygenated. We then compared the nitrite reaction for fully deoxygenated T and R state Hb. The results shown in Figure 4 indicate that while the formation of the intermediate is faster in the R-state, the release of NO from the intermediate is faster in the T-state.

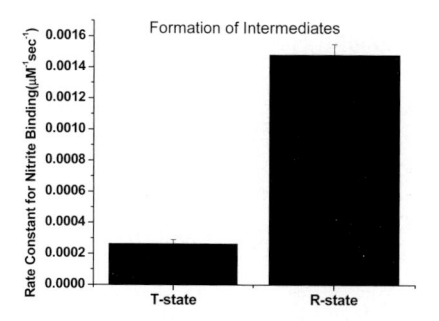
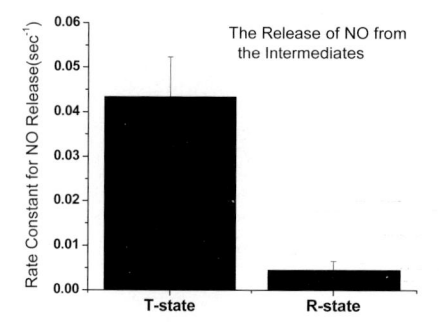

Figure 4. The effect of the quaternary T and R states on the formation of the intermediates and the release of NO from the intermediates.

4. DISCUSSION

The rat perfusion studies demonstrate that nitrite has a vasodilatory effect [6]. A role for RBC nitrite reduction in this process is suggested by the dramatic increase in RBC-NO after nitrite infusion. The correlation of the changes in MAP and CVR for each rat studied and the level of RBC-NO accumulation further support this contention. In fact correlation of basal RBC-NO with levels of MAP suggest that, even under normal conditions, RBC reduction of the plasma nitrite generated from NO oxidation also plays a role in regulating vascular tone [6].

A proposal that the RBC nitrite reduction to NO regulates vascular tone requires that the reduced nitrite is not immediately scavenged by the high concentrations of oxyHb and/or deoxyHb always present in the RBC. To explain this process we have studied the formation of the metastable delocalized species that retains reduced nitrite in a form that does not react with Hb. This intermediate represents a pool of potentially bioactive NO that can accumulate in the RBC. Our studies of R and T state Hb and the associated changes in the uptake and release of NO (Fig. 4) provide a model to explain the role of the RBC in delivering NO to the vasculature. As the blood enters the arterioles and begins to release oxygen, Hb retains the R-state but has unliganded subunits that can react with nitrite. This results in the accumulation of the bioactive intermediate. As more oxygen is removed and the quaternary conformation shifts to the T-state NO is released from this intermediate. Since T-state hemoglobin has a higher affinity for the RBC membrane [14], some fraction of this NO is released while Hb is bound to the membrane. This NO can diffuse through the membrane before being scavenged by Hb. Once out of the cell it can diffuse to the vasculature and react with guanylate cyclase causing vasodilation.

Additional studies are necessary to demonstrate to what extent this mechanism contributes to the nitrite induced vasodilatory. However, the contribution of the nitrite induced vasodilatory effects and their coupling with RBC-NO strongly suggest that the pathway described for the transfer of NO-bioactivity from the RBC to the vasculature needs to be considered.

5. ACKNOWLEDGMENTS

This research was supported in part by the Intramural Research Program of the NIH, National Institute on Aging. The help of Joy G. Mohanty in preparing this manuscript is acknowledged.

6. REFERENCES

1. L. J. Ignarro, G. Cirino, A. Casini and C. Napoli, Nitric oxide as a signaling molecule in the vascular system: an overview, J Cardiovasc Pharmacol, 34(6), 879-886 (1999).
2. E. Nagababu, S. Ramasamy, D. R. Abernethy and J. M. Rifkind, Active nitric oxide produced in the red cell under hypoxic conditions by deoxyhemoglobin-mediated nitrite reduction, J Biol Chem, 278 (47), 46349-46356 (2003).
3. K. Cosby, K. S. Partovi, J. H. Crawford, R. P. Patel, C. D. Reiter, S. Martyr, B. K. Yang, M. A. Waclawiw, G. Zalos, X. Xu, K. T. Huang, H. Shields, D. B. Kim-Shapiro, A. N. Schechter, R. O. 3. Cannon and M. T. Gladwin, Nitrite reduction to nitric oxide by deoxyhemoglobin vasodilates the human circulation, Nat Med, 9(12), 1498-1505 (2003).

4.   R. M. Pluta, A. Dejam, G. Grimes, M. T. Gladwin and E. H. Oldfield, Nitrite infusions to prevent delayed cerebral vasospasm in a primate model of subarachnoid hemorrhage, Jama, 293(12), 1477-1484 (2005).

5.   J. H. Crawford, T. S. Isbell, Z. Huang, S. Shiva, B. K. Chacko, A. N. Schechter, V. M. Darley-Usmar, J. D. Kerby, J. D. Lang, Jr., D. Kraus, C. Ho, M. T. Gladwin and R. P. Patel, Hypoxia, red blood cells, and nitrite regulate NO-dependent hypoxic vasodilation, Blood, 107(2), 566-574 (2006).

6.   J. M. Rifkind, E. Nagababu, E. Barbiro-Michaely, S. Ramasamy, R. M. Pluta and A. Mayevsky, Nitrite infusion increases cerebral blood flow and decreases mean arterial blood pressure in rats: a role for red cell NO, Nitric Oxide, 16(4), 448-456 (2007).

7.   E. Nagababu, S. Ramasamy and J. M. Rifkind, S-nitrosohemoglobin: A mechanism for its formation in conjuction with nitrite reduction by deoxyhemoglobin., Nitric Oxide, (In press) (2006).

8.   K. R. S. Wadhwani, Blood flow in the central and peripheral nervous systems., In: Lase Doppler Blood Flowmetry, edited by Shephard AP and Oberg PA. Boston: Kulvar Academic Pub., 265-304 (1996).

9.   A. Mayevsky and B. Chance, Intracellular oxidation-reduction state measured in situ by a multichannel fiber-optic surface fluorometer, Science, 217(4559), 537-540 (1982).

10.   A. Mayevsky, Brain NADH redox state monitored in vivo by fiber optic surface fluorometry, Brain Res, 319(1), 49-68 (1984).

11.   K. Chida, M. Miyagawa, W. Usui, H. Kawamura, T. Takasu and K. Kanmatsuse, Effects of chemical stimulation of the rostral and caudal ventrolateral medulla on cerebral and renal microcirculation in rats, J Auton Nerv Syst, 51(1), 77-84 (1995).

12.   M. Brunori, E. Antonini, J. Wyman, R. Zito, J. F. Taylor and A. Rossi-Fanelli, Studies on the Oxidation-Reduction Potentials of Heme Proteins. Ii. Carboxypeptidase Digests of Human Hemoglobin, J Biol Chem, 239, 2340-2344 (1964).

13.   R. F. Eich, T. Li, D. D. Lemon, D. H. Doherty, S. R. Curry, J. F. Aitken, A. J. Mathews, K. A. Johnson, R. D. Smith, G. N. Phillips, Jr. and J. S. Olson, Mechanism of NO-induced oxidation of myoglobin and hemoglobin, Biochemistry, 35 (22), 6976-6983 (1996).

14.   A. Tsuneshige, K. Imai and I. Tyuma, The binding of hemoglobin to red cell membrane lowers its oxygen affinity, J Biochem (Tokyo), 101(3), 695-704 (1987).

THE CONTROL OF OXIDATIVE PHOSPHORYLATION IN THE ADRENAL GLAND (Y1) CELL LINE

James E.J. Murphy[1] and Richard K. Porter [1]*

Abstract: We determined the proportion of oxygen consumption due to oxidative phosphorylation by mitochondria in an adrenal gland cell line (Y1 cells). In addition we determined the relative proportion of in situ mitochondrial oxygen consumption attributable to (i) proton leak and (ii) ATP turnover in these cells. This approach allowed use of top-down elasticity analysis to determine control of oxidative phosphorylation by mitochondrial (a) proton leak flux (b) substrate oxidation flux and (c) ATP turnover flux, as a function of changes in *in situ* mitochondrial membrane potential. Our data show that resting oxygen consumptions rates of Y1 cells to be 87 ± 7 nmolO/min/10^7 cells of which $38\pm 3\%$ was not due to oxidative phosphorylation. We demonstrated that mitochondrial proton leak accounted for $7\pm3\%$ of total cellular oxygen consumption or $12\pm6\%$ of resting mitochondrial oxygen consumption, with ATP turnover accounting for $55\pm3\%$ of total cellular oxygen consumption or $78\pm6\%$ of mitochondrial oxygen consumption. Control of resting mitochondrial oxygen consumption in Y1 cells was shared by (a) substrate oxidation flux ($37\pm8\%$), (b) proton leak flux ($15\pm8\%$) and (c) ATP turnover ($56\pm8\%$). Our data demonstrate, for the first time, that the majority of oxygen consumption by resting Y1 cells is due to oxidative phosphorylation.

1. INTRODUCTION

The majority of oxygen consumed in mammalian cells is due to the synthesis of ATP. Oxidative phosphorylation is the process by which the majority of ATP is synthesised and occurs within and across the mitochondrial inner membrane. A proton electrochemical gradient (Δp), is established across the mitochondrial inner membrane by proton translocation as a result of reducing equivalents (*e.g.* NADH$_2$) delivering electrons to the proton pumps/translocators within the inner membrane. The major component of Δp is the electrochemical gradient ($\Delta\Psi$) and the ultimate acceptor of the

[1] School of Biochemistry and Immunology, Trinity College Dublin, Dublin 2. Ireland.   * rkporter@tcd.ie

P. Liss et al. (eds.), *Oxygen Transport to Tissue XXX*, DOI 10.1007/978-0-387-85998-9_6,
© Springer Science+Business Media, LLC 2009

electrons is oxygen which is converted to water[1,2]. However, ATP synthesis is not perfectly coupled to oxygen consumption due to a phenomenon called proton leak across the inner membrane. Mitochondrial proton leak accounts for ~20% of primary hepatocyte[3-5] and thymocyte[6] oxygen consumption rates.

Cell lines and cancerous cells are highly glycolytic even when present in an oxygen replete environment (reviewed by Pedersen[7]). The role of mitochondria in ATP supply in cancer cells has never been quantified although one would presume variability depending on cancer cell type. In addition, cancer cells are known to have increased expression and activity of monooxygenase activity (reviewed in Bruno and Najr[8]). We set out to determine (a) the proportion of resting oxygen consumption rate that was due to oxidative phosphorylation by mitochondria in an adrenal gland cell line (Y1 cells). In addition, (b) we set out to determine the relative proportion of in situ mitochondrial oxygen consumption attributable to (i) proton leak and (ii) ATP turnover in these cells. This approach allowed use of top-down elasticity analysis to determine control of oxidative phosphorylation by mitochondrial (i) proton leak flux (ii) substrate oxidation flux and (iii) ATP turnover flux, as a result of changes in situ mitochondrial membrane potential.

2. MATERIALS & METHODS

2.1. Maintenance and Storage of Cells

The cell line used in this study was Y1 adrenocortical cells (Y1 cells). Y1 cells are derived from a mouse adrenal tumour (Cuprak and Sato[9]). Y1 cells were grown in "complete medium" which contained Dulbecco's Modified Eagles' medium (DMEM) containing 10% (w/v) fetal calf serum (FCS), gentamycin (100mg/ml) and supplemented with L-glutamine (2mM). Stock cultures were maintained at 37^0C in a humidified atmosphere with 5% CO_2, in $175cm^2$ tissue culture flasks. The cells were passaged twice weekly and removed from the surface of the flasks by incubation with 0.025% (w/v) trypsin in DMEM for 1-5 min. Once adherent cells were removed, 10ml of complete medium was added to dilute the trypsin. The cells were triturated with a 10ml pipette to obtain a single cell suspension and an aliquot ($1-2 \times 10^6$ cells) reseeded into tissue culture flasks with a total volume of 30ml of complete medium. Experiments were performed on cells of passage number between 21 and 36. Cells were stored in liquid nitrogen in cryostat tubes (1.8ml capacity) in 1ml aliquots of approximately 2×10^6 cells in complete medium containing 10% (v/v) dimethylsulphoxide.

2.2. Preparation of Y1 Cells for Oxygen Electrode

Cell suspensions were pelleted following centrifugation for 5min at 1000xg. Cells were resuspended in 3ml DMEM without FCS, gentamycin or L-glutamine. Cell concentration was determined using a haemocytometer and only suspensions with viability (as determined using trypan blue exclusion) of >90% were used. Suspensions were diluted with DMEM to a concentration of 4.5×10^6 viable cells per ml and cells were maintained on ice at 5% CO_2 and 95% air.

2.3. Trypan Blue Viability Stain

An aliquot of freshly trypsinized cells, triturated to obtain a single cell suspension, was mixed with and equal volume of stain. The stain consisted of freshly prepared 4.25% (w/v) saline and 0.2% (w/v) trypan blue. Cell viability (*i.e.* the ability to exclude trypan blue) was determined for each preparation on a haemocytometer and a Leica MDIL light microscope.

2.4. Measurement of Cell Oxygen Consumption and Mitochondrial Membrane Potential [4, 5, 10, 11]

Y1 cells (1.35×10^7) were incubated at $37°C$ in 3ml of DMEM containing $0.1\mu M$ methyltriphenylphosphonium ($TPMP^+$) and $1\mu Ci/ml$ [^3H]- $TPMP^+$ (without FCS, gentamycin or L-glutamine). Cells were incubated for 15min in a shaking water-bath prior to any additions, to allow re-establishment of ion gradients. Additions were made after 15min and cells were incubated for a further 40min to allow $TPMP^+$ to come to equilibrium with the electrochemical gradient across membranes. Substrate oxidation kinetics were determined by simultaneous measurement of oxygen consumption and membrane potential across the inner membrane in the presence of increasing amounts of oligomycin ($0-0.3\mu g/ml$). Proton leak kinetics were determined by simultaneous measurement of oxygen consumption and membrane potential in the presence of excess oligomycin ($0.3\mu g/ml$) and increasing amounts of myxothiazol ($0-0.2\mu g/ml$). ATP turnover kinetics were determined by simultaneous measurement of oxygen consumption and membrane potential in the presence of increasing amounts of myxothiazol ($0-0.01\mu g/ml$), and then correcting for oxygen consumption due to proton leak. Oxygen consumption was measured using a Clark-type oxygen electrode. Membrane potential across the mitochondrial inner membrane was measured as described in Porter and Brand[5] and Joyce et al.[11], except plasma membrane potential was not determined directly. [^3H]-$TPMP^+$ accumulation due to the plasma membrane potential and non-specific binding was defined as the $TPMP^+$ accumulation in the presence of excess oligomycin ($0.3\mu g/ml$), myxothiazol ($0.2\mu g/ml$) and FCCP ($1\mu M$).

2. 5. Top-Down Elasticity Analysis

Top-down elasticity analysis was performed as described by Brown et al.[4], Porter and Brand[5], and Joyce et al.[11], Hafner et al.[12], Brown et al.[13]. Essentially we determined the effect of varying the mitochondrial proton electrochemical gradient (in this instance measuring the mitochondrial membrane potential) on the flux through each of the three defined systems, namely, (a) substrate oxidation flux, (b) mitochondrial proton leak flux and (c) mitochondrial ATP turnover for resting (non-stimulated) Y1 cells.

2.6. Materials

[^{3}H]-TPMP$^+$ was purchased from GE-Amersham International (Amersham, Bucks., UK). All other chemicals were purchased from Sigma-Aldrich (Tallaght, Dublin) unless stated in the text. Y1 cells were from the European Collection of Cell Cultures [ECACC 85062807].

3. RESULTS

Figure 1A shows the classic raw data showing kinetic profiles for in situ mitochondrial (a) substrate oxidation, (b) proton leak and (c) ATP turnover (uncorrected for proton leak) previously observed in primary cells[4,5,14]. Calculations from Figure 1A show that oxygen consumption rates for Y1 cells were 87±7 nmolO/min/10^7 cells of which 33.5±2.5 nmolO/min/10^7 (38±3%) was a result of oxygen consumption not due to oxidative phosphorylation, the remaining 54.2±8 nmolO/min/10^7 (62±15%) clearly being due to mitochondrial oxidative phosphorylation.

Figure 1B shows the kinetics of the (a) substrate oxidation system (b) proton leak and (c) ATP turnover (corrected for proton leak), all of which have been corrected for oxygen consumption not due to oxidative phosphorylation. The kinetics of the substrate oxidation system was determined by titrating the resting Y1 cells with oligomycin and hence by decreasing ATP-turnover we could measure the effect of increasing mitochondrial membrane potential on oxygen consumption by in situ mitochondria (Fig. 1B). Consequently, the resting mitochondrial membrane potential increases, as expected, from 110±5 to 120±2.5mV, while resting oxygen consumption rates decrease, as expected, from 54.2±8 to 11.5±2.5 nmolO/min/10^7 in non-phosphorylating mitochondria following titration with oligomycin.

The kinetics of the proton leak was determined by titration of the non-phosphorylating mitochondria with myxothiazol until oxygen consumption due to mitochondria was zero. It was then possible to estimate the proportion of resting oxygen consumption due to proton leak by extrapolation to the oxygen consumption axis at the resting membrane potential. We estimated an in situ oxygen consumption rate due to proton leak of 6.5±2.5 nmolO/min/10^7 which is ~12±6% of resting mitochondrial oxygen consumption, with ATP turnover [48±8 nmolO/min/10^7] accounting for the remaining 78±6% of mitochondrial oxygen consumption.

The data in Figure 1B are amenable to the application of top-down elasticity analysis defined by (a) substrate oxidation flux, (b) proton leak flux and (c) ATP turnover, a branched pathway with the proton electrochemical gradient (or in this instance mitochondrial membrane potential) as the central intermediate. This approach has previously been applied to determine the control of oxygen consumption by mitochondria in situ in primary cells[4,5,11]. This is the first time such an approach has been applied to cancer cells.

Control of resting mitochondrial oxygen consumption (and resting mitochondrial membrane potential) in Y1 cells was shared by (a) substrate oxidation flux (37±8%), (b) proton leak flux (15±8%) and (c) ATP turnover (56±8%).

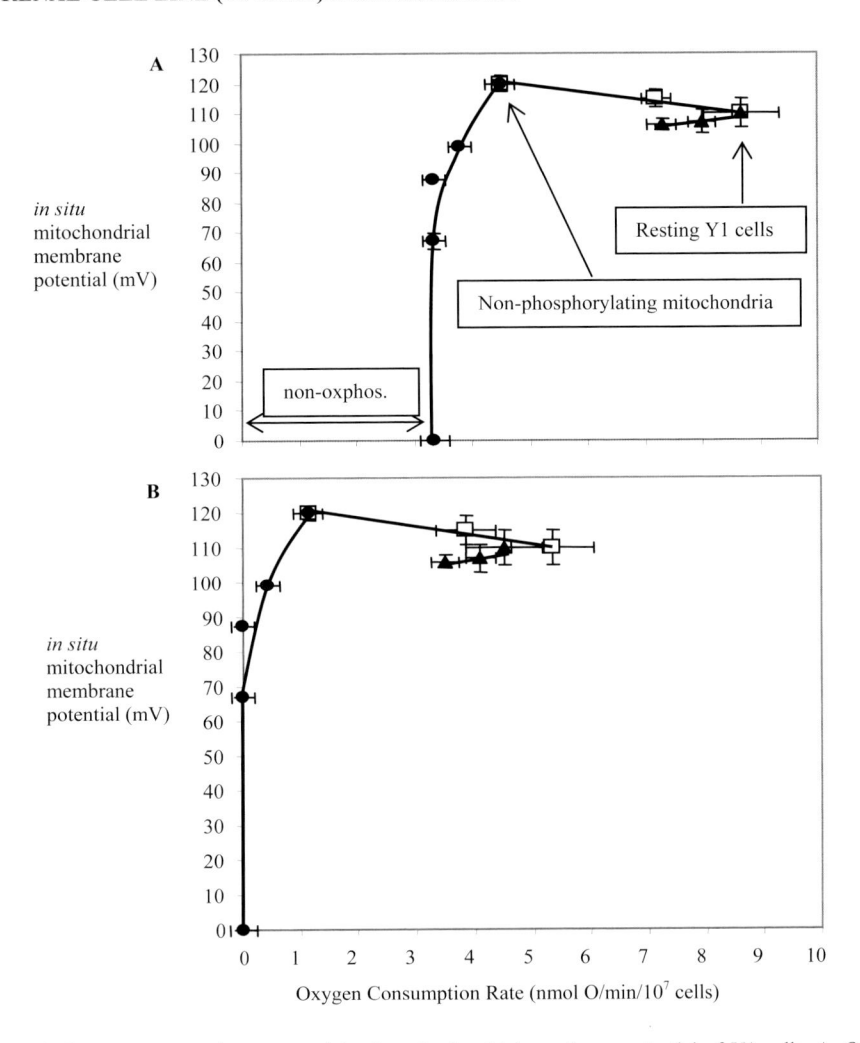

Figure 1. Oxygen consumption rates and *in situ* mitochondrial membrane potential of Y1 cells. **A.** Oxygen consumption rates (uncorrected for oxygen consumption not due to oxidative phosphorylation) and mitochondrial membrane potential of resting Y1 cells were (a) titrated with oligomycin [(0-0.3μg/ml)], until in situ mitochondria were non-phosphorylating, to give substrate oxidation kinetics (□), (b) titrated with myxothiazol (0-0.2μg/ml) in the presence of oligomycin (0.3μg/ml) to give the kinetics of the proton leak (●) and (c) titrated with myxothiazol (0-0.01μg/ml) alone to give mitochondrial ATP-turnover kinetics (▲), uncorrected for proton leak. **B:** Oxygen consumption rates (corrected for oxygen consumption not due to oxidative phosphorylation) and mitochondrial membrane potential of resting Y1 cells were (a) titrated with oligomycin [(0-0.3μg/ml)], until in situ mitochondria were non-phosphorylating, to give substrate oxidation kinetics (□), (b) titrated with myxothiazol (0-0.2μg/ml) in the presence of oligomycin (0.3μg/ml) to give the kinetics of the proton leak (●) and (c) titrated with myxothiazol (0-0.01μg/ml), corrected for proton leak, to give mitochondrial ATP-turnover kinetics (▲). The data represent three separate experiments each performed in triplicate (mean ± sem (n)), where n is the number of experiments. Zero mitochondrial membrane potential and zero mitochondrial oxygen consuming capacity were assumed under conditions of excess oligomycin (0.3μg/ml), myxothiazol (0.2μg/ml) and FCCP (1μM).

Table 1. Flux control coefficients over *in situ* mitochondrial oxygen consumption rates by (i) substrate oxidation, (ii) mitochondrial proton leak and (iii) mitochondrial ATP turnover in resting (non-stimulated) Y1 cells.

System	Flux control coefficient (%)[*]
Substrate oxidation	56 ± 8(3)
Proton Leak	15 ± 8(3)
ATP turnover	37 ± 8(3)

[*] The results represent data from three separate experiments each performed in triplicate (mean ± sem (n)), where n is the number of experiments

4. DISCUSSION

These results are the first reported data to dissect out the bioenergetic components of cancerous cells. Our data show that 62% of resting oxygen consumptions rate of Y1 cells is due to oxidative phosphorylation. The remaining 38% of the oxygen consumption in resting Y1 cells is most probably due to monooygenase reactions. Primary candidates are the cytochrome P450 enzymes which are known to have increased expression and activity in cancer cells, and are considered anti-cancer targets[8]. It is interesting and of metabolic significance that in primary cells these latter oxygen consuming reactions only account for 2% (thymocytes) to 5% hepatocytes of the total cellular oxygen consumption rates[4-6]. Our attention was also focused on dissecting out the relative proportion of oxygen consumption, by *in situ* mitochondria, due to proton leak and ATP turnover. We were able to demonstrate that mitochondrial proton leak accounted for 12% of resting mitochondrial oxygen consumption, with ATP turnover accounting for 78% of mitochondrial oxygen consumption. These data compare well with values for resting hepatocytes (20% for proton leak; 80% for ATP turnover)[4,5] and differ for those of rat brain synaptosomes (85% proton leak; 15% ATP turnover)[11]. Control of resting mitochondrial oxygen consumption in Y1 cells was shared by substrate oxidation flux (37%), proton leak flux (15%) and ATP turnover (56%). Again these data are not too dissimilar from those for hepatocytes (substrate oxidation flux (29%), proton leak flux (22%) and ATP turnover (49%))[4,5] but differ considerably from those for synaptosomes (substrate oxidation flux (71%), proton leak flux (2) and ATP turnover (27%))[11]. Our data demonstrate, for the first time, that the majority of oxygen consumption by resting Y1 cells is due to oxidative phosphorylation.

4.1 Acknowledgements

Funding for this research has come from BioResearch Ireland, the Human Frontiers Scientific Programme and Science Foundation Ireland (06/IN.1/B67).

5. REFERENCES

1. D.F.S. Rolfe, A.J. Hulbert and M.D. Brand, Characteristics of mitochondrial proton leak and control of oxidative phosphorylation in the major oxygen-consuming tissues of the rat, *Biochim. Biophys. Acta* **1118,** 405-416 (1994).

2. D.F.S. Rolfe and M.D. Brand, The physiological significance of mitochondrial proton leak in animal cells and tissues, *Biosci. Rep.* **17**, 9-16 (1997).

3. M.D. Brand, L-F. Chien, E.K. Ainscow, D.F.S. Rolfe and R.K. Porter, The causes and functions of mitochondrial proton leak, *Biochim. Biophys. Acta,* **1187,** 132-139 (1994).

4. G.C. Brown, P.L. Lakin-Thomas and M.D. Brand Control of respiration and oxidative phosphorylation in isolated rat liver cells, *Eur. J. Biochem.* **192**, 355-362 (1990).

5. R.K. Porter and M.D. Brand, Causes of differences in respiration rate of hepatocytes from mammals of different body mass, *Am. J. Physiol.* **269,** R1213-R1224 (1995).

6. C.D. Nobes, G.C. Brown, Olive, P.N. and M.D. Brand, Non-ohmic proton conductance of the mitochondrial inner membrane in hepatocytes. *J. Biol. Chem.* **265**, 12903-12909 (1990).

7. P.L. Pedersen, The cancer cell's "power plants" as promising therapeutic targets: an overview. *J. Bioenerg. Biomembr.* **39**, 1-12 (2007).

8. R.D. Bruno and V.C Najr, Targeting cytochrome P450 enzymes: a new approach in anti-cancer drug development, *Bioorg. Med. Chem.* **15**, 5047-5060 (2007).

9. L.J. Cuprak and G. Sato, Nutritional requirements of mouse adrenal cortex tumor cells in culture. *Exp. Cell Res.* **52**, 632-645 (1968).

10. M.D. Brand, Measurement of mitochondrial protonmotive force, in *Bioenergetics: A Practical Approach*, edited by G.C. Brown and C.E. Cooper (IRL Press, Oxford,1995).

11. O.J.P. Joyce, M.K. Farmer, K.F. Tipton, and R.K. Porter, Oxidative Phosphorylation by *in situ* Synaptosomal Mitochondria from Whole Brain of Young and Old Rats. *J. Neurochem.* **86**, 1032-1041 (2003)

12. R.P. Hafner, G.C. Brand and M.D. Brand, Analysis of the control of respiration rate, phosphorylation rate, proton leak rate and protonmotive force in isolated mitochondria using the 'top-down' approach of metabolic control theory, *Eur. J. Biochem.* **188**, 313-319(1990).

13. G.C. Brown, R.P. Hafner and M.D. Brand A 'top-down' approach to the determination of control coefficients in metabolic control theory, *Eur. J. Biochem.* **188**, 321-325 (1990).

14. M.D. Brand, The proton leak across the mitochondrial inner membrane, *Biochim. Biophys. Acta,* **1018,** 128-133 (1990).

REPLICATION OF MURINE MITOCHONDRIAL DNA FOLLOWING IRRADIATION

Hengshan Zhang, David Maguire, Steven Swarts, Weimin Sun, Shanmin Yang, Wei Wang, Chaomei Liu, Mei Zhang, Di Zhang, Louie Zhang, Kunzhong Zhang, Peter Keng, Lurong Zhang, and Paul Okunieff[*]

Abstract: The effect of radiation on the mitochondrial genome *in vivo* is largely unknown. Though mitochondrial DNA (mtDNA) is vital for cellular survival and proliferation, it has little DNA repair machinery compared with nuclear DNA (nDNA). A better understanding of how radiation affects mtDNA should lead to new approaches for radiation protection. We have developed a new system using real-time PCR that sensitively detects the change in copy number of mtDNA compared with nDNA. In each sample, the DNA sequence coding 18S rRNA served as the nDNA reference in a run simultaneously with a mtDNA sequence. Small bowel collected 24 hours after 2 Gy or 4 Gy total body irradiation (TBI) exhibited increased levels of mtDNA compared with control mice. A 4 Gy dose produced a greater effect than 2 Gy. Similarly, in bone marrow collected 24 hours after 4 Gy or 7 Gy TBI, 7 Gy produced a greater response than 4 Gy. As a function of time, a greater effect was seen at 48 hours compared with 24 hours. In conclusion, we found that radiation increased the ratio of mtDNA:nDNA and that this effect seems to be tissue independent and seems to increase with radiation dose and duration following radiation exposure.

1. INTRODUCTION

Breakage of cellular DNA following radiation is a dose dependent phenomenon and occurs in both the nuclear and extra-nuclear DNA. Thus, while the vast majority of DNA is located in the nucleus (nDNA), on a mass basis, mitochondrial DNA (mtDNA) is equally affected. After radiation damage, the nuclear DNA has many advantages over

---
[*] Department of Radiation Oncology, University of Rochester Medical Center, 601 Elmwood Avenue, Rochester, NY 14642 USA

P. Liss et al. (eds.), *Oxygen Transport to Tissue XXX*, DOI 10.1007/978-0-387-85998-9_7, © Springer Science+Business Media, LLC 2009

mtDNA. These have been reviewed previously[1] and include 1) rapid and efficient double strand break repair,[1, 2] and 2) rapid and near perfect single strand break repair.[3] Following irradiation, mtDNA has few repair mechanisms and continued mitochondrial function is preserved primarily due to its high copy number.[4, 5]

The cellular response to radiation with regard to the protection of mtDNA has rarely been studied. It is predictable that unchecked mtDNA damage will prevent the replication of properly functioning mitochondria over time, leading to mitochondrial depletion, and then late cell dysfunction.[6] The latter would be most pronounced for long-lived cell types such as myocardial, renal, or neuronal cells. Consistent with this hypothesis, these tissues begin to demonstrate serious dysfunction 2 to 8 years after radiation exposure.[7] Thus, as a means of self-protection, cells will want to respond to radiation to reduce the frequency and delay the onset of lethal radiation damage to the mitochondria. One such mechanism is enhanced replication of mtDNA.

In this study, we examined the early effects of radiation on the mtDNA copy number of different cell types using doses similar to those that might be experienced by bioterror victims or patients undergoing a bone marrow transplant. Two sets of primers were designed: one set coding for the 12S rRNA gene of the murine mitochondrial genome, and the other set coding for the 18S rRNA gene, an abundant housekeeping gene in the nDNA. The ratio of these two gene frequencies was compared at different times following different radiation doses to look for evidence of enhanced replication of mtDNA following irradiation.

2. MATERIALS AND METHODS

2.1. Animals and Treatments

Male Balb/c mice, 6 to 8 weeks old, were divided into groups of 5 animals. Control mice were not irradiated, while other groups received total body irradiation (TBI) via a J. L. Shepherd Irradiator with a ^{137}Cs γ-ray source. The dose rate was 1.84 Gy/min and homogeneity was better than ±6.5%. The mice were then sacrificed at 24 or 48 hours post-exposure. Mice were housed in a pathogen-free barrier facility. All protocols were in accordance with USPHS guidelines and approved by the University of Rochester Committee on Animal Research.

2.2. DNA Extraction from Mouse Tissues

The small bowel and bone marrow tissues were collected and frozen immediately at -70°C until use. Total DNA was extracted from these tissues by standard proteolytic digestion followed by phenol/chloroform/isoamyl alcohol purification. Absorbance at 260 nm was used for quantification of the extracted DNA.

2.3. Real-time Quantitative PCR to Estimate mtDNA

The PCR primers for detecting the 12S rRNA gene of murine mitochondrial genome were designed on the basis of the GenBank nucleotide sequence (NC_006914, Mus musculus domesticus mitochondrion). The sequences for the primers were 5'-ACC GCG GTC ATA CGA TTA AC-3' (forward) and 5'-CCC AGT TTG GGT CTT AGC TG-3' (reverse), producing a 177 bp amplicon. The primer sequences amplifying mouse nuclear 18S rRNA gene are 5'-CGC GGT TCT ATT TTG TTG GT-3' (forward) and 5'-AGT CGG CAT CGT TTA TGG TC-3' (reverse), which produce a 219 bp product. The real-time quantitative PCR was carried out using separate tubes for mtDNA and nDNA amplification. Each tube contained 30 ng of total DNA from the same extract as well as 20 μl of reaction mixture consisting of 2 x iQ^{TM} SYBR Green Supermix (1 x final), and one pair of the primers directed against either mtDNA or nDNA, with each primer being 0.5 μM. Each reaction was subjected to an initial denaturation of 5 min at 95°C followed by 40 amplification cycles of denaturation at 95°C for 30 s, 51°C to anneal for 20 s, and 1 min of extension at 72°C. A Bio-Rad i-cycler iQ real-time PCR detection system was employed to run the reactions. A conventional PCR was first run to develop the cycling conditions using the same primers. PCR assays were performed in duplicate or triplicate for each DNA sample. The cycle number (Ct) at which the fluorescent signal of a given reaction crossed the threshold value was used as a basis for quantification of mtDNA and nDNA copy numbers. The difference in Ct and the assumption of a doubling of DNA per cycle were used to calculate the mtDNA:nDNA ratio.

3. RESULTS AND DISCUSSION

3.1. Optimization of Real-Time Quantitative PCR

Conventional PCR was first performed to optimize the real-time PCR conditions. Duplicate PCR were run using the mtDNA and nDNA primers added to purified mouse small bowel DNA (Figure 1). As predicted, the size of the amplicon derived from the mtDNA was 177 bp, and from the nDNA was 219 bp. The band size and intensity were visualized with ethidium bromide after electrophoresis in 4% agarose gel; analysis indicated a high mtDNA:nDNA ratio.

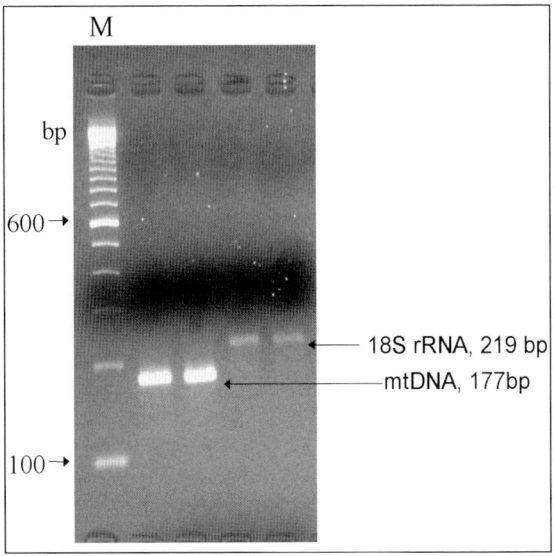

Figure 1. Conventional PCR for mtDNA and nDNA. PCR conditions were optimized to produce amplicons of the expected size and intensity.

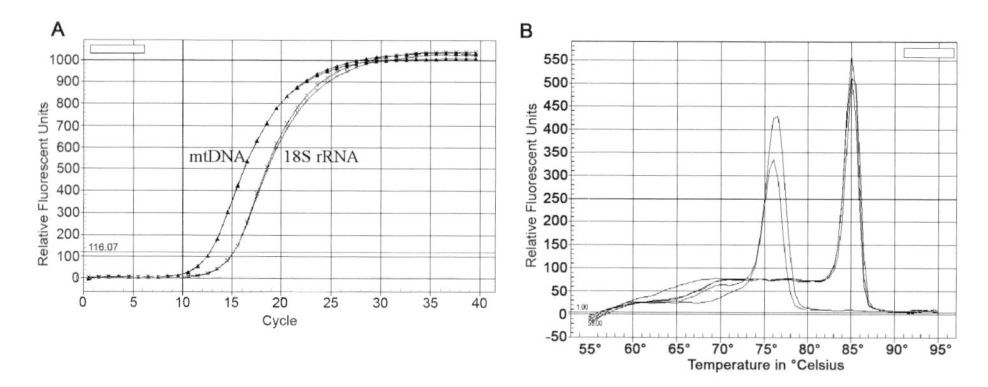

Figure 2. Optimized real-time quantitative PCR. **A**. Sample amplification curves obtained using the real-time quantitative PCR system for mtDNA and nDNA genes. **B**. Sample melt-curve analysis of mtDNA and nDNA products from a SYBR Green I assay.

The real-time PCR amplification plot in Figure 2A shows the typical two phases, an exponential phase followed by a plateau phase. A threshold is set during the early exponential phase, as shown. The lower number of cycles needed to reach the threshold level indicates a greater copy number of the mtDNA compared with the nDNA. The 18S rRNA gene is present in multiple copies in the nuclear genome, and therefore the ratios obtained by this assay are not meant to be equivalent to the average copies of mtDNA per cell.[8] Melt curves are shown in Figure 2B. The expected single, narrow peaks indicate that the mtDNA and nDNA products are sufficiently pure and specific. Taken together, the data confirm that the real-time PCR conditions are satisfactorily optimized and can be used to determine relative changes in copy numbers of the respective transcripts.

3.2. Dose-Dependent Increase of mtDNA Copy Number in Small Bowel

The small bowel mucosa is a rapidly renewing tissue. Within a few days of radiation exposure, there is complete replacement of the entire luminal epithelium.[9] Thus loss of mtDNA could seriously impact recovery given the potential for hemi-depletion with each cell division. It was therefore interesting to see how mtDNA might respond to low dose radiation. To test this, radiation doses that have no obvious histological effects were used. The small bowel tissues were collected from mice 24 hr after 0 Gy, 2 Gy, or 4 Gy irradiation. The total DNA was extracted from each sample and subjected to real-time PCR. Each cycle leads to almost a doubling in DNA copies, and thus the difference between cycle numbers measured at the threshold (designated ß) can be used to calculate the relative ratio of mtDNA:nDNA using the equation: 2^β. At 24 hours after irradiation, we observed an increase in the mtDNA:nDNA ratio compared with non-irradiated tissues (Figures 3 and 4A). Tissues irradiated to 4 Gy indicated a clear increase in copy ratio compared with 2 Gy, consistent with a dose response. This result is also consistent with a compensatory replication of mtDNA to replace lost or damaged mtDNA. Other explanations include a lowering of the copy number of chromosomes per cell by redistribution of the cell cycle; however, this is very unlikely since the proportion of cells in S-phase at 24 hrs after irradiation is stable or moderately increased, which would lead to a decrease rather than the observed increase in mtDNA:nDNA ratio.

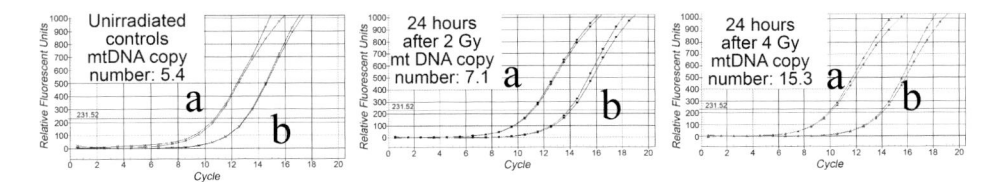

Figure 3. Representative real-time PCR of mtDNA and nDNA extracted from small bowel of mice 24 hours after 2 or 4 Gy TBI. **(a)** Denotes change in relative copy number of mtDNA; **(b)** Denotes change of relative copy number of nDNA.

3.3. Dose- and Time-Dependent Increase in mtDNA Copy Number in Bone Marrow

Bone marrow is probably the most sensitive organ to radiation. Within a few days of TBI, the bone marrow has only a few microscopically detectable hematopoietic cells, yet within two or three weeks of sublethal exposure, the bone marrow repopulates. This extraordinary cellular replication must include a matching mtDNA replacement. Damage to bone marrow mtDNA could otherwise lead to devastating replication-related hemi-depletion. To examine for changes in mtDNA of the bone marrow following irradiation, animals were given sublethal TBI (4 Gy) or near $LD_{50/30}$ doses (7 Gy). Marrow was harvested at 24 or 48 hours after exposure and analyzed as above. Again the copy ratio of mtDNA in bone marrow of mice increased after 4 Gy.

As in the previous small bowel studies, the amplification of mtDNA seems to be dose dependent; here, 7 Gy produced a higher mtDNA:nDNA ratio. The ratio seems to increase further at 48 hrs compared with 24 hrs (Figure 4B).

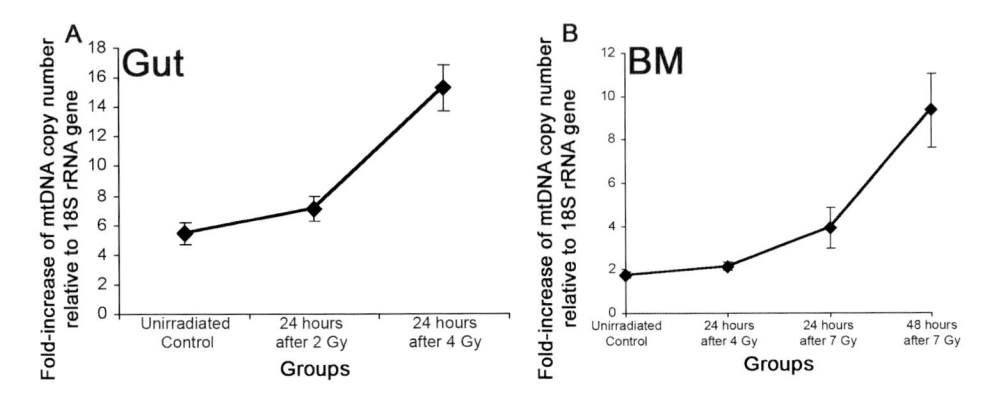

Figure 4. A: Increased mtDNA in the irradiated gut is dose- and time-dependent. Mice were exposed to 2 or 4 Gy TBI. The small bowel was harvested at 24 hours. **B:** Increased mtDNA in the irradiated bone marrow is dose- and time-dependent. Mice were exposed to 4 or 7 Gy TBI. The bone marrow was harvested at either 24 or 48 hours. Values are mean ± 1 SD.

While the nDNA is a traditional standard to normalize the number of cells, radiation can cause cell cycle synchrony. However, we did not observe any change in either the S or the G2 phase, and the expected increase in nDNA, if synchrony occurred, would have lowered, not raised, the mtDNA/nDNA ratio.

4. CONCLUSIONS

We report a set of primers and procedures that can be used to measure relative copy number of mtDNA to nDNA, and which can be used on DNA purified from murine tissues. Sequence modification would make this technology applicable to other animal tissues. Further, we have demonstrated that small bowel and bone marrow collected from irradiated mice have an increase in mtDNA copy number after irradiation when measured at 24 or 48 hours and that the copy number may increase with radiation dose. These results are consistent with our previous hypothesis that some radiation effects are due to damage of mitochondrial DNA offset in part by adaptive replication of mtDNA.[1, 10]

5. ACKNOWLEDGEMENTS

This research was supported in part by the Center for Medical Countermeasures Against Radiation Program, U19-AI067733, National Institute of Allergy and Infectious Diseases. The authors wish to thank Amy K. Huser for editorial assistance.

REFERENCES

1. P. Okunieff, S. Swarts, P. Keng, W. Sun, W. Wang, J. Kim, S. Yang, H. Zhang, C. Liu, J. P. Williams, A. K. Huser, and L. Zhang, Antioxidants reduce consequences of radiation exposure, *Adv. Exp. Med. Biol.* in press, (2007).
2. G. Iliakis, H. Wang, A. R. Perrault, W. Boecker, B. Rosidi, F. Windhofer, W. Wu, J. Guan, G. Terzoudi, and G. Pantelias, Mechanisms of DNA double strand break repair and chromosome aberration formation, *Cytogenet. Genome Res.* **104**(1-4), 14-20 (2004).
3. S. Purkayastha, J. R. Milligan, and W. A. Bernhard, On the chemical yield of base lesions, strand breaks, and clustered damage generated in plasmid DNA by the direct effect of x rays, *Radiat. Res.* **168**(3), 357-366 (2007).
4. N. B. Larsen, M. Rasmussen, and L. J. Rasmussen, Nuclear and mitochondrial DNA repair: similar pathways? *Mitochondrion.* **5**(2), 89-108 (2005).
5. D. L. Croteau, R. H. Stierum, and V. A. Bohr, Mitochondrial DNA repair pathways, *Mutat. Res.* **434**(3), 137-148 (1999).
6. L. Weissman, N. C. de Souza-Pinto, T. Stevnsner, and V. A. Bohr, DNA repair, mitochondria, and neurodegeneration, *Neuroscience.* **145**(4), 1318-1329 (2007).
7. A. M. Gaya, and R. F. Ashford, Cardiac complications of radiation therapy, *Clin. Oncol. (R. Coll. Radiol.)* **17**(3), 153-159 (2005).
8. K. Bross, and W. Krone, On the number of ribosomal RNA genes in man, *Humangenetik.* **14**(2), 137-141 (1972).
9. J. W. Denham, M. Hauer-Jensen, T. Kron, and C. W. Langberg, Treatment-time-dependence models of early and delayed radiation injury in rat small intestine, *Int. J. Radiat. Oncol. Biol. Phys.* **48**(3), 871-887 (2000).
10. S. Wolff, The adaptive response in radiobiology: evolving insights and implications, *Environ. Health. Perspect.* **106**(Suppl 1), 277-283 (1998).

EFFECTS OF ANESTHESIA ON BRAIN MITOCHONDRIAL FUNCTION, BLOOD FLOW, IONIC AND ELECTRICAL ACTIVITY MONITORED IN VIVO

Nava Dekel[1], Judith Sonn[1], Efrat Barbiro-Michaely[1], Eugene Ornstein[2] and Avraham Mayevsky[1]*.

Abstract: Thiopental, a well-known barbiturate, is often used in patients who are at high risk of developing cerebral ischemia, especially during brain surgery. Although barbiturates are known to affect a variety of processes in the cerebral cortex, including oxygen consumption by the mitochondria, the interrelation between mitochondrial function and anesthetics has not been investigated in detail under in vivo conditions.

The aim of this study was to examine the effects of thiopental on brain functions in normoxia and under partial or complete ischemia. The use of the multiparametric monitoring system permitted simultaneous measurements of microcirculatory blood flow, NADH fluorescence, tissue reflectance, and ionic and electrical activities of the cerebral cortex. Thiopental caused a significant, dose-dependent decrease in blood flow and a significant decrease in extracellular levels of potassium, with no significant changes in NADH levels in normoxic and ischemic rats. Following complete ischemia (death), the increase in the reflectance was significantly smaller in the anesthetized normoxic group versus the awake normoxic group. The time until the secondary increase in reflectance, seen in death, was significantly shorter in the anesthetized ischemic group.

In conclusion, it seems that the protective effect of thiopental occurs only under partial ischemia and not under complete ischemia.

1. INTRODUCTION

In experimental practice, general anesthesia can be achieved by using a variety of drugs and routes of administration (inhalation or intravenous). Knowledge of the effect of anesthetics on the cerebral circulation, metabolism and intracranial pressure, in both normal and pathological conditions, is crucial for neurobiological purposes [1]. Since mitochondria play a major role in cellular energy metabolism, it is extremely important to verify the interaction between mitochondrial function and anesthetics [2, 3]. Most of the anesthetics inhibit the cerebral metabolic rate (CMR) probably through the electrophysiological effects.

---

* [1] The Mina & Everard Goodman Faculty of life sciences and The Leslie and Susan Gonda Multidisciplinary Brain Research Center, Bar-Ilan University, Ramat-Gan, Israel.
[2] The Department of Anesthesiology, College of physicians and Surgeons of Columbia University, New-York.

P. Liss et al. (eds.), *Oxygen Transport to Tissue XXX*, DOI 10.1007/978-0-387-85998-9_8,
© Springer Science+Business Media, LLC 2009

Namely, the spontaneous electrical activity is inhibited and this is associated with a decrease in CMR [4] and in cerebral blood flow (CBF), and a decline in the consumption of glucose and oxygen in the cerebral tissue [5]. During neurosurgical procedures, the issue of anesthesia is crucial with reference to adequate cerebral blood flow. In other words, it is important to use the types of anesthetics and the doses that do not impair cerebral blood supply. Preoperative brain protection includes strategies to prevent or reduce ischemic damage and to improve neurological outcome. The protective effect of anesthesia, compared to the awake state, has been demonstrated in animals but remains to be validated in clinical practice [6]. From this point of view, the vasoconstrictor effect of pentobarbital in the healthy brain may be clinically important, since it may partly explain the protective properties of these drugs against cerebral ischemia [7]. In the present study, we evaluated the effect of the anesthetic thiopental on the hemodynamic, metabolic and electrical activity of the brain cortex in normoxic and ischemic rats.

2. METHODS

2.1 Brain Function Multiprobe (BFM)

The monitoring of brain functions was performed using the Brain Function Multiprobe assembly (BFM) developed in our laboratory [8]. The BFM includes a bundle of optical fibers for NADH redox state fluorometry and for CBF (Cerebral Blood Flow) monitoring. We used a mini-electrode for the monitoring of extracellular K^+ levels, surrounded with a DC-potential electrode. The intracranial pressure (ICP) was monitored by a Camino transducer (Camino Laboratories, San Diego, CA, USA). A needle thermistor was used for brain temperature measurements and two needle platinum electrodes were used for electrocorticogram ECoG (Fig.1).

2.2 Monitoring Techniques

2.2.1. Mitochondrial NADH Redox State (NADH)

Mitochondrial NADH redox state was evaluated by the measurement of NADH fluorometry from the surface of the brain [9]. Excitation light at the wavelength of 366 nm passes from the fluorometer to the tissue via a bundle of quartz optical fibers. The emitted fluorescent light at 450 nm, together with the reflected light at the excitation wavelength (366 nm), is transferred to the fluorometer via another bundle of optical fibers and through appropriate filters. The changes in the reflected light are correlated to changes in cerebral blood volume and are used to correct for hemodynamic artifacts, which also affect the fluorescent signal at 450 nm.

2.2.2. Cerebral Blood Flow (CBF)

Cerebral Blood Flow (CBF) was monitored by laser Doppler flowmetry (LDF) (Perimed, Inc, Sweden; model PF2B) at 632.8 nm, which measures relative changes in microcirculatory blood flow, where the basal level is considered as 100% and the level monitored on death is 0% [10, 11].

2.2.3. Extracellular Levels of K^+

Extracellular levels of K^+ were measured by ion selective mini-electrodes (WPI Inc. USA) with a sensitivity near the values of Nernst equation [12].

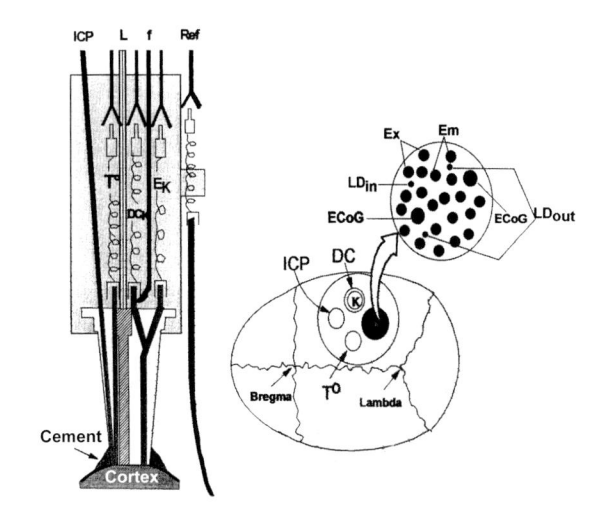

Figure 1: A schematic representation of the Brain Function Mutiprobe (BFM) assembly and its location on the cerebral cortex. The probe includes a bundle of optical fibers for the monitoring of mitochondrial NADH reflectance/fluorescence (Ex-Excitation, Em-Emission), laser Doppler flowmetry (LD in/out), fibers for intracranial pressure monitoring (ICP), brain temperature probe (T°), mini-electrode for extracellular K+ level monitoring and DC-potential and EcoG electrodes.

2.3. Animal Preparation

All experiments were performed in accordance with the Guidelines of the Animal Care Committee of Bar-Ilan University. Wistar male rats (250 – 300 g) were anesthetized with an IP injection (0.3 ml/100gw) of equithesin (each ml contains: pentobarbital 9.72 mg, chloral hydrate 42.51 mg, magnesium sulfate 21.25 mg, propylene glycol 44.34 % w/v, alcohol 11.5 % and distilled water).

The rat's head was fixated in a head holder and a midline incision of the skin was made exposing the rat's skull. A hole of 6mm in diameter was drilled in the right parietal bone and the Dura mater was gently removed. In addition, four stainless steel screws were fastened in the parietal and frontal bones to ensure a better fixation of the monitoring probe to the cortical surface. The BFM was placed on the cerebral cortex and fixated by acrylic cement. When the ECoG showed full wakefulness, the specific protocol was conducted. At the end of the experiment, the rats were sacrificed by the injection of a lethal dose of thiopental to the femoral vein.

2.3.1. Experimental Protocols:

The following protocols were performed in the awake animals:
a)   Two doses of thiopental were tested: 2 injections of 30 mg/kg I.V. at an interval of 15 minutes (Low dose) or gradual infusion of thiopental 70-100mg/kg I.V. during a period of 30 minutes to induce burst suppression in the ECoG (High dose). Fifteen minutes later, a lethal dose of thiopental was injected to induce cardiac arrest. In the control groups, saline solution was injected and death was induced by the injection of a lethal dose of thiopental.
b)   Partial ischemia was induced by bilateral carotid arteries occlusion 24 hours before monitoring. Thereafter, the section was sutured and the rat was returned to the cage with a free access to food and water. One day later, a high dose of thiopental was injected (70-100mg/kg I.V.).

3. RESULTS

The injection of thiopental caused an immediate decrease in CBF in both groups (Figure 2). However, while this decrease continued in the high dose group for most of the experimental period, in the low dose group, 4 minutes after thiopental injection CBF fully recovered and showed the same pattern as the control rats. The changes in CBF were associated with changes in the reflectance. As seen, the injection of thiopental to normoxic rats under both doses (high and low) caused an initial significant increase in the reflectance ($p < 0.05$) as compared to the control rats. Thereafter, the reflectance showed a trend of increase in the control group as well, hence the differences between the treated rats and the control rats disappeared.

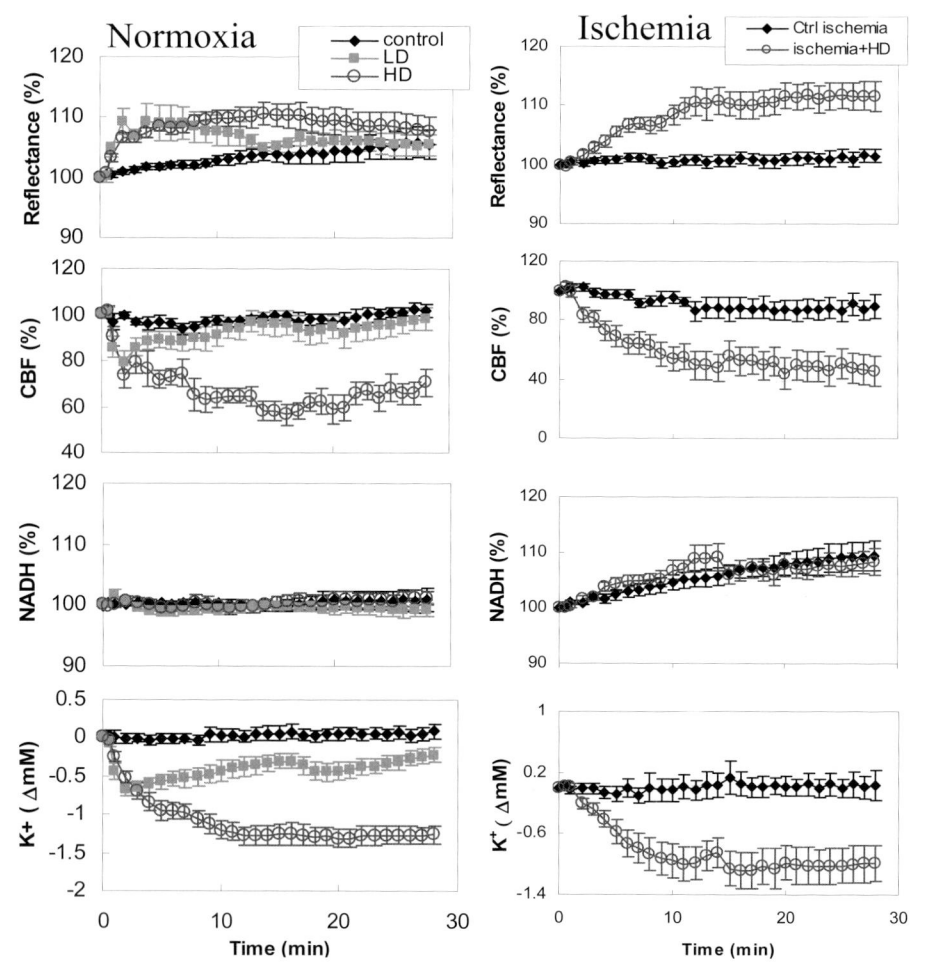

Figure 2: The effects of two doses of thiopental (LD - low dose and HD - high dose) and Saline (control, n=8) on the hemodynamic and metabolic state of the cerebral cortex in normoxic rats and in ischemic rats. For the groups of normoxic rats: low dose (n=9), high dose (n=7). For the ischemic rats: control (n=11), thiopental (n=9).

Nevertheless, the reflectance was significantly elevated ($p < 0.05$) after the high dose exposure versus the control group, through most of the experiment.

Mitochondrial function, evaluated by the level of reduced NADH, showed no differences in all three groups. Regarding the changes in the extracellular level of K^+, three types of response were observed. In the control group, K^+ level remained stable through the entire experimental period. In the low dose group, a significant decrease was seen in the first 4 minutes followed by a partial recovery of approximately 0.4mM change and biphasic decrease/increase of 0.1mM. In the high dose group, the extracellular level of K^+ significantly decreased by 1.3mM ($p < 0.01$), and remained low through the entire experimental period. The effects of the high dose of thiopental on the ischemic brain are presented in Figure 2. Following thiopental administration, CBF decreased by 50%, whereas the control ischemic rats showed a decrease of 10% in CBF. Subsequently, the reflectance increased in the ischemic thiopental group by approximately 12%, while in the control group the reflectance remained stable. Concerning NADH levels, a significant increase of approximately 8% was observed in both groups ($p < 0.05$). The ionic changes showed a massive decrease in the level of extracellular K^+ (1mM) with no changes in the control group. The effects of the high dose of thiopental on the monitored parameters, in a typical experiment, are presented in Figure 3. After thiopental was injected, CBF decreased to very low levels (near 0) leading to a rapid increase in the reflectance and NADH levels, exhibiting a further secondary increase to above 100%. Approximately 1 minute after the injection of thiopental, the extracellular K^+ levels showed a gradual increase, followed by a rapid increase to above 30mM. This was associated with a decrease of 10mV in the DC potential around the K^+ electrode. Immediately after thiopental injection, ECoG was completely depressed or disappeared.

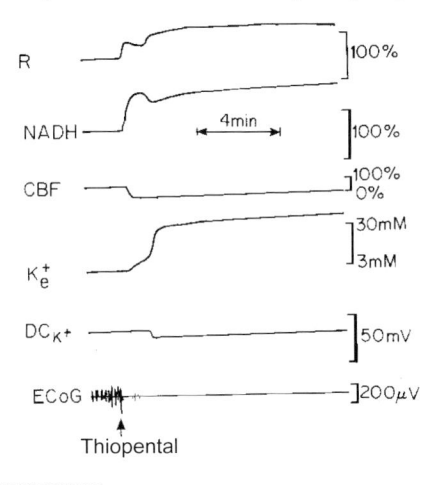

Figure 3: The responses of the cerebral cortex to the injection of a lethal dose of thiopental. R - Reflectance, NADH - mitochondrial NADH, K^+_e – extracellular level of K+, DC $_{K^+}$ - DC potential around the electrode of K^+, ECoG – Electrocorticogram.

4. DISCUSSION

In selecting an anesthetic agent to be used for neurosurgical procedures, the agent's effects on the intracranial pressure (ICP), cerebral blood flow (CBF), and cerebral metabolic rate of oxygen consumption ($CMRO_2$) must be considered. In this regard, barbiturates are commonly used, especially in patients who have undegone traumatic brain injury and suffer from increased ICP, or in patients after stroke. The reason for using barbiturates in such cases is mainly based on their vascular activity. In the present study, we showed that the injection of thiopental yielded a dose-dependent decrease in

CBF, as was previously reported [13]. This decrease was seen both in the normoxic and ischemic rats. With the low dose, the decrease in CBF continued only for a short period; while with the high dose, the decrease lasted for a long time. Consequently, the reflectance increased due to the decrease in the cerebral blood volume. Although blood supply to the tissue significantly declined, there were no significant changes in NADH level. This fact indicates that the cerebral tissue was not in a state of oxygen deficit, implying a protective effect on the cerebral tissue for thiopental. In other words, thiopental decreases blood supply to the brain but also decreases the demand in the cerebral tissue, due to its effect of decreasing cerebral metabolic rate ($CMRO_2$) [14]. The ionic homeostasis was altered, as seen by the reduction in the extracellular K^+ levels, probably due to the decrease in the permeability for K^+ [15]. The latter effect may be also due to a decrease in the activity of Na^+/K^+ ATPase, as was previously reported for insulin secreting cells [16].

5. ACKNOWLEDGMENTS

This research was supported by the Israel Science Foundation, Grant No. 358/04

6. REFERENCES

1. Z. Karwacki, P. Kowianski, and J. Morys, General anaesthesia in rats undergoing experiments on the central nervous system, *Folia Morphol. (Warsz.)* **60**, 235-242 (2001)
2. S. Muravchick and R. J. Levy, Clinical implications of mitochondrial dysfunction, *Anesthesiology.* **105**, 819-837 (2006)
3. K. Nouette-Gaulain, A. Quinart, T. Letellier, and F. Sztark, Mitochondria in anaesthesia and intensive care, *Ann. Fr. Anesth. Reanim.* **26**, 319-333 (2007)
4. J. C. Drummond and H. M. Shapiro, Cerebral physiology, in: *Anesthesia.* edited by. R.D. Miller (Churchill Livingstone, New York, 1990), pp. 621-658.
5. B. K. Siesjo, Anaesthesia, analgesia and sedation. in: *Brain Energy Metabolism.* Edited by B.K. Siesjo. (John Wiley & Sons, New York, 1978), pp. 233-255.
6. P. Hans and V. Bonhomme, The rationale for perioperative brain protection, *Eur. J. Anaesthesiol.* **21**, 1-5 (2004)
7. R. Hemmingsen, D. I. Barry, and M. M. Hertz, Cerebrovascular effects of central depressants: a study of nitrous oxide, halothane, pentobarbital and ethanol during normocapnia and hypercapnia in the rat, *Acta Pharmacol. Toxicol. (Copenh)* **45**, 287-295 (1979)
8. A. Mayevsky, R. Nakache, H. Merhav, M. Luger-Hamer, and J. Sonn, Real time monitoring of intraoperative allograft vitality, *Transplant. Proc.* **32**, 684-685 (2000)
9. A. Mayevsky and G. Rogatsky, Mitochondrial function *in vivo* evaluated by NADH fluorescence: From animal models to human studies, *Am. J. Physiol Cell Physiol.* **292**, C615-C640 (2007)
10. D. Arvidsson, H. Svensson, and U. Haglund, Laser-Doppler flowmetry for estimating liver blood flow, *Am J Physiol* **254**, G471-G476 (1988)
11. K.C. Wadhwani and S. I. Rapoport, Blood flow in the central and peripheral nervous systems. in: *Laser Doppler Blood Flowmetry,* edited by A.P. Shepherd and P.A Oberg, (Kluwer Academic Pub., Boston, 1990), pp. 265-304.
12. C. M. Friedli, D. S. Sclarsky, and A. Mayevsky, Multiprobe monitoring of ionic, metabolic and electrical activities in the awake brain, *Am. J. Physiol.* **243**, R462-R469 (1982)
13. J. Donegan, Effect of anesthesia on cerebral physiology and metabolism. in: *Neuroanesthesia: Handbook of Clinical and Physiologic Essentials.* Edited by P. Newfield, J.E Cottrell and C.B Wilson, (Little Brown and Company, New York, 1991), pp. 17-29.
14. E. M. Nemoto, R. Klementavicius, J. A. Melick, and H. Yonas, Suppression of cerebral metabolic rate for oxygen ($CMRO_2$) by mild hypothermia compared with thiopental, *J Neurosurg Anesthesiol.* **8**, 52-59 (1996)
15. P. Arhem and H. Kristbjarnarson, A barbiturate-induced potassium permeability increase in the myelinated nerve membrane, *Acta Physiol Scand.* **113**, 387-392 (1981)
16. R. Z. Kozlowski, C. N. Hales, and M. L. Ashford, Dual effects of diazoxide on ATP-K^+ currents recorded from an insulin-secreting cell line, *Br. J Pharmacol.* **97**, 1039-1050 (1989)

THE INFLUENCE OF FLOW REDISTRIBUTION ON WORKING RAT MUSCLE OXYGENATION

Louis Hoofd, and Hans Degens[*]

Abstract: We applied a theoretical model of muscle tissue O_2 transport to investigate the effects of flow redistribution on rat soleus muscle oxygenation. The situation chosen was the anaerobic threshold where redistribution of flow is expected to have the largest impact. In the basic situation all capillaries received an equal proportion of the total flow through the tissue, resulting in 4.7% anoxic tissue and a mean tissue $PO_2 = 3.62$ kPa. Both a redistribution of flow where 1) capillaries in blocks of tissue receiving 50% of the basic flow alternated with tissue blocks with capillaries receiving 150% of the basic flow (6.8% anoxic tissue; mean tissue $PO_2 = 3.32$ kPa) and 2) matching flow to O_2 consumption (3.3% anoxic tissue; mean tissue $PO_2 = 3.60$ kPa) had little effect. When overall flow was decreased by 20%, the anoxic tissue increased to 7.6% and the mean tissue PO_2 decreased to 3.22 kPa. The conclusion from these model calculations is, that flow redistribution has little impact on skeletal muscle oxygenation, which is in line with earlier findings for rat heart.

1. INTRODUCTION

Tissue oxygenation is the result of a balance between O_2 supply, by the capillary blood, and O_2 consumption, in the tissue cells. In between, there is an important role for O_2 diffusion. For a muscle with a varying O_2 consumption depending on the work it performs, the question arises how blood flow and O_2 consumption are related and, in particular, if flow can be matched to consumption to maintain adequate oxygenation. To maintain an adequate oxygenation, the muscle has two mechanisms available; capillary recruitment and increasing blood flow. During maximal recruitment all, that is 100%, of the capillaries are open and available for O_2 exchange with the surrounding tissue. It is unlikely, however, that each individual capillary receives an equal proportion of the total

[*] Louis Hoofd, 143 Department of Physiology, Radboud University Nijmegen Medical Centre, P.O. Box 9101; 6500 HB Nijmegen, e-mail: l.hoofd@fysiol.umcn.nl. Hans Degens, Institute for Biophysical and Clinical Research into Human Movement, Manchester Metropolitan University, Alsager Campus, Alsager, Cheshire ST7 2HL, UK, e-mail: h.degens@mmu.ac.uk

P. Liss et al. (eds.), *Oxygen Transport to Tissue XXX*, DOI 10.1007/978-0-387-85998-9_9,
© Springer Science+Business Media, LLC 2009

blood flow through the muscle, since capillaries are distributed inhomogeneously in the tissue. Indeed, the flow distribution over capillaries may vary with varying (working) conditions. Yet, in an earlier investigation on rat heart[1], we found that flow redistribution has little influence on tissue oxygenation. It should be noted that this does not automatically apply to skeletal muscle, as skeletal muscle differs from heart in a number of ways, in particular in the much wider range of working states including lactate production at maximum work.

Heterogeneity of capillary spacing is the most important factor in muscle tissue oxygenation[2]. Consequently, reliable data of capillary localisation must be available. Average tissue values of capillary density and heterogeneity of capillary spacing allow to select a representative tissue portion where calculations can be based on. Here we use data of rat soleus muscles that were obtained in a previous study[3].

2.   MATERIALS AND METHODS

The muscle tissue considered here is rat soleus skeletal muscle. Since muscle working state can be very different, from rest to maximum work, we had to select a state where the relation between flow and consumption will be the most relevant. At rest, only few capillaries will be open. At maximum work, the muscle produces a significant amount of lactate from anaerobic energy production strongly suggesting inadequate O_2 supply at least locally. Thus, the anaerobic threshold, at the verge of lactate production, seemed the most relevant state for our investigation. According to textbooks on work physiology, we assumed this threshold to be at ⅔ of the maximum oxygen consumption and ⅔ of the maximum flow.

2.1. Mathematical model

The mathematical treatment is based on oxygen diffusion from a number of point-source capillaries into a surrounding plane, coordinate $\vec{r}$ [4]:

$$PO_2 + P_F S_{MbO_2} = \frac{Q}{4\wp}\left[\Phi(\vec{r}) - \sum_{i=1}^{N} \frac{A_i}{\pi} \ln\left(\frac{|\vec{r} - \vec{r_i}|^2}{r_{ci}^2} \right) \right]$$

where PO_2 is oxygen partial pressure, P_F is facilitation pressure[5] of the tissue myoglobin, Mb, with saturation S_{MbO_2}, Q and $\wp$ are the tissue's oxygen consumption and oxygen permeability respectively, N is the number of capillaries, and A_i, $\vec{r_i}$ and r_{ci} are supply area, location and radius of the ith capillary respectively. The term $\Phi(\vec{r})$ accounts for the distribution of oxygen consumption and can be calculated according to the cited paper; here, the solution for a homogeneous rectangle was taken r_4. The N supply areas can be calculated from the N capillary rim pressures P_{ri} which in turn were calculated from the capillary O_2 pressures P_{ci} by the method of the Extraction Pressure EP[6, 7]:

$$P_{ri} = P_{ci} - \frac{A_i}{A} EP$$

where A is the average supply area. The EP is the PO_2 gradient in and near the capillary for the average capillary and depends on a variety of local capillary data, the most important being blood velocity and hematocrit. Values were calculated by the approximate method of Bos[8]. The above equation is equivalent to a flux-dependent PO_2 difference as used by other authors[9].

The consecutive planes are coupled by subtracting the amount of oxygen delivered for each capillary k[10]:

$$\frac{d}{dz}(c_{tk}O_2) = -\frac{Q}{F_k}\left(A_k - \pi r_{ck}^2\right)$$

where z is the coordinate perpendicular to the planes, and $c_{tk}O_2$, F_k are the capillary's total O_2 content and blood flow, respectively. The $c_{tk}O_2$ incorporates both free oxygen and oxygen bound to hemoglobin, Hb. Contrary to the other equations, which are all analytical, this latter equation has to be solved numerically which was done by taking a non-infinitesimal step Δz equal to the distance between the planes instead of the infinitesimal dz.

2.2. Input data

The overall tissue and blood data were: Q = 0.092 mM, i.e., ⅔ of maximal consumption[11] assuming that the soleus contains 90% type I and 10% type II fibers[3]; Permeability = Krogh's diffusion coefficient[12] $\wp = 1.18 \ 10^{-11}$ mol·m^{-1}·kPa^{-1}·sec^{-1}; Mb $P_{50} = 0.7$ kPa and $D_{Mb}/D_{O2} = 0.075$[13] and cMb = 0.28 mM[3, 11] leading to $P_F = 2$ kPa; blood O_2 solubility $\alpha O_2 = 0.024$ L·L^{-1}·atm^{-1}[14] and Hb content[15] 17 g·dL^{-1}; ⅔ of maximum flow[16] of 276 mL·min^{-1}·(100 g)$^{-1}$; $r_c = 2.65$ μm from capillary luminal diameter of 5.3 μm[17]. Hb saturation was described by the Hill equation with $P_{50} = 4.93$ kPa and n = 2.69. For these data, EP = 0.79 kPa[8].

A representative tissue cross-section of 400 × 400 μm was selected (rat 14A[3]) containing 91 capillaries and extended to a block of 800 μm capillary length. At z = 0, capillary PO_2 was set at 13 kPa.

2.3. Situations calculated

A basic situation (BASIC) was calculated where all capillary data were identical, in particular, capillary flow. A border zone of 45 μm was excluded from the calculations to avoid border effects. Other situations were compared with this basic situation. These were: 16 adjacent regions of 80 × 80 μm with alternating 150% and 50% of the average flow in a checkerboard pattern (4 × 4); 4 adjacent regions of 160 × 160 μm alike (2 × 2); flow relative to O_2 delivery for each capillary (MATCHED); and 20% less flow uniformly distributed (80%).

3.   RESULTS

For each of the situations, tissue PO_2 was calculated at equidistant points in the tissue block with a spacing of 10 μm, except when within a capillary. The resulting 73 224 points were gathered into a histogram of class width 0.5 kPa. A region where the calculated PO_2 was below zero was considered an anoxic region where in fact $PO_2 = 0$. These regions are indicated in the leftmost bar of the histograms. Figure 1 shows the results for four of the situations; the 4 × 4 case was only slightly different from the 2 × 2 case and not shown.

As expected, the basic situation (upper left panel in Figure 1) showed some anoxic tissue, i.e. 4.7%, consistent with the emergence of lactate around the anaerobic threshold. Most of the tissue PO_2 was not low, a mean PO_2 of 3.62 kPa with a standard deviation (SD) of 2.46 kPa indicating quite a heterogeneous tissue PO_2 distribution.

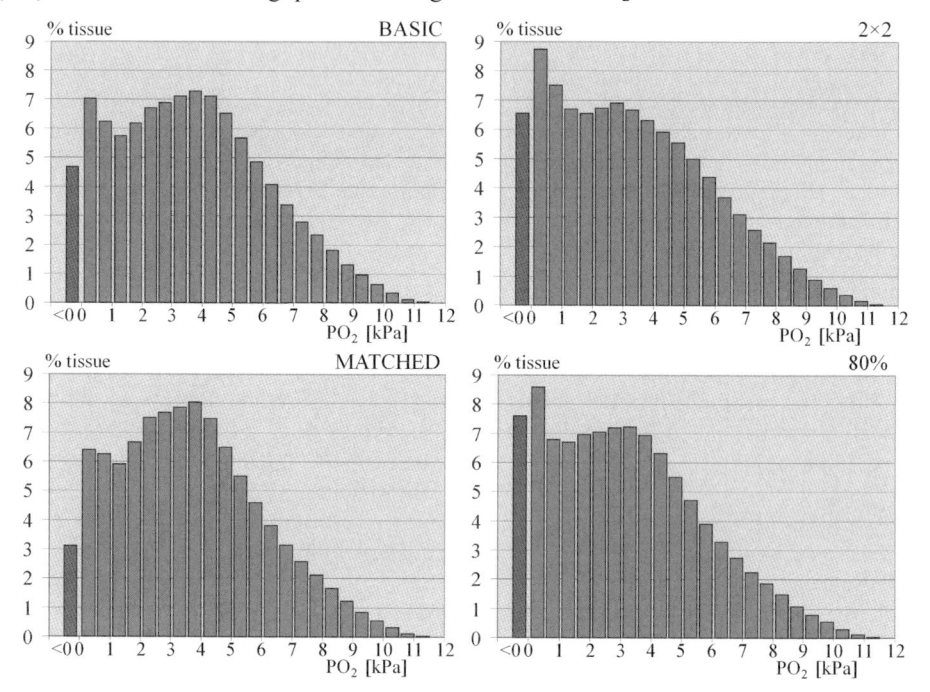

Figure 1. PO_2 histograms, %tissue with PO_2 within the class boundaries of 0.5 kPa, calculated for the basic situation (upper left), the 4 flow regions (upper right), matched capillary flow (lower left), and 20% overall decreased flow (lower right).

During the conditions of a heterogeneous flow distribution, regions of high (150%) and low (50%) flow, there was no large change in the tissue PO_2 distribution. For the 2 × 2 case (upper right panel in Figure 1), where two large regions of 160 × 160 μm suffer from halved flow, the anoxic tissue increased with only 1.9% to 6.6%, and mean PO_2 decreased no further than to 3.33 kPa. For the 4 × 4 case, these figures were 6.8% and 3.32 kPa. The standard deviation was virtually unaltered in both situations; 2.51 and 2.46 kPa, respectively vs. 2.46 kPa in the BASIC situation

As expected, matched flow (lower left panel in Figure 1) resulted in a marginally better oxygenation; there remained anoxic regions, now 3.1%, whereas mean PO_2 was unchanged, 3.60 kPa and the histogram became slightly narrower, SD = 2.35 kPa.

Also as expected, decreased overall flow (lower right panel in Figure 1) worsened oxygenation, but only moderately. The anoxic tissue portion increased to 7.6% and mean tissue PO_2 decreased to 3.22 kPa; histogram SD was unchanged, 2.44 kPa.

4. DISCUSSION

Tissue is supplied with oxygen by the blood in the capillary network and consequently capillary distribution within the tissue is a major determinant of tissue PO_2. Beside capillarisation also the blood flow through the capillaries is important for tissue oxygenation. Here we addressed the question how the distribution of blood flow over the various capillaries affects the oxygenation of the tissue. To do so, it is important, that the capillary geometry is adequately accounted for. Since the mathematical model allows for individually independent capillary data[6], it is suitable for theoretical predictions of the impact of alterations of flow distribution on tissue oxygenation.

The most remarkable finding is, that flow redistribution only marginally affects tissue oxygenation. This is best appreciated when comparing this outcome with the impact of, for instance, a homogeneous distribution of capillaries on tissue oxygenation. We calculated a case where all capillaries were equidistant in a rectangular grid of 42 μm spacing, a capillary density identical to the current situation. Then, the mean tissue PO_2 becomes 4.45 kPa vs. 3.62 kPa in the basic situation, the histogram narrows and there are no PO_2s below 1.2 kPa. The flow redistributions we considered here have much less effect. This is in line with earlier findings, that heterogeneity of capillary spacing is by far the most important factor in tissue oxygenation[2, 5, 18].

We also calculated situations where the redistribution of flow was more marked than presented here. A 40% overall flow decrease gave 13% anoxic tissue and a 190% – 10% flow redistribution in the checkerboard cases resulted in 10% anoxic tissue. These calculations, however, were considered less reliable; the model will overestimate the anoxic portion, because the term $\Phi(\overset{u}{r})$ was applied for uniform O_2 consumption and anoxic regions do not consume oxygen. So in fact, the anoxic portion in the present calculations will also be somewhat too high. Thus, considering that the amount of anoxic tissue is overestimated with our model it is even more remarkable that the 190% – 10% 2 × 2 case, making a region of 160 × 160 μm virtually devoid of blood flow, had such a limited effect on tissue PO_2.

The current calculations can only be compared with results also for a heterogeneous tissue layout. Recently, Goldman et al.[18] considered a situation of 24 capillaries in a 165 × 155 μm rectangular field in a numerical calculation. Because of their different objective, results are not well comparable. They found only a moderate PO_2 decrease for a flow reduction of baseline to 75%, but a sudden and significant increase in anoxic tissue for a reduction from 50% to 40%. The current model is not numerical but semi-analytical. Both methods have their advantages and disadvantages. The current semi-analytical model runs very fast (a few seconds on a standard PC) even for the 91 capillaries and it has no trouble with boundary conditions, as numerical models do. However, the term $\Phi(\overset{u}{r})$ in the model will have to be adapted for anoxic regions if a better analysis of such situations is to be done.

5. REFERENCES

1. L. Hoofd and Z. Turek, in: *Adv. Exper. Med. Biol., Vol. 345*, edited by P. Vaupel, R. Zander, and D.F. Bruley (Plenum Press, New York and London, 1994), pp. 275–282.
2. Z. Turek, L. Hoofd, S. Batra, and K. Rakusan, The effect of realistic geometry of capillary networks on tissue Po₂ in hypertrophied rat heart. in: *Adv. Exper. Med. Biol., Vol. 317,* edited by W. Erdmann and D.F. Bruley (Plenum Press, New York and London, 1992) pp. 567–572.

3.   H. Degens, D. Deveci, A. Botto-van Bemden, L.J.C. Hoofd, and S. Egginton, Maintenance of heterogeneity of capillary spacing is essential for adequate oxygenation in the soleus muscle of the growing rat. *Microcirc.* **13**, 467–476 (2006)

4.   L. Hoofd, Calculation of oxygen pressures in tissue with anisotropic capillary orientation. I: Two-dimensional analytical solution for arbitrary capillary characteristics. *Math. Biosci.* **129**, 1–23 (1995)

5.   L. Hoofd, Updating the Krogh model - assumptions and extensions. in: *Oxygen Transport in Biological Systems, Soc. Exper. Biol. Seminar Series 51,* edited by S. Egginton and H.F. Ross (Cambridge University Press, Cambridge, 1992) pp. 197–229.

6.   C. Bos, L. Hoofd, and T. Oostendorp, The effect of separate red blood cells on capillary tissue oxygenation calculated with a numerical model. *IMA J. Math. Appl. Med. Biol.* **13**, 259–274 (1996)

7   L. Hoofd and C. Bos, Extraction pressures calculated for rat heart and dog skeletal muscle and application in models of tissue oxygenation. in: *Adv. Exper. Med. Biol., Vol. 428*, edited by D.K. Harrison and D.T. Delpy (Plenum Press, New York, 1998) pp. 679–685.

8.   C.G. Bos, *Mathematical modelling of oxygen transport from capillaries to tissue,* (Dissertation Thesis, University of Nijmegen, the Netherlands, 1997) pp. 78-83

9.   D. Goldman and A.S. Popel, A computational study of the effect of capillary network anastomoses and tortuosity on oxygen transport, *J. Theor. Biol.* **206**, 181–194 (2000)

10. L. Hoofd, Calculation of oxygen pressures in tissue with anisotropic capillary orientation. II: Coupling of two-dimensional planes. *Math. Biosci.* **129**, 25–39 (1995)

11. B.J. van Beek-Harmsen, M.A. Bekedam, H.M. Feenstra, F.C. Visser, and W.J. van der Laarse, Determination of myoglobin concentration and oxidative capacity in cryostat sections of human and rat skeletal muscle fibres and rat cardiomyocytes. *Hist. Cell Biol.* **121**, 335-342 (2004).

12. A. Krogh, The number and distribution of capillaries in muscles with calculations of the oxygen pressure head necessary for supplying the tissue. *J. Physiol.* **52**, 409–415 (1919)

13. W.J. Federspiel, A model study of intracellular oxygen gradients in a myoglobin-containing skeletal muscle fiber. *Biophys. J.* **49**, 857–868 (1986)

14. P.L. Altman, J.F. Gibson, and C.C. Wang, in: *Handbook of Respiration,* edited by D.S. Dittmer and R.M. Grebe (Saunders, Philadelphia & London, 1958) pp. 6–9.

15. L.F.M. van Zutphen, V. Baumans, and A.C. Beynen (eds.), *Proefdieren en dierproeven.* Bunge Scientific Publishers, Utrecht, 1991 (1st ed), pp. 1-365.

16. R.B. Armstrong, and M.H. Laughlin, Blood flows within and among rat muscles as a function of time during high speed treadmill exercise. *J. Physiol.* **344**, 189-208 (1983)

17. Y. Kano, S. Shimegi, H. Furukawa, H. Matsudo, and T. Mizuta, Effects of aging on capillary number and luminal size in rat soleus and plantaris muscles. *J. Gerontol.* **57**, B422-B427 (2002)

18. D. Goldman, R.M. Bateman, and C.G. Ellis, Effect of decreased O_2 supply on skeletal muscle oxygenation and O_2 consumption during sepsis: role of heterogeneous capillary spacing and blood flow. *Am. J. Physiol. Heart Circ. Physiol.* **290**, 2277–2285 (2006).

HETEROGENEITY OF CAPILLARY SPACING IN THE HYPERTROPHIED PLANTARIS MUSCLE FROM YOUNG-ADULT AND OLD RATS

Hans Degens[1], Christopher I. Morse, and Maria T.E. Hopman

Abstract: Heterogeneity of capillary spacing may affect tissue oxygenation. The determinants of heterogeneity of capillary spacing are, however, unknown. To investigate whether 1) impaired angiogenesis and increased heterogeneity of capillary spacing delays development of hypertrophy during aging and 2) heterogeneity of capillary spacing is determined by variations in fiber size we overloaded the left m. plantaris in young-adult (5-month-old) and old (25-month-old) rats for 1, 2 or 4 weeks by denervation of synergists, while the right leg served as an internal control. Fiber size, capillary density and capillary to fiber ratio were similar in control young-adult and old muscles. The time course and degree of hypertrophy were similar at both ages, indicating that in rats up to the age of 25 months the hypertrophic response is maintained. The variation in fiber size and heterogeneity of capillary spacing were, however, larger in old than young-adult muscles, larger in the superficial than the deep region of the muscle, and correlated significantly (R = 0.558; P < 0.001). This suggests that part of the heterogeneity of capillary spacing is due to heterogeneity in fiber size and may reflect that morphological constraints for positioning of capillaries partly determines heterogeneity of capillary spacing in muscle.

1. INTRODUCTION

The increase in fiber size and capillary proliferation during compensatory hypertrophy of the plantaris muscle follow a similar time course[1]. This suggests that growth of muscle fibers and angiogenesis are tightly linked. During ageing, however, the

[1]Hans Degens, Institutte for biophysical and Clinical Research into Human Movement, Manchester Metropolitan University, Alsager, ST7 2HL, UK Phone: +44.161.2475686, Fax: +44.161.2476375, E-mail: h.degens@mmu.ac.uk. Chris I. Morse Manchester Metropolitan University, Alsager, UK. Maria T.E. Hopman, Radboud University Medical Centre, Nijmegen, The Netherlands.

P. Liss et al. (eds.), *Oxygen Transport to Tissue XXX*, DOI 10.1007/978-0-387-85998-9_10,
© Springer Science+Business Media, LLC 2009

capillary to fiber ratio and capillary density are decreased[2,3]. This, one may expect, could impair the hypertrophic response and even more so if one considers that the angiogenic potential is reduced at old age[4]. Further evidence for attenuated muscle plasticity at old age are the delayed response to chronic electrical stimulation[5] and an impaired regeneration after denervation[6]. Therefore, in the present study we will compare the time course of hypertrophy in young-adult and old rats.

Not only a reduced number of capillaries, but also an increased heterogeneity of capillary spacing will negatively affect tissue oxygenation[7-10]. Little is known, however, about the determinants of capillary spacing. As capillaries are never found within a muscle fiber the positioning of capillaries is limited and we propose that the heterogeneity of capillary spacing is at least partly determined by the heterogeneity of fiber sizes. Aging[11] and hypertrophy[12] have been reported to be accompanied by an increased variability in fiber size and are therefore potentially interesting models to investigate whether heterogeneity of fiber size is a determinant of the heterogeneity of capillary spacing.

2. MATERIALS AND METHODS

2.1 Animals

Young-adult and old male Wistar rats were housed two to a cage and had free access to food and water. The environment was maintained at 22^0C at a 2 h light 12 h dark cycle. The rats were randomly assigned to groups in which the left plantaris muscle was subjected to 1, 2 or 4 weeks of overload, while the right plantaris muscle served as an internal control, as described previously[3,13]. Briefly, the branches of the n. ischiadicus to the gastrocnemius and soleus muscles were cut and sewed into the biceps femoris to prevent reinnervation. As a result of this procedure the plantaris muscle is overloaded and develops compensatory hypertrophy. After surgery rats received a subcutaneous injection of Rimadyl (0.5 mg·kg^{-1}) as an analgesic. The surgeries to induce overload of the plantaris muscle were staggered so that all animals of the young-adult group were 5- and those of the old group were 25-months old at the time of the terminal experiment, irrespective of their muscles being subjected to 1, 2 or 4 weeks of overload. The animals were killed by an intraperitoneal injection of an overdose of pentobarbital sodium and the plantaris muscles excised, blotted dry, weighed, frozen in liquid N_2 and stored at -80^{0}C until analysis.

2.2 Determination of capillarization

Transverse 12-μm sections of the plantaris muscle were cut at -20^{0}C in a cryostat, and stained for alkaline phosphatase as described previously[3,8,13] to reveal all anatomically present capillaries. The capillarization was analyzed with the method of capillary domains[3,7,8,13,14]. In addition to conventional measures of muscle capillarization, such as the numerical capillary density (CD) and capillary to fiber ratio (C:F), this method gives the oxygen supply area of a capillary and an indication of the heterogeneity of capillary spacing. Heterogeneity of capillary spacing may have a pronounced effect on muscle oxygenation[7,8]. Briefly, sections were photographed and capillary coordinates read with a digitizing tablet (Model MMII 1201, Summagraphics, TX, USA).

Table 1. Indices of hypertrophy and capillary supply in the deep and superficial region of the plantaris muscle of 5-month- and 25-month-old male Wistar rats in which the right plantaris muscle was overloaded for 1, 2 or 4 weeks by denervation of the gastrocnemius and soleus muscles. The contralateral leg served as control (C)

	5-month-old				25-month-old			
	C $n=15$	1 week $n=5$	2 week $n=4$	4 week $n=5$	C $n=10$	1 week $n=3$	2 week $n=3$	4 week $n=3$
MM (mg)	429 ± 8	480 ± 12[1]	513 ± 20[1]	511 ± 17[1]	428 ± 17	461 ± 15[1]	496 ± 19[1]	543 ± 52[1]
MM:BM (mg·g⁻¹)	0.97 ± 0.02	1.08 ± 0.03[1]	1.16 ± 0.03[1]	1.17 ± 0.05[1]	0.71 ± 0.02[2]	0.79 ± 0.04[1,2]	0.82 ± 0.03[1,2]	0.87 ± 0.07[1,2]
Deep								
FCSA (μm²)	2508 ± 101	2880 ± 112	3045 ± 358	3107 ± 276	2572 ± 143	2864 ± 379	2856 ± 163	3722 ± 213
FCSA SD (μm²)	761 ± 63	760 ± 56	726 ± 88	814 ± 108	899 ± 75[2]	1205 ± 275[2]	1138 ± 166[2]	1611 ± 237[1,2]
C:F	2.26 ± 0.04	2.22 ± 0.13	2.26 ± 0.11	2.52 ± 0.19	2.44 ± 0.10	2.41 ± 0.36	2.40 ± 0.15	2.43 ± 0.14
CD (mm⁻²)	975 ± 49	836 ± 63	857 ± 105	895 ± 90	1007 ± 70	855 ± 62	890 ± 30	727 ± 95
R (μm)	18.20 ± 0.41	19.57 ± 0.74	19.82 ± 1.52	18.90 ± 1.00[1]	18.04 ± 0.58	19.23 ± 0.66	19.18 ± 0.36	21.24 ± 1.21[1]
DOM (μm²)	1056 ± 46	1210 ± 89	1255 ± 200	1135 ± 119[1]	1032 ± 63	1164 ± 81	1156 ± 44	1427 ± 158[1]
Log$_D$SD	0.166 ± 0.003	0.165 ± 0.007	0.158 ± 0.008	0.168 ± 0.005[1]	0.180 ± 0.009[2]	0.187 ± 0.009[2]	0.176 ± 0.006[2]	0.185 ± 0.004[2]
Superficial								
FCSA (μm²)	2999 ± 116	3817 ± 492	3658 ± 292	4247 ± 387	3165 ± 252	3501 ± 267	4118 ± 1004	4797 ± 270
FCSA SD (μm²)	972 ± 59	975 ± 159	1029 ± 76	1453 ± 349	1315 ± 126[2]	1717 ± 185[2]	1603 ± 451[2]	1700 ± 201[2]
C:F	1.50 ± 0.11	1.51 ± 0.27	1.87 ± 0.23	1.62 ± 0.25	1.57 ± 0.16	1.72 ± 0.12	1.99 ± 0.41	1.82 ± 0.10
CD (mm⁻²)	548 ± 36	433 ± 79	592 ± 73	423 ± 53	525 ± 38	510 ± 20	555 ± 96	417 ± 90
R (μm)	24.33 ± 0.90	27.36 ± 2.16	23.77 ± 1.43	27.85 ± 1.77	24.36 ± 0.72	24.83 ± 0.53	24.46 ± 1.63	27.42 ± 1.96
DOM (μm²)	1901 ± 151	2412 ± 361	1795 ± 214	2476 ± 311	1879 ± 109	1939 ± 84	1896 ± 245	2386 ± 350
Log$_D$SD	0.185 ± 0.005	0.193 ± 0.013	0.161 ± 0.008	0.199 ± 0.014	0.202 ± 0.006[2]	0.208 ± 0.010[2]	0.180 ± 0.006[2]	0.211 ± 0.014[2]

MM: muscle mass; BM: body mass; FCSA: fiber cross-sectional area; FCSA SD: standard deviation of the FCSA; C:F: capillary to fibre ratio; CD: capillary density; R: radius of capillary domain area; DOM: capillary domain area; Log$_D$SD: logarithmic standard deviation of the capillary domain area; Deep and superficial region differ for each parameter at $p < 0.001$; [1]: significantly different from control at $p < 0.05$; [2]: significantly different from young-adult.

Domains were constructed, defined as the area surrounded by a capillary delineated by equidistant boundaries from adjacent capillaries. The area of the capillary domain provides an estimate of the oxygen supply area of a particular capillary. The radius (R) of a domain was calculated from a circle with the same area, giving an estimate for the maximal diffusion distance from the capillary. The $\log_D SD$ of the domain areas gives an estimate of the heterogeneity of capillary spacing. The fiber cross-sectional area (FCSA) was determined by tracing the outlines of the fibers on the digitizing tablet.

2.3 Statistics

As the deep and superficial regions differ from each other in capillary supply and FCSA the two regions were analyzed separately with a two-way ANOVA for age and duration of overload. If a significant duration effect was found a Dunnett 1-sided post-hoc test was performed to determine which time-points differed from the control situation. Effects, interactions and differences were considered significant at $P \leq 0.05$. Data are presented as mean $\pm$ SEM.

3. RESULTS

3.1 Hypertrophy

The mass of the control plantaris muscles were similar in young-adult and old rats. As the old were heavier than young-adult rats (body mass at end of experiment: 442 ± 5 g, n = 15 vs. 608 ± 21 g, n = 9; p < 0.001), the muscle:body mass (MM:BM) ratio was lower in old than young-adult animals. Hypertrophy in terms of change in muscle mass or MM:BM was already evident after 1 week overload (Table 1) and was also reflected by an increase in FCSA after 2 and 4 weeks of overload in both the deep and superficial regions of the muscle (Table 1). The absence of age*duration interactions indicates that the pattern of hypertrophy was similar in young-adult and old rats.

3.2 Capillarization

There were no age*duration interactions for any of the parameters of capillary supply, indicating that changes in capillary supply during hypertrophy followed a similar pattern in young-adult and old rats. Despite significant hypertrophy the C:F ratio was not significantly changed in either the deep or superficial region of the muscle during 4 weeks of overload (Table 1). In the deep region this resulted in significantly increased diffusion distances after 4 weeks overload, as reflected by increased R and domain area, and a reduction in the CD, while in the superficial region these changes were not significant (Table 1). The latter may be due to the relatively low number of animals or the relatively short duration of overload, as previously we observed after 6 weeks overload significant alterations also in the superficial region of the muscle[3,13].

3.3 Heterogeneity of capillary spacing

$\text{Log}_D SD$ was significantly larger in the superficial than the deep region of the plantaris muscle (Table 1). In both regions $\text{Log}_D SD$ was significantly larger in old than young-adult rats, but was not significantly changed during hypertrophy (Table 1). A similar pattern was observed for the heterogeneity of fiber size, which was larger in the superficial than deep region of the muscle and increased in both regions with age (Table 1). Figure 1A illustrates the difference in the distribution of fiber sizes in young-adult and old rats for the muscle as a whole (deep and superficial fibers pooled). Figure 1B shows a significant correlation between the heterogeneity of fiber sizes (FCSASD) and the heterogeneity of capillary spacing (Log_DSD).

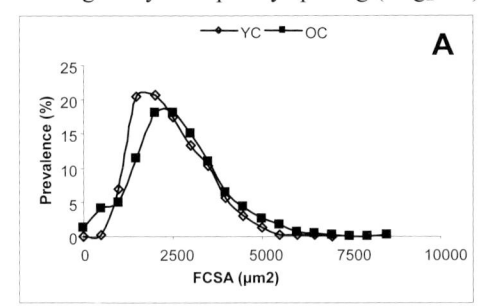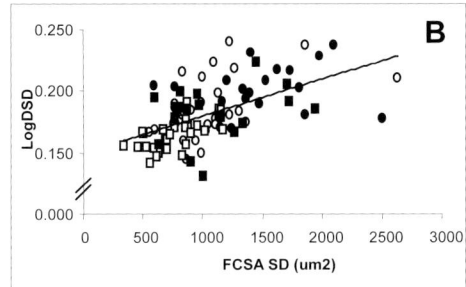

Figure 1. A) The distribution of fiber sizes (FCSA) and **B)** the relation between the FCSASD (heterogeneity of fiber size) and heterogeneity of capillary spacing (Log_DSD) in the deep (Y: □; O: ■) and superficial (Y:○; O: ●) regions of the plantaris muscle from young-adult (YC: 5-month-old) and old (O: 25-month-old) male rats. The regression line is for all data pooled; R = 0.558, n = 96, P < 0.001.

4. DISCUSSION

We observed, that the quantitative capillary supply and the response to hypertrophy of the rat plantaris muscle are similar in young-adult and old (25-month-old) rats. The plantaris muscle of old rats, however, exhibited a larger variation of fiber sizes and heterogeneity of capillary spacing. Also the larger heterogeneity of capillary spacing in the superficial than the deep region of the muscle was associated with a larger variability in fiber sizes in the superficial region. This and the significant correlation between the variability of fiber size and heterogeneity of capillary spacing suggest that the fiber size heterogeneity is one of the determinants of capillary distribution.

In line with previous observations[3] the plantaris muscle from young-adult and old rats developed a similar degree, and followed a similar time-course, of hypertrophy. In young-adult rats the capillary density was decreased only after 3 weeks of overload[1] and here we extend this observation by showing that also in old rats the capillary density is diminished 4 weeks after induction of hypertrophy, while before that the capillary density is maintained. Thus, while in 24-month-old rats the response to chronic electrical stimulation[5] and regeneration from denervation[6] appear to be attenuated, we obtained no evidence for an impaired hypertrophic response. It should be noted, however, that in rats older than the old rats in the present study the hypertrophic response is impaired[15].

A heterogeneous distribution of capillaries may cause an impaired tissue oxygenation[7-10]. Since the capillary supply is dependent on oxidative capacity and fiber size[2] one might expect that the larger heterogeneity of capillary spacing in the superficial than the deep region is explained by the smaller fibers and higher oxidative capacity in the latter region. It is unlikely, however, that these factors explain the capillary distribution within muscle as while the capillary density in slow and fast eel muscles differs 35-fold, the heterogeneity in capillary spacing is similar in both muscles[16] and despite a 5-fold increase in fiber size during normal muscle growth[7], or during hypertrophy as observed in the present study, no change in the heterogeneity of capillary spacing was observed. Here we propose that the heterogeneity of fiber size is a determinant of the capillary distribution within the muscle. In support of this, we observed that the larger heterogeneity of capillary spacing in 1. old than young-adult muscles and 2. the superficial than the deep region of the muscle were associated with a larger variability in fiber size. Moreover, there appeared to be a significant correlation between the variability in fiber size and the heterogeneity of capillary spacing. This may indicate that physical constraints to positioning of capillaries at least partly determine the distribution of capillaries in muscle.

5. ACKNOWLEDGEMENTS

We appreciate the help of J.A.M Evers and M. de Jong with the surgery and digitizing.

6. REFERENCES

1. M. J. Plyley, B. J. Olmstead, and E. G. Noble, Time course of changes in capillarization in hypertrophied rat plantaris muscle, J Appl Physiol 84, 902-907 (1998).
2. H. Degens, Age-related changes in the microcirculation of skeletal muscle, Adv Exp Med Biol 454, 343-348 (1998).
3. H. Degens, Z. Turek, L. J. Hoofd, and R. A. Binkhorst, Capillary proliferation related to fibre types in hypertrophied aging rat M. plantaris, Adv Exp Med Biol 345, 669-676 (1994).
4. A. Rivard, J. E. Fabre, M. Silver, D. Chen, T. Murohara, M. Kearney, M. Magner, T. Asahara, and J. M. Isner, Age-dependent impairment of angiogenesis, Circulation 99, 111-120 (1999).
5. T. J. Walters, H. L. Sweeney, and R. P. Farrar, Influence of electrical stimulation on a fast-twitch muscle in aging rats, J Appl Physiol 71, 1921-1928. (1991).
6. K. K. White, and D. W. Vaughan, The effects of age on atrophy and recovery in denervated fiber types of the rat nasolabialis muscle, Anat Rec 229, 149-158 (1991).
7. H. Degens, D. Deveci, A. Botto-van Bemden, L. J. Hoofd, and S. Egginton, Maintenance of heterogeneity of capillary spacing is essential for adequate oxygenation in the soleus muscle of the growing rat, Microcirculation 13, 467-476 (2006).
8. H. Degens, B. E. Ringnalda, and L. J. Hoofd, Capillarisation, fibre types and myoglobin content of the dog gracilis muscle, Adv Exp Med Biol 361, 533-539 (1994).
9. D. Goldman, R. M. Bateman, and C. G. Ellis, Effect of decreased O2 supply on skeletal muscle oxygenation and O2 consumption during sepsis: role of heterogeneous capillary spacing and blood flow, Am J Physiol Heart Circ Physiol 290, H2277-2285 (2006).
10. J. Piiper, and P. Scheid, Diffusion limitation of O2 supply to tissue in homogeneous and heterogeneous models, Respir Physiol 85, 127-136 (1991).
11. L. M. Snow, L. K. McLoon, and L. V. Thompson, Adult and developmental myosin heavy chain isoforms in soleus muscle of aging Fischer Brown Norway rat, Anat Rec A Discov Mol Cell Evol Biol 286, 866-873 (2005).
12. G. R. Chalmers, R. R. Roy, and V. R. Edgerton, Variation and limitations in fiber enzymatic and size responses in hypertrophied muscle, J Appl Physiol 73, 631-641 (1992).
13. H. Degens, Z. Turek, L. J. Hoofd, M. A. Van't Hof, and R. A. Binkhorst, The relationship between capillarisation and fibre types during compensatory hypertrophy of the plantaris muscle in the rat, J Anat 180, 455-463 (1992).
14. L. Hoofd, Z. Turek, K. Kubat, B. E. Ringnalda, and S. Kazda, Variability of intercapillary distance estimated on histological sections of rat heart, Adv Exp Med Biol 191, 239-247 (1985).
15. H. Degens, and S. E. Alway, Skeletal muscle function and hypertrophy are diminished in old age, Muscle Nerve 27, 339-347. (2003).
16. S. Egginton, Z. Turek, and L. J. Hoofd, Differing patterns of capillary distribution in fish and mammalian skeletal muscle, Respir Physiol 74, 383-396 (1988).

MICROVASCULARITY OF THE LUMBAR ERECTOR SPINAE MUSCLE DURING SUSTAINED PRONE TRUNK EXTENSION TEST

Rammohan V. Maikala,[1] and Yagesh N. Bhambhani.

Abstract: This study evaluated the reliability of oxygenation and blood volume responses, from the right erector spinae in twenty two healthy men and women, during static prone trunk extension on two separate days. Near-infrared spectroscopy (NIRS)-derived *physiological change* for oxygenation was calculated as the difference between the 'baseline' before the start of the trunk extension and 'minimum' at the point of volitional exhaustion. The *physiological change* for blood volume was calculated as the difference between the 'baseline' value and 'maximum' at the point of volitional exhaustion. Test-retest reliability, based on the intraclass correlation coefficients for the *physiological change* were: oxygenation - men: +0.60 versus women: +0.37; blood volume - men: +0.93 versus women: +0.59, respectively. Results suggest that NIRS-derived blood volume measurements were more reliable than the oxygenation responses. The most interesting observation of the study was the hyperemia in blood volume responses with a parallel decrease in oxygenation as participants continued the test until volitional exhaustion. Such an increase in muscle blood volume contradicts the theory that sufficient occlusion of blood flow to the lumbar muscle region is possible with static trunk extension resulting in muscle fatigue.

1. INTRODUCTION

There is a growing body of literature on the importance of applying Near-infrared spectroscopy (NIRS) in the evaluation of lumbar extensor muscle haemodynamics during a variety of static and dynamic activities.[1-7] The prone trunk extension method,[8] also known as the "Sørensen test" or "couch method" has been widely utilized to evaluate static back muscle endurance and fatigue in healthy and low-back pain populations. In

---

[1] Rammohan V. Maikala, PhD, Liberty Mutual Research Institute for Safety, 71 Frankland Road, Hopkinton, MA 01748, USA. Yagesh N. Bhambhani, PhD Faculty of Rehabilitation Medicine, University of Alberta, Edmonton, Alberta, T6G 2G4, Canada.

P. Liss et al. (eds.), *Oxygen Transport to Tissue XXX*, DOI 10.1007/978-0-387-85998-9_11,
© Springer Science+Business Media, LLC 2009

the trunk musculature, women have greater percentages of slow twitch fibers,[9] but have 10-14% less hemoglobin than men, resulting in lower oxygen carrying capacity.[10] However, it is not clear if oxygen saturation in the lumbar muscle of women differs from men during the Sørensen test. Therefore, the present investigation was intended to examine the lumbar muscle oxygenation and blood volume trends and their reliability within both men and women during static endurance activity until volitional exhaustion. It was hypothesized that the physiological change within the NIRS region of interest in women would differ from men.

2.   METHODS

Written informed consent that conformed to the Declaration of Helsinki was obtained from healthy 11 men (mean ± SD: age 23.7 ± 4.5 years) and 11 women (age 24.5 ± 4.0 years). They all completed the Revised Physical Activity Readiness Questionnaire of the Canadian Society for Exercise Physiology to identify contraindications for exercise. The human research ethics board at the University approved the test protocol.

2.1 Static Back Muscle Endurance Test

Each participant completed the Sørensen test on two separate days but at the same time of the day. These tests were separated by at least one week to avoid delayed onset muscle soreness. The participant was asked to lie prone on an adjustable plinth, with the anterior border of the iliac crest adjusted to the edge of a marker on the plinth. One pillow was placed on the mid-thigh and another beneath the tibialis region with two elastic straps positioned at mid-thigh and mid-calf region. These straps were tightened so that the participant's lower body was completely restrained at those positions. An adjustable rope with a weight attached at its edge was lowered to the position between the participant's scapula to serve as a reference point for the trunk alignment. The participant was rested for two minutes before beginning of the test. Just five seconds before the endurance test began, a countdown was started and the section of the plinth supporting only the upper body was lowered at the final countdown. The participant was verbally instructed to place the arms across their chest and minimize trunk rotations throughout the endurance session, and maintain contact with the reference point for as long as possible. When the participant was unable to maintain contact with this reference point the test was terminated, and the plinth support to the upper body was immediately restored. Endurance time to the nearest second was recorded. Thereafter, the participant was allowed to recover for four minutes in the prone position, at the end of which the session was terminated.

2.2 Oxygenation And Blood Volume Measurements

A continuous dual wave NIRS unit (MicroRunman, NIM Inc., PA, USA) was used to evaluate relative changes in the muscle oxygenation and blood volume. Before the test session, a sensor was placed on the right erector spinae muscle at the 3rd lumbar vertebra, approximately 3 cm from the midline of the spine.[7] A piece of clear plastic was wrapped around the sensor to prevent sweat from distorting the absorbency signal. A dark elastic bandage was also wrapped around the sensor to secure it in place and minimize any loss of light. Muscle skinfold thickness at the sensor location was measured twice using a Lange skinfold caliper (Cambridge Scientific Industries, Inc.

Maryland, USA) before placing the sensor on the muscle belly. Position of the sensor at the 3rd lumbar region from the mid line of the spine was recorded to the nearest millimeter and identified with a marker, thus ensuring the correct placement of the sensor on each subject for both testing sessions. As per the manufacturer's specifications, the sensor was calibrated at 760 nm and 850 nm wavelengths before the test. The light source-detector separation was set to 4 cm. The difference in absorbency between these two wavelengths indicated the change in oxygenation whereas the sum of these absorbencies indicated the change in total muscle blood volume. Both oxygenation and blood volume were measured in terms of optical density (od) where 'od' is defined as the logarithmic ratio of intensity of light calibrated at each wavelength to the intensity of measured light at the same wavelength. Real time data were recorded using the software provided by the manufacturer.

3. DATA ANALYSIS

All the real time NIRS-derived responses were averaged every 5 seconds using a customized Microsoft Excel™ macro written for this study. Baseline values were recorded during the initial rest. The *physiological change* for oxygenation was calculated as the difference between the 'baseline' before the start of the trunk extension and 'minimum' at the point of volitional exhaustion. The *physiological change* for blood volume was the difference between the 'baseline' value and 'maximum' at the point of volitional exhaustion. Adipose tissue thickness (a sum of fat and skin layer) at the site of NIRS measurements was calculated as the mean value of skinfold thickness (in mean $\pm$ SD) - men: 5.4 ± 1.5 mm; and women: 6.0 ± 2.3 mm.

4. STATISTICAL ANALYSIS

Independent 't' tests were used to compare the physical characteristics of the men and women. A two-way analysis of covariance, using the adipose tissue thickness of the lumbar muscle as the covariate with gender as a between-subjects factor, and day of the protocol (day 1 and day 2) as a repeated measure factor was used to compare the physiological change in oxygenation and blood volume responses. The statistical significance was considered at $P < 0.05$. Significant F ratios were further analyzed with the Scheffe post hoc multiple comparison test. Intraclass correlations were calculated to establish the test-retest reliability of NIRS-derived measurements. A nonparametric Spearman correlation test was used to examine the relationship between logarithmic value of adipose tissue thickness and oxygenation and blood volume responses during three protocols. The Statistical Package for Social Sciences SPSS (version 15) was used for all statistical analyses (SPSS Inc., Chicago, USA).

5. RESULTS

The men were heavier than the women (mean $\pm$ SD in kg: 77.5 ± 13.8 versus 57.6 ± 13.8, respectively; $P < 0.01$) and taller (in cm: 176 ± 8 versus 163 ± 6, respectively; $P < 0.05$). The oxygenation and blood volume trends in a typical male and female participant are shown in the Figures 1 and 2. No significant interaction between the day of the protocol and gender was observed in either the oxygenation or blood volume responses (Table 1, P>0.05).

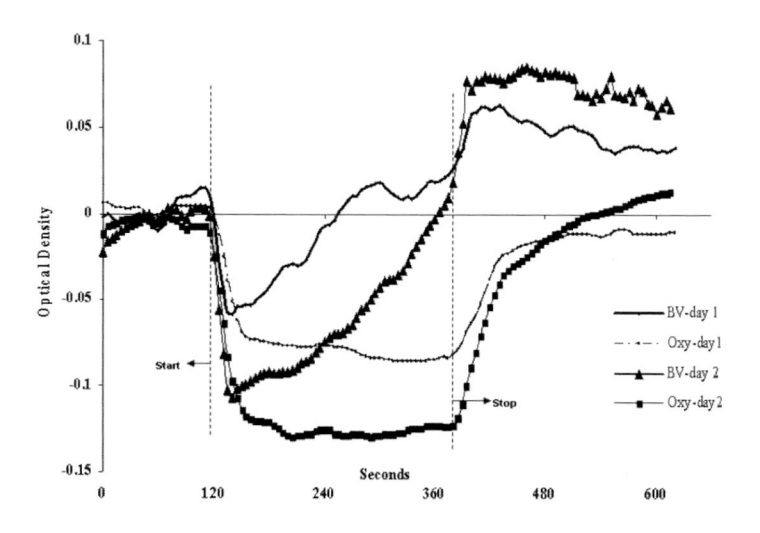

Figure 1, NIRS-determined oxygenation (Oxy) and blood volume (BV) trends of a typical female on day1 and day 2 in the lumbar muscle during sustained prone trunk extension until volitional exhaustion.

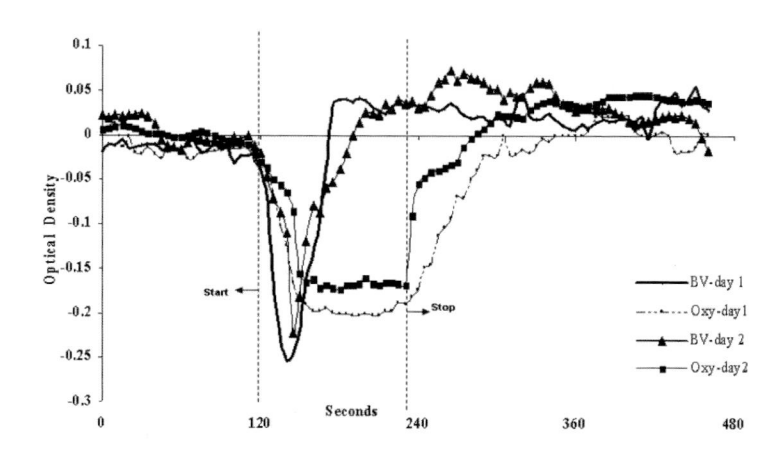

Figure 2, NIRS-determined oxygenation (Oxy) and blood volume (BV) trends of a typical male on day1 and day 2 in the lumbar muscle during sustained prone trunk extension until volitional exhaustion.

Table 1. Physiological change during the Sørensen test until volitional exhaustion.

		Endurance time in seconds	Oxygenation in od	Blood volume in od
Men	Day 1	139 ± 53	0.100 ± 0.10	0.110 ± 0.26
	Day 2	135 ± 54	0.088 ± 0.18	0.075 ± 0.10
Women	Day 1	203 ± 72	0.084 ± 0.06	0.066 ± 0.11
	Day 2	205 ± 74	0.061 ± 0.06	0.080 ± 0.05

Test-retest reliability, based on the intraclass correlation coefficients for the physiological change were: oxygenation - men: +0.60 *versus* women: +0.37; blood volume - men: +0.93 *versus* women: +0.59, respectively. In terms of endurance time, intraclass correlations were: men: +0.97 *versus* women: +0.98. Adipose tissue thickness calculated from the right side of erector spinae skinfolds at the 3rd lumbar region for men and women was not significantly different from each other ($P>0.05$).

6. DISCUSSION

It is evident from the low to high test-retest reliability results that physiological changes in oxygenation and blood volume measurements were highly reproducible during both days. At the beginning of the contraction period (Figures 1 and 2), the majority of participants demonstrated a rapid muscle desaturation (as evidenced by the decrease in oxygenation) resulting in restriction of blood flow due to a greater intramuscular pressure in the lumbar muscle. As the exhaustion point was reached, there was a plateau in oxygenation trends, suggesting a local homeostatic adjustment over a period may result in constant capillary oxygen tension.[1] This phenomenon might further prevent decrease in haemodynamics in these paraspinal muscles. Concomitantly, as the trunk extension continued, a systematic increase in blood volume trend was observed (Figures 1 and 2). At final stages, the majority of participants demonstrated a recovery in blood volume to pretest levels. Such hyperemia in blood volume, with a parallel decrease in oxygenation during trunk extension until volitional exhaustion, suggests that complete occlusion of arterial flow to the lumbar muscle was not possible with this static protocol.

An out-of-phase haemodynamic trend in the lumbar muscle (Figures 1 and 2) is similar to insufficient cuff occlusion trends of some leg muscles. Rundell et al.[11] demonstrated hyperemia in vastus medialis and biceps femoris muscles, with a simultaneous decrease in the vastus laterlais and rectus femoris muscles, suggesting a complete arterial occlusion is not feasible for all the four leg muscles. Since erector spinae is a collection of three muscles: illiocostalis, longissimus, and spinalis, the static test chosen for the present study might not have achieved complete reduced conditions, thus demonstrating hyperemic response in blood volume but decrease in oxygenation (Figures 1 and 2). This important finding is in contradiction to the report that the Sørensen test induces at least 40% of maximal voluntary contractions to lumbar muscles,[9] and such levels of contraction might result in sufficient occlusion of blood flow to the paraspinal extensor muscles.[12] Therefore, based on the physiological change (Table 1), the static test was not able to establish vascular occlusion for the lumbar muscle. However, one should be cautious in extrapolating the blood volume changes observed directly to the changes in blood flow. This is due to the fact that, in theory, an increase or decrease in the optically-derived muscle blood volume reflects the balance between local muscle blood flow (at the NIRS regions of interest), the effect of muscular contraction on vascular hemoglobin volume, metabolic vasodilation, hemoconcentration, and capillary recruitment.[13] However, an increase or decrease in NIRS-derived blood volume changes most likely reflects greater or lesser amounts of blood entering the tissue, respectively.[14]

Endurance time during the static test was significantly greater in women (by 32%, P<0.01) than in men (Table 1, Figures 1 and 2), and is in agreement with several studies.[8,9,15] It is known that greater endurance time in women might be attributed to their lower upper-body weight, and greater percentages of slow twitch fibers in the trunk musculature compared to men.[9] Although a greater decrease in deoxygenation in men is attributed to their lower oxidative capacity, gender did not influence any of the oxygenation and blood volume responses. Therefore, similar NIRS responses at the point of volitional exhaustion imply that, irrespective of the differences in endurance times, venous oxygen saturation in men and women are similar.

7. CONCLUSIONS

NIRS-derived blood volume measurements were more reliable than the oxygenation responses. The most interesting finding of the present study is that the Sørensen test is not able to occlude arterial blood flow to the lumbar muscle. Although endurance of the trunk muscles in women was much greater, their peripheral limitations at the erector spinae level are similar to those of men. It should be noted that the small sample size of the present study might have played a prominent role in not detecting significant gender differences in physiological responses.

8. ACKNOWLEDGEMENTS

This work was supported in part by the Small Faculties Grant, University of Alberta. We are indebted to Martha Roxburgh and Sharon Brintnell of the Occupational Performance and Analysis Unit, University of Alberta for laboratory space. The first author is also thankful for the manuscript assistance of Debra Larnis of the Liberty Mutual Research Institute for Safety.

9. REFERENCES

1. B. R. Jensen, K. Jorgensen, A. R. Hargens, P. K. Nielsen, and T. Nicolaisen, Physiological response to submaximal isometric contractions of the paravertebral muscles, *Spine* **24**, 2332-2338 (1999).
2. R. V. Maikala, M. Farag, and Y. N. Bhambhani, Low-back muscle oxygenation trends during simple postural variations – Intersession reliability using NIRS. *Proceedings of International Ergonomics Conference*, San Diego, USA: pp 5.9-5.12 (2000).
3. S. M. McGill, R. L. Hughson, and K. Parks, Lumbar erector spinae oxygenation during prolonged contractions: implications for prolonged work, *Ergonomics* **43**, 486-493 (2000).
4. Y. Yoshitake, H. Ue, M. Miyazaki, and T. Moritani, Assessment of lower-back muscle fatigue using electromyography, mechanomyography, and near-infrared spectroscopy, *Eur J Appl Physiol* **84**, 174-179 (2001).
5. W. J. Albert, G. G. Sleivert, J. P. Neary, and Y. Bhambhani, Monitoring individual erector spinae fatigue responses using electromyography and near infrared spectroscopy, *Can J Appl Physiol* **29**, 363-378 (2004).

6. M. Kankaanpää, W. N. Colier, S. Taimela, C. Anders, O. Airaksinen, S. M. Kokko-Aro, and O. Hanninen, Back extensor muscle oxygenation and fatigability in healthy subjects and low back pain patients during dynamic back extension exertion, *Pathophysiology* **12**, 267-273 (2005).
7. R. V. Maikala, and Y. N. Bhambhani, In vivo lumbar erector spinae oxygenation and blood volume measurements in healthy men during seated whole-body vibration, *Exp Physiol* **91**, 853-866 (2006).
8. F. Biering-Sørensen, Physical measurements as risk indicators for low back trouble over a one-year period, *Spine* **9**, 106-119 (1984).
9. K. Jorgensen, Human trunk extensor muscles – physiology and ergonomics, *Acta Physiol Scand* **160** (S637), 1-58 (1997).
10. C. L. Wells, *Women, Sport and Performance: A Physiological Perspective* (Human Kinetics, Champaign, Illinois, 1991).
11. K. W. Rundell, S. Nioka, and B. Chance, Hemoglobin/myoglobin desaturation during speed skating, *Med Sci Sports Exerc* **29**, 248-258 (1997).
12. F. Bonde-Petersen, A. L. Mork, and E. Nielsen, Local muscle blood flow and sustained contractions of human arm and back muscles, *Eur J Appl Physiol* **34**, 43-50 (1975).
13. D. S. DeLorey, J. M. Kowalchuk, and D. H. Paterson, Relationship between pulmonary O_2 uptake kinetics and muscle deoxygenation during moderate-intensity exercise, *J Appl Physiol* **95**, 113-120 (2003).
14. C. J. McNeil, B. J. Murray, and C. L. Rice, Differential changes in muscle oxygenation between voluntary and stimulated isometric fatigue of human dorsiflexors, *J Appl Physiol* **100**, 890-895 (2006).
15. T. Mayer, R. Gatchel, J. Betancur, and E. Bovasso, Trunk muscle endurance measurement – isometric contrasted to isokinetic testing in normal subjects, *Spine* **20**, 920-927 (1995).

FIBER CAPILLARY SUPPLY RELATED TO FIBER SIZE AND OXIDATIVE CAPACITY IN HUMAN AND RAT SKELETAL MUSCLE

Rob C.I. Wüst[1], Sarah L. Gibbings and Hans Degens

Abstract: The capillary supply of a muscle fiber is thought to be determined by its type, oxidative capacity, size and metabolic surrounding. Size and oxidative capacity, however, differ between fiber types. To investigate which of these factors determines the capillary supply of a myofiber most we analysed in sections from human vastus lateralis (n = 11) and rat plantaris muscle (n = 8) the type, succinate dehydrogenase activity (SDH), reflecting oxidative capacity, and capillary supply of individual fibers. Capillary fiber density differed between fiber types in rat (P < 0.03) but not in human muscle. In human muscle only, the local capillary to fiber ratio (LCFR) correlated with the integrated SDH (fiber cross-sectional area · SDH) of a fiber (R = 0.62; P < 0.001). Backward multiple regression revealed, however, that the LCFR was primarily determined by fiber size, type (R = 0.71, human) and surrounding of the fiber (R = 0.62; rat plantaris muscle), i.e. whether it came from the deep or superficial region of the muscle (all P < 0.001) and not SDH. In conclusion, size, type and metabolic surrounding rather than mitochondrial activity determine the capillary supply to a muscle fiber.

1. INTRODUCTION

Since the pioneering work of Krogh[1], many scientists have tried to understand the factors underlying the distribution of capillaries in skeletal muscle[2-5]. The blood flow to a muscle is largely dependent on the oxidative capacity of a muscle with red muscles having a higher maximal blood flow than white muscles and red muscles having a more dense capillary network than glycolytic muscles[6-8]. The capillary density of skeletal muscles varies widely between different species, but also between muscles and even

[1]Rob C.I. Wüst, Instititute for Biophysical and Clinical Research into Human Movement, Manchester Metropolitan University, Alsager, ST7 2HL, UK. Phone: +44 161 2475314, Fax: +44 161 2476375, E-mail: r.wust@mmu.ac.uk. Sarah L. Gibbings and Hans Degens, Manchester Metropolitan University, Alsager, UK.

P. Liss et al. (eds.), *Oxygen Transport to Tissue XXX*, DOI 10.1007/978-0-387-85998-9_12,
© Springer Science+Business Media, LLC 2009

within a single muscle[4]. Within the plantaris muscle for instance the deep, oxidative region has a denser capillary network than the superficial, glycolytic region[6]. The capillary supply to individual muscle fibers is thought to be determined by its oxidative capacity[6,9], size[2,6] and type[6,8]. Yet, to what extent the capillary supply of an individual fiber is determined by these factors and how these relations may vary between human and rat is not systematically investigated.

Therefore, the primary aim of the present study is to quantify to what extent fiber type, oxidative capacity, fiber size and metabolic surrounding determine the capillary supply of an individual fiber and whether or not these relations are qualitatively the same in rats and humans. As a measure for the oxidative capacity of a fiber we used sections stained for succinate dehydrogenase (SDH). It has been shown that the staining intensity of a muscle fiber for this enzyme from the citric acid cycle correlates very well with the maximum cellular oxygen uptake[10], and hence the maximum demand for oxygen that has to be supplied by the surrounding capillaries.

2. MATERIALS AND METHODS

2.1 Sampling

2.1.1 Rat Biopsies

Eight male Wistar rats were killed by an intraperitoneal injection of an overdose of pentobarbital sodium and the plantaris muscles excised, frozen in liquid N_2 and stored at -80^0C until analysis.

2.1.2 Human Biopsies

After local anesthesia with 2% lidocaine, a percutaneous biopsy of the vastus lateralis muscle (at ~50% of the muscle length) was obtained from 5 men and 6 women using a conchotome, placed on cork, frozen in liquid N_2 and stored at -80^0C.

2.2 Histochemistry

Transverse 10-μm sections were cut in a cryostat at -20^0C and stained with alkaline phosphatase (rat)[6] or lectin (human)[2] to depict capillaries. In a serial section fibers were classified as type I, IIa, or IIb with myofibrillar ATPase (pH 4.55)[11]. Another serial section was stained for SDH[10]. SDH activity was calculated as the absorbance at 660 nm per μm section thickness per second of staining time ($\Delta A660 \; \mu m^{-1} \; s^{-1}$).

2.3 Determination of capillarization

The capillarization was analyzed with the method of capillary domains[6,12]. In short, domains were constructed around capillaries, defined as the area surrounded by a capillary delineated by equidistant boundaries from adjacent capillaries. The fiber cross-sectional area (FCSA) was determined by tracing the outlines of the fibers on the digitizing tablet. The capillary supply to a fiber was given as the local capillary to fiber

Table 1. Indices of fiber size and capillary supply in the human vastus lateralis and the deep and superficial region of the plantaris muscle of male Wistar rats.

	%	FCSA (μm^2)	LCFR	CFD (mm^{-2})	SDH activity ($\Delta A660 \; \mu m^{-1} \; s^{-1}$)
Human VL					
I	52.2 (5.4) [#+†]	4068 (1300) [#+]	1.51 (0.68) [+†]	362 (145) [#]	1.64 (0.12) 10^{-5} [+†]
IIa	45.3 (5.3) [#+†]	3754 (1555) [#+]	1.33 (0.70) [+]	325 (138) [#]	1.34 (0.22) 10^{-5} [+†]
IIb	2.5 (1.9) [#+]	3669 (1382)	0.90 (0.69) [#]	254 (105) [#]	1.28 (0.12) 10^{-5} [+]
Total (n = 426)	N/A	3897 (1387) [#]	1.33 (0.67)	349 (140) [#]	1.42 (0.10) 10^{-5} [#]
Rat plantaris, deep region					
I	8.8 (4.7) [+Δ†]	1694 (218) [†]	1.28 (0.35) [+†]	765 (245) [+Δ†]	1.26 (0.55) 10^{-5} [+†]
IIa	24.3 (5.3) [†]	1998 (452) [†]	1.24 (0.25) [†]	640 (148) [+†]	1.57 (0.69) 10^{-5} [+†]
IIb	66.8 (8.3)	3116 (641)	1.81 (0.37)	613 (174) [+]	0.87 (0.49) 10^{-5} [+]
Total (n = 571)	N/A	2581 (523) [+]	1.55 (0.31) [+]	638 (170) [+]	1.14 (0.49) 10^{-5} [+]
Rat plantaris, superficial region					
I (n=5)	0.8 (1.0) [Δ†]	1271 (266) [†]	0.56 (0.26) [+†]	494 (268) [†]	1.04 (0.49) 10^{-5} [†]
IIa	14.4 (20.0) [†]	2174 (1103) [†]	0.76 (0.33) [†]	395 (119)	0.88 (0.48) 10^{-5} [†]
IIb	83.8 (20.2)	3605 (739)	1.09 (0.23)	329 (58)	0.50 (0.38) 10^{-5}
Total (n = 347)	N/A	3453 (449)	1.07 (0.16)	334 (55)	0.50 (0.23) 10^{-5}

VL: vastus lateralis; FCSA: fiber cross-sectional area; LCFR: local capillary to fiber ratio; CFD: capillary fiber density. #: significantly different from rat plantaris deep region at $P < 0.02$; +: significantly different from rat plantaris superficial region at $P < 0.02$; Δ: different from type IIa of same origin at $P < 0.05$; †: different from type IIb of same origin at $P < 0.05$.

ratio (LCFR), calculated as the sum of the fractions of the capillary domains overlapping a given fiber.

2.4 Statistics

A 3x3 ANOVA was performed with as factors fiber type and muscle origin. If interactions were found, a repeated measures ANOVA (with Bonferroni post-hoc tests) was done to detect differences between fiber types within a muscle or muscle region, and an ANOVA (with Bonferroni post-hoc tests) for muscle origin was done to detect whether a certain fiber type differed between human and rat deep or superficial region. To test which factor predicted the capillary supply of the muscle most, backwards multiple regression analysis was performed. Correlations and differences were considered significant at $P \leq 0.05$. Data are presented as mean ± SD.

3. RESULTS

In line with previous observations[13] we found a significant correlation between the LCFR and SDH·FCSA in human (R = 0.62; $P < 0.001$), but not in rat muscle. However, backward multiple regression revealed that the local capillary to fiber ratio was not determined by SDH, but determined by fiber size and type in the human vastus lateralis muscle (R = 0.71) and also surrounding of the fiber (R = 0.62; rat plantaris muscle); i.e. whether it came from the deep or superficial region of the muscle (all $P < 0.001$). Figure 1A shows the positive correlation between LCFR and FCSA for each muscle. There was

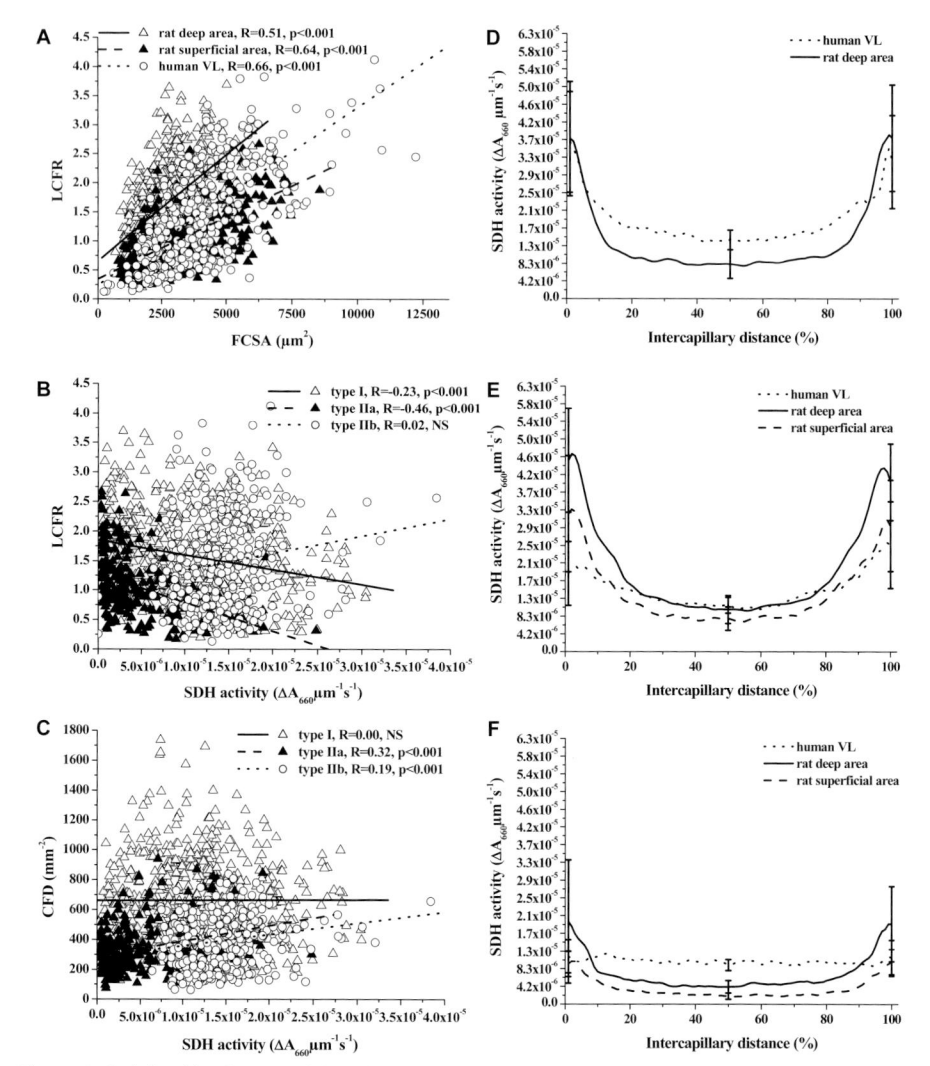

Figure 1. Relationships between (**A**) FCSA and LCFR for rat deep region, rat superficial region and human vastus lateralis, (**B**) SDH activity and LCFR and (**C**) SDH activity and CFD for the deep region only. SDH activity between two capillaries shown in **D-F** (**D**: type I, **E**: type IIa, **F**: type IIb).

even a negative relation within fiber types between LCFR and capillary fiber density (CFD = LCFR/FCSA) with SDH activity of a fiber, as shown in Fig. 1B and 1C for the deep region of the rat plantaris muscle. However, if we controlled for FCSA and fiber type, the correlations between SDH activity and capillarization disappeared ($R < 0.1$, NS). As mitochondria are not homogenously distributed within a muscle cell we plotted the mitochondrial distribution against the fraction of the intercapillary distance, where capillaries were located at opposite sites of the muscle cell. The mitochondrial density was highest in close proximity of the capillaries (subsarcolemmal mitochondria) and exhibited a steep decline to the center of the myofiber (interfibrillar mitochondria) in the oxidative type I and IIa fibers. Furthermore, the profile of the mitochondrial density was flatter in glycolytic type IIb than in the more oxidative type I and IIa fibers (Fig. 1 D-F).

Where in human muscle type IIb fibers have a higher LCFR than type I and IIa fibers, the opposite was found in rat muscle. This apparent discrepancy is largely related to fiber size (Table 1). To take this into account we also determined the CFD for each fiber type and it appeared that type I fibers had a higher CFD than type IIb fibers in both the deep and superficial region of the rat plantaris muscle (Table 1), but not in human muscle.

Table 1 and Fig. 1A show that the relation between LCFR and FCSA is not fixed, but differs between human and rat muscle and also between the deep and superficial region of the rat plantaris muscle. In terms of CFD, the capillary supply to a fiber is highest in the deep region of the plantaris muscle and no significant difference was observed between the CFD of fibers from the human vastus lateralis muscle and the superficial region of the rat muscle, even though the SDH activity of the human fibers was higher.

4. DISCUSSION

The main observation of the present study is that in both human and rat muscle the capillary supply to a fiber is primarily determined by its type and size, while the oxidative capacity plays no significant role.

Many studies have shown a significant relation between capillarization and oxidative capacity in muscle tissue[8]. Also the capillary supply to individual fibers has been reported to be related to mitochondrial volume, or integrated SDH activity, in hummingbirds[5] and humans[13]. As there is a hyperbolic relationship between fiber size and SDH activity[14], correlations between capillary supply and oxidative capacity may be confounded by fiber size. Indeed, it has been observed in both rat[3] and human muscle[2] that size is an important determinant of capillary supply to a fiber. Here we took into account both fiber size and SDH activity and found that not SDH, but fiber size was an important determinant of capillary supply to a fiber in both rat and human muscle (Fig. 1A). Moreover, in rat muscle (integrated) SDH did not correlate significantly with capillary supply to a fiber. Thus, taking into account both fiber size and oxidative capacity it appeared that size, rather than oxidative capacity is a significant predictor of the local capillary to fiber ratio.

Although Ahmed et al.[2] suggested that there was no effect of fiber type we did observe with backward regression that also the type of the fiber had an impact on the capillary supply of a fiber. The discrepancy may be related to the type of analysis, as they did not take the fiber type into account in their regression analysis. Nevertheless, in line with what we found they observed that the average LCFR was type I > IIa > IIb. The effect of type is, however, rather small, as both Ahmed et al.[2] and we observed that the CFD, which takes into account the size of the fiber, was similar in each fiber type. In the rat plantaris muscle, on the other hand, the CFD of type IIb fibers was lower than that of type I and IIa fibers, but this difference was not explicable by differences in SDH activity between fiber types as revealed with backward regression for type and SDH. Even the FCSA·SDH, to obtain a measure for the total demand for O_2 of the myofiber[13], did not significantly correlate with the capillary supply to a fiber in the rat plantaris muscle.

As the mitochondria ultimately consume O_2 one would have expected that the oxidative capacity is the main determinant of the capillary supply to a fiber. It should be noted, however, that mitochondria are not homogenously distributed within the fiber, where the density is highest subsarcolemmal, decreasing steeply to the interior of the cell[4]. Here we confirm that observation and show that the density distribution shows a flatter profile in human than rat fibers and in type IIb than type I and IIa fibers. Such a profile would considerably shorten the diffusion distance to the majority of the mitochondria, and hence result in an adequate O_2 supply to the mitochondria. This would, at least partly, explain why there is no significant relation between the oxidative capacity of a fiber and its capillary supply.

Finally, in line with previous observations[6] we observed that the capillary supply to a fiber is modulated by the metabolic surrounding of the fiber. This is reflected by the different slope of the FCSA-LCFR relation and the lower CFD for each fiber type in the superficial than the deep region of the muscle.

In conclusion, in both rat and human muscle size and type rather than mitochondrial activity is the primary determinant of capillary supply to a muscle fiber.

5. REFERENCES

1. A. Krogh, The number and distribution of capillaries in muscles with calculations of the oxygen pressure head necessary for supplying the tissue, J. Physiol. 52(6), 409-415 (1919).
2. S. K. Ahmed, S. Egginton, P. M. Jakeman, A. F. Mannion, and H. F. Ross, Is human skeletal muscle capillary supply modelled according to fibre size or fibre type?, Exp. Physiol. 82(1), 231-234 (1997).
3. H. Degens, B. E. Ringnalda, and L. J. Hoofd, Capillarisation, fibre types and myoglobin content of the dog gracilis muscle, Adv. Exp. Med. Biol. 361, 533-539 (1994).
4. H. Hoppeler, O. Mathieu, R. Krauer, H. Claassen, R. B. Armstrong, and E. R. Weibel, Design of the mammalian respiratory system. VI Distribution of mitochondria and capillaries in various muscles, Respir. Physiol. 44(1), 87-111 (1981).
5. O. Mathieu-Costello, R. K. Suarez, and P. W. Hochachka, Capillary-to-fiber geometry and mitochondrial density in hummingbird flight muscle, Respir. Physiol. 89(1), 113-132 (1992).
6. H. Degens, Z. Turek, L. J. Hoofd, M. A. Van't Hof, and R. A. Binkhorst, The relationship between capillarisation and fibre types during compensatory hypertrophy of the plantaris muscle in the rat, J. Anat. 180 (Pt 3), 455-463 (1992).
7. S. Egginton, Numerical and areal density estimates of fibre type composition in a skeletal muscle (rat extensor digitorum longus), J. Anat. 168(Feb), 73-80 (1990).
8. B. Saltin, and P. D. Gollnick, Skeletal muscle adaptability: significance for metabolism and performance, in: Handbook of Physiology, Skeletal Muscle, edited by L.D. Peachy (Williams and Wilkins, Bethesda, MD, 1983), pp. 555-631.
9. O. Mathieu-Costello, and R. T. Hepple, Muscle structural capacity for oxygen flux from capillary to fiber mitochondria, Exerc. Sport Sci. Rev. 30(2), 80-84 (2002).
10. W. J. van der Laarse, P. C. Diegenbach, and G. Elzinga, Maximum rate of oxygen consumption and quantitative histochemistry of succinate dehydrogenase in single muscle fibres of Xenopus laevis, J. Muscle Res. Cell Motil. 10(3), 221-228 (1989).
11. M. H. Brooke, and K. K. Kaiser, Muscle fiber types: how many and what kind? Arch. Neurol. 23(4), 369-379 (1970).
12. L. Hoofd, Z. Turek, K. Kubat, B. E. Ringnalda, and S. Kazda, Variability of intercapillary distance estimated on histological sections of rat heart, Adv. Exp. Med. Biol. 191, 239-247 (1985).
13. M. A. Bekedam, B. J. van Beek-Harmsen, A. Boonstra, W. van Mechelen, F. C. Visser, and W. J. van der Laarse, Maximum rate of oxygen consumption related to succinate dehydrogenase activity in skeletal muscle fibres of chronic heart failure patients and controls, Clin. Physiol. Funct. Imaging 23(6), 337-343 (2003).
14. W. J. van der Laarse, A. L. Des Tombe, M. B. Lee-de Groot, and P. C. Diegenbach, Size principle of striated muscle cells, Netherlands Journal of Zoology 48(3), 213-223 (1998).

NOVEL THERAPEUTIC APPROACH TARGETING THE HIF-HRE SYSTEM IN THE KIDNEY

Masaomi Nangaku[1]

Abstract: Recent studies emphasize the role of chronic hypoxia in the tubulo-interstitium as a final common pathway to end-stage renal disease. Therefore, therapeutic approaches which target the chronic hypoxia should prove effective against a broad range of renal diseases.

Many of hypoxia-triggered protective mechanisms are hypoxia inducible factor (HIF)-dependent. Although HIF-1α and HIF-2α share both structural and functional similarity, they have different localization and can contribute in a non-redundant manner. While gene transfer of constitutively active HIF has been shown effective, pharmacological approaches to activate HIF are more desirable. Oxygen-dependent activation of prolyl hydroxylases (PHD) regulates the amount of HIF by degradation of this transcription factor. Therefore, PHD inhibitors have been the focus of recent studies on novel strategies to stabilize HIF. Cobalt is one of the inhibitors of PHD, and stimulation of HIF with cobalt is effective in a variety of kidney disease models. Furthermore, crystal structures of the catalytic domain of human prolyl hydroxylase 2 have been clarified recently. The structure aids in the design of PHD selective inhibitors for the treatment of hypoxic tissue injury.

Current advance has elucidated the detailed mechanism of hypoxia-induced transcription, giving hope for the development of novel therapeutic approaches against hypoxia.

---

[1] Masaomi Nangaku, Division of Nephrology and Endocrinology, University of Tokyo School of Medicine, 7-3-1 Hongo, Bunkyo-ku, Tokyo 113-8655, Japan

P. Liss et al. (eds.), *Oxygen Transport to Tissue XXX*, DOI 10.1007/978-0-387-85998-9_13,
© Springer Science+Business Media, LLC 2009

1. INTRODUCTION

Recent studies emphasize the role of chronic hypoxia in the tubulointerstitium as a final common pathway to end-stage renal failure[1]. Hypoxia in the kidney has been demonstrated in a variety of disease models utilizing pimonidazole staining, a Clark-type electrode, blood oxygen level dependent (BOLD)-MRI, and transgenic animals expressing a hypoxia-sensing reporter vector[2].

Chronic hypoxia of the kidney occurs via several mechanisms acting in concert. When advanced, tubulointerstitial damage is associated with the loss of peritubular capillaries. Associated interstitial fibrosis impairs oxygen diffusion and supply to tubular and interstitial cells. In addition, a number of mechanisms that induce tubulointerstitial hypoxia at an early stage have been identified. Glomerular injury and vasoconstriction of efferent arterioles due to imbalances in vasoactive substances decrease post-glomerular peritubular capillary blood flow. Oxidative stress also hampers the efficient utilization of oxygen in tubular cells, leading to reduced renal oxygen tension[3]. Relative hypoxia in the kidney also results from increased metabolic demand in tubular cells. Further, renal anemia hinders oxygen delivery. These factors can affect the kidney before the appearance of significant pathological changes in the vasculature and predispose the kidney to tubulointerstitial injury.

Therefore, therapeutic approaches which target the chronic hypoxia should prove effective against a broad range of renal diseases. At the center of the cellular response to hypoxia is hypoxia-inducible factor, HIF, and activation of this "master gene" switch results in a broad and coordinated downstream reaction to protect organs against hypoxia.

2. HIF IN THE KIDNEY

HIF is composed of two subunits, an oxygen-sensitive HIF-α subunit and a constitutively expressed HIF-β subunit (also known as ARNT, the aryl hydrocarbon receptor nuclear translocator). Both HIF-1α and HIF-1β are members of the basic helix-loop-helix PER/ARNT/SIM (HLH-PAS) family of transcription factors. HIF binds to the HRE in the cis-regulatory regions of its target genes, and transcriptionally activates various genes encoding proteins that mediate adaptive responses to reduced oxygen availability. Under normoxic conditions, two conserved proline residues within the central oxygen-dependent degradation domains of the HIF proteins are hydroxylated by the protein products "*prolyl hydroxylase domain* containing" (PHDs). This promotes binding of the von Hippel Lindau tumor suppressor protein (pVHL), part of a ubiquitin ligase complex, resulting in polyubiquitylation and rapid degradation. Similarly, a conserved asparagine residue in the carboxyl-terminal transactivation domain (CAD) of the HIF proteins is hydroxylated in normoxia by factor inhibiting HIF (FIH), preventing recruitment of the p300/CBP transcriptional co-activators and thus leading to transcriptional repression. Under hypoxia, oxygen is lacking as an essential substrate for the hydroxylation reaction, and the unmodified HIF proteins avoid degradation but rather heterodimerize with HIF-β and up-regulate the transcription of target genes.

HIF-α subunits have different isoforms, and a biological role of each isoform remains to be elucidated. HIF-1α is expressed in most cell types, whereas HIF-2α shows a more restricted pattern of expression. In the adult kidney, HIF-2α is expressed in peritubular endothelial cells and fibroblasts as well as glomerular endothelial cells, whereas HIF-1α is predominantly localized in tubular cells[4,5]. Up-regulation of the two HIF-α isoforms in the kidney by hypoxia was demonstrated in models of segmental renal infarction and radio contrast nephropathy[6,7]. While cell-type specificity of HIF isoforms in these models was consistent with previous findings, temporal and spatial profiles of HIF activation were relatively complex, suggesting an important but complicated role of HIF in tissue preservation as a response to regional renal hypoxia.

Because mice with complete deficiency of HIF are embryonic or perinatal lethal, analysis of a biologic function of a HIF isoform after birth has been hampered. One possible way to overcome this problem is to utilize heterozygous knockout mice. Studies using $Hif1a^{+/-}$ mice showed no benefit of preconditioning of the heart by hypoxia, demonstrating that cardiac protection against ischemia-reperfusion injury by preconditioning is critically dependent on $Hif1a$ gene dosage[8]. We employed HIF-2α knockdown mice and showed that these mice were more susceptible to ischemia-reperfusion injury model of the kidney[9]. Our studies utilizing the Cre-loxP system to rescue HIF-2α specifically in the endothelium demonstrated that HIF-2α in the kidney endothelium is responsible for regulation of oxidative stress and subsequent tubulointerstitial injury.

3. PHD IN THE KIDNEY

Under normoxia, hydroxylation of HIF-α-subunits by HIF prolyl hydroxylases (PHD) is required for binding to the pVHL-E3-ubiquitin ligase complex. After polyubiquitination, HIF-α is degraded by the proteasome. The enzymatic activity of PHD depends on iron as the activating metal, 2-oxoglutarate as a co-substrate, and ascorbic acid as a cofactor. Three PHD with the potential to catalyze this reaction have been identified, and these proteins, termed PHD1, PHD2, and PHD3, appear to have arisen by gene duplication. In the kidney all three isoforms of PHD are expressed, and all PHD exhibited much higher levels in renal medulla than cortex[10]. PHD2 is the most abundant isoform in various organs including the kidney[11]. The contribution of each to the physiological regulation of HIF remains uncertain.

Recent experiments using suppression by small interference RNA showed that each of the three PHD isoforms contributes in a non-redundant manner to the regulation of both HIF-1α and HIF-2α subunits, and that the contribution of each PHD is strongly dependent on its relative abundance[12]. Although Phd1(-/-) and Phd3(-/-) mice did not display apparent angiogenic defects, conditional knockout of Phd2 led to hyperactive angiogenesis and angiectasia, demonstrating a major role of PHD2 as a negative regulator for vascular growth in adult mice[13].

4. GENE TRANSFER TO ACTIVATE HIF IN THE KIDNEY

We previously showed that *in vivo* gene transfer of DNA expressing constitutively active fusion protein of HIF protects the kidney against ischemic injury[14]. Recently phase I trial of adenoviral delivery of a constitutively active form of HIF was conducted in patients with critical limb ischemia. HIF gene transfer therapy was well tolerated in the patients, and at 1 year, limb status observations in HIF-1α patients included complete rest pain resolution in 14 of 32 patients and complete ulcer healing in 5 of 18 patients[15].

5. PHD INHIBITORS

PHD inhibitors have been the focus of recent studies on novel strategies to stabilize HIF. More than half a century ago, oral administration of cobaltous chloride was employed to treat anemia associated with chronic renal disease and led to a transient but significant erythropoietic response[16]. Although the mechanism of erythropoiesis was unknown at that time, cobalt is now recognized as an inhibitor of PHD, and thereby serves as a stimulator of HIF. We demonstrated the renoprotective effects of chemical pre-conditioning with cobaltous chloride in an ischemic model of renal injury[17]. Administration of cobalt induced up-regulation of HIF-regulated genes, such as VEGF and EPO, and subsequently protected the kidney against the tubulointerstitial damage induced by hypoxia. Pretreatment of experimental animals with cobalt also ameliorated disease manifestations of a new model of glomerulonephritis induced by co-administration of angiotensin II and Habu snake venom[18]. Furthermore, we showed that cobalt treatment was effective when given after the initial insult in a chronic progressive glomerulonephritis model, a model of cyclosporin nephrotoxicity, and a model of chronic renal failure with glomerular hypertension, demonstrating not only its preventive but also its therapeutic potential[19-21]. HIF activation in these models restored the peritubular capillary network, reduced the number of apoptotic cells, and improved renal functions as well as histological damages.

Although cobalt administration has been somewhat effective in experimental animals, long-term administration to humans is hindered by various side effects. Less toxic and more potent PHD inhibitors are desirable, and a variety of new candidates are now under development. Three distinct PHD inhibitors, l-mimosine (L-Mim), ethyl 3,4-dihydroxybenzoate (3,4-DHB), and 6-chlor-3-hydroxychinolin-2-carbonic acid-N-carboxymethylamid (S956711), induced HIF-α protein in human and rodent cells and enhanced angiogenesis in the sponge assays[22]. Systemic administration of L-Mim and S956711 in rats led to HIF-α induction in the kidney. Systemic treatment of mice with 3,4-DHB displayed significantly increased viability and enhanced exercise performance in severe hypoxia[23]. Pretreatment with the novel PHD inhibitor FG-4487 strongly induced the accumulation of HIF-1α and HIF-2α in tubular and peritubular cells, respectively, with significant amelioration of ischemic renal injury[24].

Technological advances in computer-based drug design will be useful to develop specific PHD inhibitors. Recent studies by Schofield's group clarified crystal structures of the catalytic domain of human PHD2[25], and this information enables us to perform docking simulation based on the three-dimensional structure of PHD2 for development of novel compounds.

6. OTHER TARGETS OF HIF ACTIVATION THERAPY

FIH hydroxylates regulates HIF activation via hydroxylation of an asparagine residue in the COOH-terminal transactivation domain of HIF-α. During hypoxia, asparagine hydroxylation is blocked and CBP/p300 recruitment is facilitated, enabling increased levels of transcription of the target genes. In the kidney, FIH is expressed in tubular epithelial cells and glomeruli, as shown by immunohistochemical methods[26]. Thus, inhibition of FIH can lead to increased HIF target gene expression.

The phosphoinositide 3-kinase (PI3K)/Akt pathway and the protein kinase C signaling have also been implicated in the regulation of HIF-α. While the dogma proposes that stability is the rate-limiting factor to determine the protein levels of HIF-α, some inputs including activation of these pathways function at the level of transcription and translation.

Other proteins that govern HIF activity include Siah2, OS-9, ING4 (inhibitor of growth 4), Hsp90 (heat-shock protein 90), ARD1 (arrest at start of cell cycle defective 1), and p53 among others. Whether these pathways can be a good target for therapeutic approaches in kidney disease *in vivo* remains to be elucidated.

7.   CONCLUSION

Chronic hypoxia is the final common pathway to end-stage renal failure, and accumulating evidence suggests the pathogenic role of hypoxia from an early stage of kidney disease. Therapeutic approaches against this final common pathway should be effective in a broad range of renal diseases. Although a fine line between therapeutic benefit and harmful side effects must be drawn, HIF activation is promising, as it facilitates a variety of defensive mechanisms in a coordinated manner.

8.   REFERENCES

1. M. Nangaku, Chronic hypoxia and tubulointerstitial injury: a final common pathway to end-stage renal failure, J. Am. Soc. Nephrol. 17(1), 17-25 (2006)
2. T. Tanaka, T. Miyata, R. Inagi, T. Fujita, and M. Nangaku, Hypoxia in renal disease with proteinuria and/or glomerular hypertension. Am. J. Pathol. 165(6), 1979-92 (2004)
3. F. Palm, J. Cederberg, P. Hansell, P. Liss, and P.O. Carlsson, Reactive oxygen species cause diabetes-induced decrease in renal oxygen tension, Diabetologia. 46(8), 1153-60 (2003)
4. C. Rosenberger, S. Mandriota, J.S. Jurgensen, M.S. Wiesener, J.H. Horstrup, U. Frei, P.J. Ratcliffe, P.H. Maxwell, S. Bachmann, and K.U. Eckardt, Expression of hypoxia-inducible factor-1 and -2 in hypoxic and ischemic rat kidneys, J. Am. Soc. Nephrol. 13(7), 1721-1732 (2002)
5. M.S. Wiesener, J.S. Jurgensen, C. Rosenberger, C.K. Scholze, J.H. Horstrup, C. Warnecke, S. Mandriota, I. Bechmann, U.A. Frei, C.W. Pugh, P.J. Ratcliffe, S. Bachmann, P.H. Maxwell, and K.U. Eckardt, Widespread hypoxia-inducible expression of HIF-2α in distinct cell populations of different organs, FASEB J. 17, 271-273 (2003)
6. C. Rosenberger, W. Griethe, G. Gruber, M. Wiesener, U. Frei, S. Bachmann, and K.U. Eckardt, Cellular responses to hypoxia after renal segmental infarction, Kidney Int. 64(3), 874-886 (2003)
7. C. Rosenberger, S.N. Heyman, S. Rosen, A. Shina, M. Goldfarb, W. Griethe, U. Frei, P. Reinke, S. Bachmann, and K.U. Eckardt, Up-regulation of HIF in experimental acute renal failure: evidence for a protective transcriptional response to hypoxia. Kidney Int. 67(2), 531-542 (2005)
8. Z. Cai, D.J. Manalo, G. Wei, E.R. Rodriguez, K. Fox-Talbot, H. Lu, J.L. Zweier, and G.L. Semenza GL, Hearts from rodents exposed to intermittent hypoxia or erythropoietin are protected against ischemia-reperfusion injury, Circulation. 108(1), 79-85 (2003)
9. I. Kojima, T. Tanaka, R. Inagi, H. Kato, T. Yamashita, A. Sakiyama, O. Ohneda, N. Takeda, M. Sata, T. Miyata, T. Fujita T, and M. Nangaku, Protective role of hypoxia-inducible factor-2alpha against ischemic damage and oxidative stress in the kidney, J. Am. Soc. Nephrol. 18(4), 1218-26 (2007)
10. N. Li, F. Yi, C.M. Sundy, L. Chen, M.L. Hilliker, D.K. Donley, D.B. Muldoon, and P.L. Li, Expression and actions of HIF prolyl-4-hydroxylase in the rat kidneys, Am. J. Physiol. Renal Physiol. 292(1), F207-16 (2007)

11. C. Willam, P.H. Maxwell, L. Nichols, C. Lygate, Y.M. Tian, W. Bernhardt, M. Wiesener, P.J. Ratcliffe, K.U. Eckardt, and C.W. Pugh, HIF prolyl hydroxylases in the rat; organ distribution and changes in expression following hypoxia and coronary artery ligation, J. Mol. Cell. Cardiol. 41, 68-77 (2006)

12. R.J. Appelhoff, Y.M. Tian, R.R. Raval, H. Turley, A.L. Harris, C.W. Pugh, P.J. Ratcliffe, and J.M. Gleadle, Differential function of the prolyl hydroxylases PHD1, PHD2, and PHD3 in the regulation of hypoxia-inducible factor, J. Biol. Chem. 279(37), 38458-38465 (2004)

13. K. Takeda, A. Cowan, and G.H. Fong, Essential role for prolyl hydroxylase domain protein 2 in oxygen homeostasis of the adult vascular system, Circulation. [Epub ahead of print] (2007)

14. K. Manotham, T. Tanaka, T. Ohse, I. Kojima, T. Miyata, R. Inagi, H. Tanaka, R. Sassa, T. Fujita, and M. Nangaku, A biological role of HIF-1 in the renal medulla, Kidney Int. 67(4), 1428-1439 (2005)

15. S. Rajagopalan, J. Olin, S. Deitcher, A. Pieczek, J. Laird, P.M. Grossman, C.K. Goldman, K. McEllin, R. Kelly, and N. Chronos, Use of a constitutively active hypoxia-inducible factor-1alpha transgene as a therapeutic strategy in no-option critical limb ischemia patients: phase I dose-escalation experience, Circulation. 115(10), 1234-43 (2007)

16. H.F. Gardner, The use of cobaltous chloride in the anemia associated with chronic renal disease. J. Lab. Clin. Med. 41(1), 56-64 (1953)

17. M. Matsumoto, Y. Makino, T. Tanaka, H. Tanaka, N. Ishizaka, E. Noiri, T. Fujita, and M. Nangaku, Induction of renoprotective gene expression by cobalt ameliorates ischemic injury of the kidney in rats, J. Am. Soc. Nephrol. 14(7), 1825-1832 (2003)

18. Y. Kudo, Y. Kakinuma, Y. Mori, N. Morimoto, T. Karashima, M. Furihata, T. Sato, T. Shuin, and T. Sugiura, Hypoxia-inducible factor-1alpha is involved in the attenuation of experimentally induced rat glomerulonephritis, Nephron Exp. Nephrol. 100(2), e95-103 (2005)

19. T. Tanaka, I. Kojima, T. Ohse, J.R. Ingelfinger, S. Adler, T. Fujita, and M. Nangaku, Cobalt promotes angiogenesis via hypoxia-inducible factor and protects tubulointerstitium in the remnant kidney model, Lab. Invest. 85(10), 1292-1307 (2005)

20. T. Tanaka, I. Kojima, T. Ohse, R. Inagi, T. Miyata, J.R. Ingelfinger, T. Fujita, and M. Nangaku, Hypoxia-inducible factor modulates tubular cell survival in cisplatin nephrotoxicity. Am. J. Physiol. Renal Physiol. 289(5), F1123-1133 (2005)

21. T. Tanaka, M. Matsumoto, R. Inagi, T. Miyata, I. Kojima, T. Ohse, T. Fujita, and M. Nangaku, Induction of protective genes by cobalt ameliorates tubulointerstitial injury in the progressive Thy1 nephritis, Kidney Int. 68(6), 2714-2725 (2005)

22. C. Warnecke, W. Griethe, A. Weidemann, J.S. Jurgensen, C. Willam, S. Bachmann, Y. Ivashchenko, I. Wagner, U. Frei, M. Wiesener, and K.U. Eckardt, Activation of the hypoxia-inducible factor-pathway and stimulation of angiogenesis by application of prolyl hydroxylase inhibitors, FASEB J. 17, 1186-1188 (2003)

23. H. Kasiganesan, V. Sridharan, and G. Wright, Prolyl hydroxylase inhibitor treatment confers whole-animal hypoxia tolerance, Acta Physiol. (Oxf). 190(2), 163-9 (2007)

24. W.M. Bernhardt, V. Campean, S. Kany, J.S. Jürgensen, A. Weidemann, C. Warnecke, M. Arend, S. Klaus, V. Günzler, K. Amann, C. Willam, M.S. Wiesener, and K.U. Eckardt, Preconditional activation of hypoxia-inducible factors ameliorates ischemic acute renal failure, J. Am. Soc. Nephrol. 17(7), 1970-1978 (2006)

25. M.A. McDonough, V. Li, E. Flashman, R. Chowdhury, C. Mohr, B.M. Liénard, J. Zondlo, N.J. Oldham, I.J. Clifton, J. Lewis, L.A. McNeill, R.J. Kurzeja, K.S. Hewitson, E. Yang, S. Jordan, R.S. Syed, and C.J. Schofield, Cellular oxygen sensing: Crystal structure of hypoxia-inducible factor prolyl hydroxylase (PHD2), Proc. Natl. Acad. Sci. 103(26), 9814-9819 (2006)

26. E.J. Soilleux, H. Turley, Y.M. Tian, C.W. Pugh, K.C. Gatter, and A.L. Harris AL, Use of novel monoclonal antibodies to determine the expression and distribution of the hypoxia regulatory factors PHD-1, PHD-2, PHD-3 and FIH in normal and neoplastic human tissues, Histopathology. 47(6), 602–610 (2005)

CAROTID BODY SENSORY DISCHARGE AND GLOMUS CELL HIF-1α ARE REGULATED BY A COMMON OXYGEN SENSOR

S. Lahiri[1], A. Roy[1], S.M. Baby[1], C. Di Giulio[2] and D.F. Wilson[3]

Abstract: The carotid body responds to both acute and more prolonged periods of lowered oxygen pressure. In the acute response, the decrease in oxygen pressure is coupled to increased afferent neural activity while the latter involves, at least in part, increase in the hypoxia inducible transcription factor HIF-1α. In this paper, we summarize evidence that both the acute changes in neural activity and the longer term adaptive changes linked to HIF-1α induction share the same oxygen sensor, mitochondrial cytochrome c oxidase.

1. INTRODUCTION

The carotid body responds to both acute and more prolonged periods of lowered oxygen pressure (for review[1]). The response to acutely lowered oxygen pressure is increased afferent neural activity, important to the regulation of breathing. Longer periods of lowered oxygen pressures result in progressive alteration in metabolism of cells in the carotid body, an important early event being induction of HIF-1α in the glomus cells[2,3]. In this paper, we will summarize evidence that the acute changes in neural activity and the longer term adaptive changes linked to HIF-1α induction share the same oxygen sensor, and this is mitochondrial cytochrome c oxidase. In our discussions, the term «oxygen sensor» is restricted to the entity responsible for the interaction with oxygen and generating a response that is correlated with oxygen pressure. This sensor response is then transcribed the sensor response into chemical messages that lead to the appropriate biological response. It is recognized that there may be a significant number of intermediates may participate in transfer of the

---

[1] Department of Physiology, University of Pennsylvania, Philadelphia, PA 19104
[2] Department of Basic and Applied Medical Sciences, University of Chieti, Italy.
[3] Department of Biochemistry and Biophysics, University of Pennsylvania, Philadelphia, PA 19104

P. Liss et al. (eds.), *Oxygen Transport to Tissue XXX*, DOI 10.1007/978-0-387-85998-9_14,   **87**
© Springer Science+Business Media, LLC 2009

information between the sensor and the final biological response. This definition of the oxygen sensor is consistent with that used for the glucose sensor, for example, where the glucose sensor of the beta cells of the pancreatic islets is glucokinase. In both cases, the intermediate steps leading from the sensor to the biological responses, afferent neural activity and insulin release, are still being worked out[5,6].

2. METHODS

Isolated perfused-superfused carotid body preparation

Isolated perfused-superfused carotid bodies were prepared from cat and rat by the method of Iturriaga and coworkers[7]. These preparations include a section of the nerve, allowing the afferent activity to be measured and it retains physiological levels of response to oxygen and carbon dioxide. As shown in Figure 1, the nerves synapse to the glomus cells, the cells presumed to be primarily responsible for oxygen sensing.

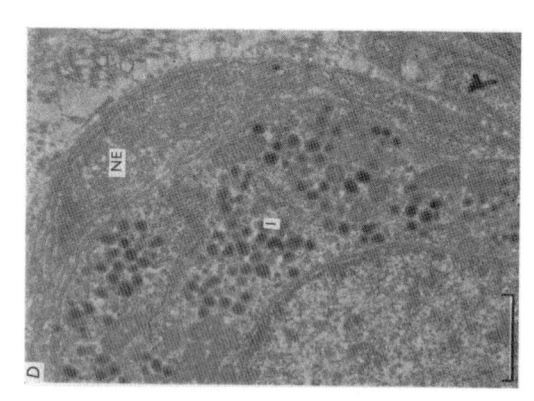

Figure 1: An electron-micrograph of a carotid body showing a nerve-ending (NE) synapsed with a glomus cell (-). These nerve endings are connected to nerve fibers that` join the carotid sinus nerve (CSN).

Inhibition of heme oxidases by carbon monoxide and its reversal by light

In hemoglobin free cells and tissue, the primary metabolic effect of CO is as an inhibitor of the terminal oxidase of mitochondria. There it competes with oxygen for binding to a reduced heme a in the active site. This competition with oxygen, combined with the unique spectral properties of complex formed between CO and heme a of cytochrome oxidase, provide very powerful tools for studying the mechanism(s) of oxygen utilization and sensing in biology. Not only do the CO complexes have a unique absorption spectrum, but many can be dissociated by light, reversing the inhibition by CO[8-11]. The latter property makes it possible to very rapidly, selectively, and reversibly switch a single metabolic pathway, such as mitochondrial oxidative phosphorylation, on and off. This greatly facilitates determining the biological roles of that pathway and its metabolic products.

3. RESULTS AND DISCUSSION

Oxygen sensing by the carotid body. The afferent neural activity of the isolated, perfused-superfused carotid body is a physiological measure of the oxygen sensing activity. Because it is an isolated, blood free preparation, experimental measurements can be made that are not possible in the intact animal. In intact animals, carbon monoxide preferentially binds to hemoglobin, suppressing oxygen delivery and thereby causing tissue hypoxia. In the isolated perfused-superfused carotid body preparation, however, there is no hemoglobin and the effects of CO are mediated primarily by competing with oxygen at the active site of cytochrome c oxidase.

When the perfusate solution for the isolated rat carotid body were switched from normoxia (PO$_2$ ~ 150 Torr) to one combined with CO (PCO ~ 350 Torr and PO$_2$ ~ 150 Torr) in the dark (Figure 2), the sensory discharge increased rapidly to a level which was about 60% of the maximal hypoxic response. This activity was sustained while the preparation was in the dark. Illuminating the carotid body with a bright white light completely reversed the CO induced increase in afferent neural activity, an effect that was fully reversible.

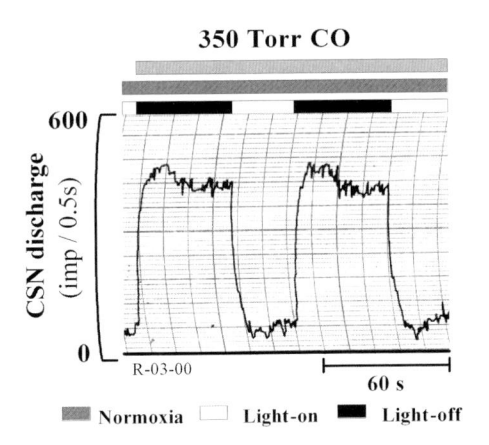

Figure 2: The effects of CO and light on the afferent neural activity of the isolated carotid body perfused with medium containing 150 Torr pO$_2$ After an initial period of perfusion with CO free medium, the perfusion medium was changed to one that also contained 350 Torr CO. The afferent activity increased quickly to near 50% of the maximal hypoxic response. Illumination of the carotid body with bright white light caused the afferent activity to rapidly, 3-5 sec, return to near the no CO levels (taken from 4).

When the CO perfused carotid body was illuminated with monochromatic light of different wavelength but similar intensities the response of the afferent neural activity was very dependent on the wavelength of the light (Figure 3). If care is taken to use light intensities that only partially reverse the CO effect, after correction to the same light intensity (number of quanta of light/sec) the response at each wavelength is directly correlated with the amount of light absorbed by CO complex at that wavelength. A plot of the response against the wavelength of light is called the "photochemical action spectrum" and is the absorption spectrum of the CO complex that is being dissociated by the light.

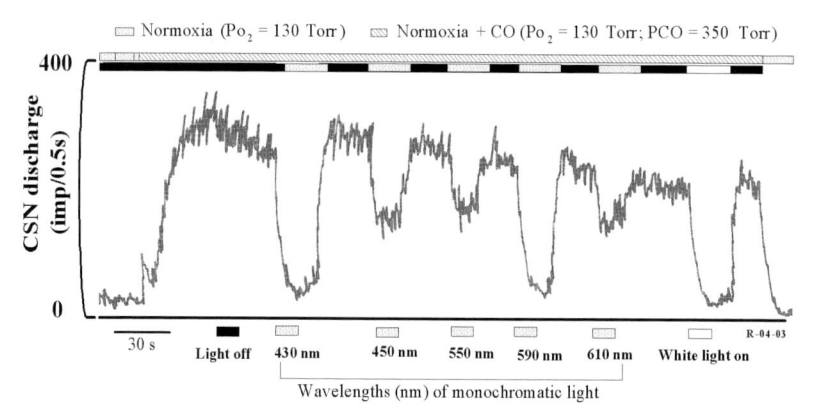

Figure 3. The efficacy of different wavelengths of monochromatic light in reversing the effect of CO (taken from 4).

A bar graph constructed from the data for 6 independent experiments like that in Figure 3 is presented in Figure 4. Error bars (± SD) are included to show the variability among measurements. Of the 5 selected wavelengths, 430, 450, 550, 590, and 610 nm, two, 430 and 590 nm, are much more effective at reversing the CO effect than the others. The intensities of the monochromatic light were not corrected to the same value and the

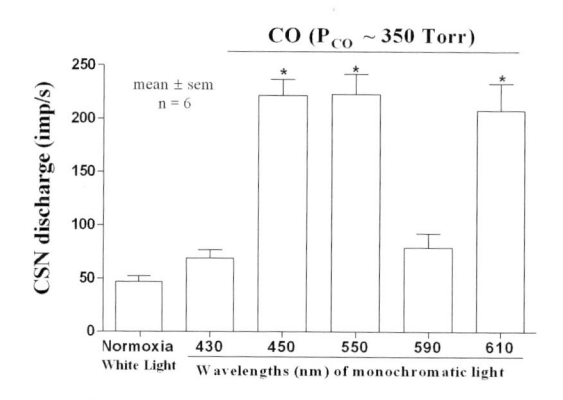

Figure 4. The relative efficacy of monochromatic light of different wavelengths in reversing the effect of CO on the afferent neural activity of the rat carotid body. Data from 6 experiments like that shown in Figure 3 has been analyzed and the data plotted ± SD for each wavelength. The values for 450, 550, and 610 nm are significantly different those at 430 and 590 nm (* P < 0.001). (Taken from 4)

intensity increased with increasing wavelength (see ref. 6 for a graph of the intensity vs wavelength for the tungsten-iodine light source). Fully corrected data for a larger number of wavelengths have been presented for the cat carotid body[6]. The photochemical action spectrum (absorption spectrum of the CO-sensor complex) obtained is that of the CO-complex of mitochondrial cytochrome c oxidase.

Early evidence had suggested the mitochondria were responsible for oxygen sensing by the carotid body, in particular the observation that addition of inhibitors of mitochondrial respiration and of oxidative phosphorylation caused stimulation of the afferent neural activity[12,13]. This stimulation observed was, however, generally much less than that for hypoxia and transient, rising to a maximum and then decreasing again. After the decrease in activity the oxygen sensitivity was lost. The transient nature of the response and the difficulty in reversing the inhibition and recovering the oxygen sensitivity after treatment resulted to the uncertainty in the interpretation. Carbon monoxide overcomes these limitations by being fully reversible and by allowing highly specific identification of the oxygen sensor as mitochondrial cytochrome oxidase.

Oxygen sensing for induction of HIF-1α: Induction of the transcription factor HIF-1α has been extensively studied (for review see Semenza[14]). It is continuously produced and degraded under normoxic conditions, the degradation occurring through hydroxylation of a proline in the molecule by a proline hydroxylase. Oxygen is a substrate for the proline hydroxylase and is essential to its activity. Decreasing availability of this substrate, oxygen, for the hydroxylation reaction is considered to limit the rate of HIF-1α degradation in an oxygen dependent manner. Although this is a reasonable mechanism for coupling the level of the transcription factor to the available oxygen pressure, it is not the only possible mechanism for the oxygen dependence. Both protein synthesis and ubiquination, an essential step in the degradation pathway, are strongly dependent on the availability of cellular energy, as expressed in the energy state ($[ATP]/[ADP][Pi]$). The cellular levels of AMP are strongly coupled to the energy state by the enzyme adenylate kinase. The latter catalyzes the reaction $2ADP \rightarrow ATP + AMP$. As a result, the level of AMP increases as the square of the increase in ADP. The intracellular concentration of AMP is known to strongly influence cellular metabolism, not only through allosteric modulation of the activity of a variety of enzymes, but also through activation of AMP dependent protein kinase, a powerful regulatory protein kinase that modulates many cellular activities.

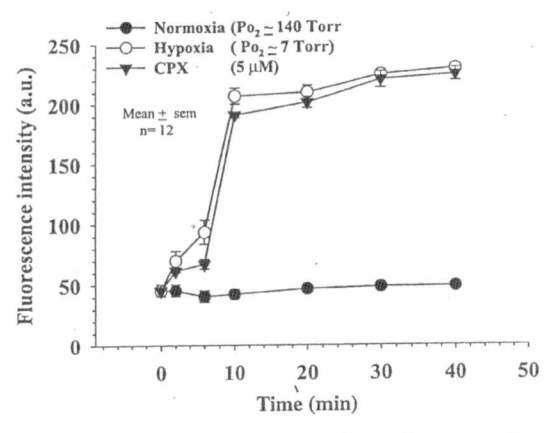

Figure 5: Increase in HIF-1α following change of the superfusion medium of isolated glomus cells to one containing either low oxygen (7 Torr) or 5 μM cyclopirox olamine (CPX). HIF-1α was fluorescently labeled with a specific antibody and the average fluorescence for individual cells measured. (taken from 15)

The increase in HIF-1α when glomus cells were subjected to lowered oxygen pressure or treated with iron chelator, cyclopirox olamine (CPX) are shown in Figure 5. The response to hypoxia (7 Torr) and to addition of 5 μM CPX was not significant until 2-3 minutes after initiation of the treatment and maximal effect was reached in about 10 min. This compares with the afferent neural activity of the carotid body, discussed earlier, where the changes begin within 1 second and are maximal in about 5 seconds. The coupling of mitochondrial function to the afferent neural activity suggests the mitochondria might also contribute to the oxygen dependence of HIF-1α levels in the cells and thereby to the responses of the carotid body to longer periods of hypoxia. The unique properties of carbon monoxide have been used to test this possibility using isolated glomus cells. The cells were superfused with medium that was normoxic (150 Torr oxygen pressure) or a medium with 150 Torr oxygen and 350 Torr CO. It was observed that even at 150 Torr oxygen pressure addition of CO led to an increase in the level of HIF-1α to near those induced by lowering the oxygen to 7 Torr as long as the cells were incubated in the dark (Figure 5). As also seen in Figure 5, comparison of the effect of CO on cells cultured in the dark with that when the cells are illuminated with bright light shows that most of this effect of CO is due to formation of CO complexes that can be dissociated by light. Thus, addition of CO can mimic hypoxia and this can be reversed by light.

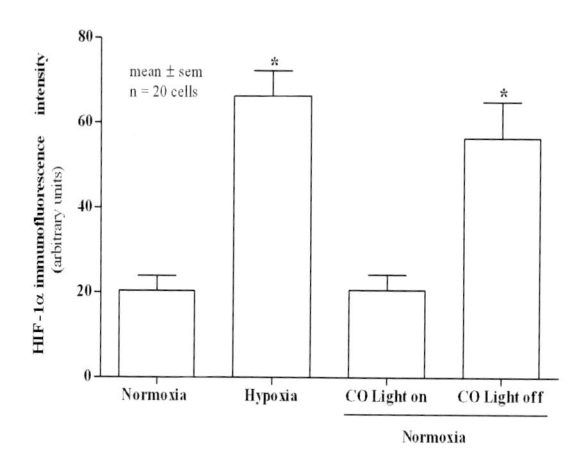

Figure 5: Effects of hypoxia (7 Torr) and of CO (350 Torr with 150 Torr oxygen) effect in normoxia) on the increase in HIF-1α and reversal of the CO effect by light (from 4). *P < 0.001 compared to either normoxic controls or illumination with white light.

Figure 6: Quantifying HIF-1α accumulation in the presence of CO when illuminated with light of 4 different wavelengths, 430, 500, 590, and 610 nm. (From 4) *P< 0.001 compared to white light illumination, **P < 0.001 compared to 430 or 590 nm illumination.

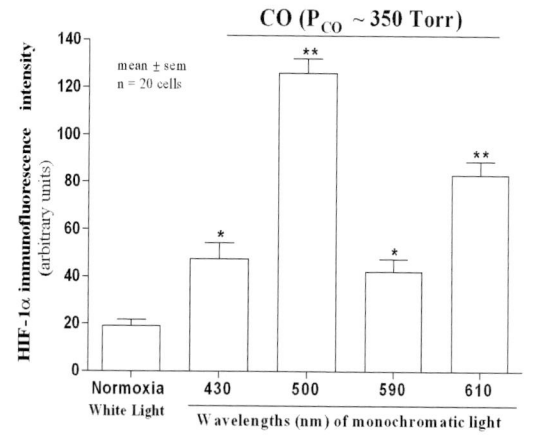

Reversal of the CO effect by light makes it possible to determine the absorption spectrum of the CO complex by determining the relative efficacy of different wavelengths of light. Measurements have been reported using 4 different wavelengths of light, 430, 500, 590, and 610 nm, selected to discriminate among the photodissociable CO complexes known to form in cells. The greatest effect of light was observed for 430 and 590 nm while the least was for 500 nm light. This is the same relationship observed for the reversal of the effect of CO on the afferent neural activity of the intact carotid body and is consistent with the effect of CO being mediated through cytochrome oxidase of the respiratory chain. Thus the induction of HIF-1α in glomus cells by addition of CO to a normoxic medium and the selective reversal by light at 430 and 590 nm is consistent with it being a response to inhibition of mitochondrial oxidative phosphorylation. It appears that the levels of HIF-1α in the glomus cells can be increased either by lowering the oxygen level available to the proline hydroxylase or by limiting the ability of mitochondrial oxidative phosphorylation to maintain the cellular energy level.

General comments on the role of mitochondria in physiological oxygen sensing and its biological relevance. The carotid and aortic bodies are specifically designed to sense the oxygen pressure in the vascular system and generate neural transmissions that communicate this information to remote locations. This oxygen dependence increases with maturation[6,] presumably in response to general metabolic control. Their oxygen sensitivity has often been called distinctive and assumed to require unique mechanism(s) for oxygen sensing. In general physiology, however, there are many examples of oxygen sensory systems that regulate tissue and cellular responses in similar ranges of oxygen pressures. Although the carotid body lies on the wall of the carotid artery and the influent blood oxygen pressures are high, within the carotid body microcirculation the oxygen pressures are not significantly different from those of many other tissues[16]. In many of these other tissues the oxygen pressure in the microcirculation is sensed and used to control local blood flow. Striking examples include autoregulation of blood flow in the brain and regulation of coronary flow in the heart. In the heart, for example, under physiological conditions the blood flow is precisely matched to oxygen extraction and the arterial-venous oxygen difference is maintained over wide ranges of work (oxygen consumption) rates[17]. Coronary flow in the isolated perfused heart, like neural activity of the isolated perfused carotid body, is stimulated by both inhibitors of mitochondrial oxidative phosphorylation and uncouplers of oxidative phosphorylation[18,19]. These agents "uncouple" flow from the oxygen pressure, as was presented for the neural activity of the carotid body, and flow becomes correlated with tissue energy state and not oxygen pressure *per se*. It appears that the mitochondria function as an oxygen sensor for a wide range of physiological functions, including regulation of blood flow in some tissues, carotid and arterial body neural activity, and adaptation to chronic hypoxia in the carotid body. It should be noted, however, that biology often uses more than one mechanism for control of critical pathways. In coronary flow, for example, there are several levels of control in addition to the energy state. These include a variety of vasoconstrictor and vasodilator agents that can be released in response to local conditions, agents controlled by the sympathetic nervous system, and modulators of contractility and work rate.

In Summary: The dependence of mitochondrial oxidative phosphorylation on oxygen pressure is appropriate for control of many oxygen dependent physiological functions. It has a primary role in the regulation of afferent neural activity of the carotid body, and by inference that of arterial bodies[20], and in mediating induction of HIF-1α in the glomus cells of the carotid body. These responses occur over different time scales (sec vs min) and are associated with the acute and chronic, respectively, effects of oxygen pressure on the carotid body.

4. ACKNOWLEDGEMENTS

Supported by the NIH grant R-37-HL-43413 (SL) and NS031465 (DFW). S.M.B is the recipient of NRSA fellowship (HL-07027-30).

5. REFERENCES

1. Wilson DF, Roy A, and Lahiri S. Immediate and long-term responses of the carotid body to high altitude. *High Alt Med Biol* 6: 97-111, 2005.
2. Baby SM, Roy A, Mokashi AM, and Lahiri S. Effects of hypoxia and intracellular iron chelation on hypoxia-inducible factor -1α and -1βin the rat carotid body and glomus cells. Histochem. Cell Biol. 120:343-352, (2003).
3. Roy A, Denys V, Baby SM, Mokashi A, Kubin L, and Lahiri S. Activation of HIF-1α mRNA by hypoxia and iron chelator in isolated rat carotid body. *Neurosci Lett* 363: 229-232, 2004.
4. Roy, A., Baby, S.M., Wilson, D.F., and Lahiri, S. Rat carotid body chemosensory discharge and glomus cell HIF-1a expression in vitro: Regulation by a common oxygen sensor. A.J.Physiol. – Reg. Integ. & Comp. Physiol.. 293(2): R829-36, 2007.
5. Matschinsky FM, Magnuson MA, Zelent D, Jetton TL, Doliba N, Han Y, Taub R, Grimsby J. The network of glucokinase-expressing cells in glucose homeostasis and the potential of glucokinase activators for diabetes therapy. *Diabetes*. 55(1):1-12, 2006. Review.
6. Wilson DF, Mokashi A, Chugh D, Vinogradov S, Osanai S, and Lahiri S. The primary oxygen sensor of the cat carotid body is cytochrome a$_3$ of the mitochondrial respiratory chain. *FEBS Lett* 351: 370-374, 1994.
7. Iturriaga R, Rumsey WL, Mokashi A, Spergel D, Wilson DF, and Lahiri S. In vitro perfused-superfused cat carotid body for physiological and pharmacological studies. *J Appl Physiol*. 70:1393-400, 1991.
8. Castor, L.N. and Chance, B. Photochemical action spectra of carbon monoxide –inhibited respiration. *J. Biol. Chem*. 217: 453-465, 1955.
9. Warburg O. Uber die Wirkung des Kohlenoxyds auf den Stoffwechsel der Hefe. *Biochem J* 177: 471-486, 1926.
10. Warburg, O. and Negelein, E. Über die Einfluss der Wellenlänge auf die Verteilung des Atmungsferments. (Absorptionsspektrum des Atmungsferments) *Biochem. Z.* 193, 339-346, 1928.
11. Wilson DF. Identifying oxygen sensors by their photochemical action spectra. *Methods in Enz* 381: 690-703, 2004.
12. Mulligan E. and Lahiri S. Separation of carotid body chemoreceptor responses to O$_2$ and CO$_2$ by oligomycin and by antimycin A. *Am J Physiol* 242: C200-206, 1982.
13. Mulligan E. Lahiri S. Storey BT. Carotid body O2 chemoreception and mitochondrial oxidative phosphorylation. J. Appl. Physiol: Resp. Env. & Exercise Physiol. 51(2): 438-46, 1981.
14. Semenza GL. HIF-1: mediator of physiological and pathological responses to hypoxia. *J Appl Physiol* 88: 1474-1480, 2000.
15. Roy A, Li J, Baby SM, Mokashi A, Buerk DG, and Lahiri S. Effects of iron-chelators on ion channels and HIF-1α in the carotid body. *Respir Physiol & Neurobiol* 141: 115-123, 2004.
16. Rumsey, WL, Iturriaga, R, Spergel, D, Lahiri, S, and Wilson, DF. Optical measurements of the dependence of chemoreception on oxygen pressure in the cat carotid body. Amer. J. Physiol. 261: C614-C622, 1991.
17. Allela, A. Williams FL, Bolene-Williams, C, and Katz, LN Interrelation between cardiac oxygen consumption and coronary blood flow. Am. J. Physiol. 183: 570-582, 1955.
18. Nuutinen EM, Nishiki K, Erecinska M, and Wilson DF. Role of mitochondrial oxidative phosphorylation in regulation of coronary blood flow. *Am J Physiol* 243: H159-H169, 1982.
19. Nuutinen EM, Nelson D, Wilson DF, and Erecinska M. Regulation of coronary blood flow: effects of 2,4-dinitrophenol and theophylline. *Amer J Physiol* 244: H396-H405, 1983.
20. Lahiri S, Mokashi E, Mulligan E, and Nishino T. Comparison of aortic and carotid chemoreceptor responses to hypercapnia and hypoxia. *J Appl Physiol* 51(1): 55-61, 1981.

IMPACT OF REACTIVE OXYGEN SPECIES ON THE EXPRESSION OF ADHESION MOLECULES *IN VIVO*

Oliver Thews[*], Christine Lambert[**], Debra K. Kelleher[*], Hans K. Biesalski[**], Peter Vaupel[*], and Juergen Frank[**]

Abstract: Many non-surgical tumor treatments induce reactive oxygen species (ROS) which result in cell damage. This study investigated the impact of ROS induction on the expression of adhesion molecules and whether α-tocopherol pre-treatment could have a protective effect. Experimental rat DS-sarcomas were treated with a combination of localized 44°C-hyperthermia, inspiratory hyperoxia and xanthine oxidase which together lead to a pronounced ROS induction. Further animals were pre-treated with α-tocopherol. The *in vivo* expression of E- and N-cadherin, α-catenin, integrins αv, $\beta 3$ and $\beta 5$ as well as of the integrin dimer $\alpha v \beta 3$ was assessed by flow cytometry. The expression of αv-, $\beta 3$-integrin, of the $\alpha v \beta 3$-integrin dimer and of E-cadherin was significantly reduced by the ROS-inducing treatment. This effect was partially reversible by α-tocopherol, indicating that ROS play a role in this process. N-cadherin, α-catenin and $\beta 5$-integrin expression were unaffected by ROS. These results indicate that the expression of several adhesion molecules is markedly reduced by ROS and may result in a decrease in the structural stability of tumor tissue. Further studies are needed to clarify the impact of ROS induction on the metastatic behavior of tumors.

1. INTRODUCTION

The integrity and stability of tumor tissue result from mechanical interaction between tumor cells and from the attachment of cells to the extracellular matrix (ECM). The inter-cellular contact is established via cadherins which are linked to cytoskeletal structures (e.g., actin filaments) by catenin molecules. At the same time, tumor cells may be attached to extracellular matrix proteins (e.g., collagen, fibronectin) by integrins which form heterodimers (α- and β-subunit) on the cell surface. These molecules play a role in attaching cells to the ECM and in intracellular signaling via RAS or PKC, which can affect the migration and invasion of tumor cells[1]. Changes in cadherin, catenin and

[*] Institute of Physiology and Pathophysiology, University of Mainz, 55099 Mainz, Germany
[**] Institute of Biological Chemistry and Nutrition, University of Hohenheim, 70593 Stuttgart, Germany

P. Liss et al. (eds.), *Oxygen Transport to Tissue XXX*, DOI 10.1007/978-0-387-85998-9_15,
© Springer Science+Business Media, LLC 2009

integrin expression may result in increased shedding of tumor cells into the blood stream and have, therefore, been thought to be responsible for an enhanced metastatic spread of tumor cells[2,3].

Besides cytokines, some parameters related to the tumor metabolic micro-environment (e.g., hypoxia) have been identified which can affect adhesion molecule expression *in vitro*[4]. The generation of reactive oxygen species (ROS) may also influence adhesion molecule expression or may interact with the intracellular signaling pathway linked to integrins resulting in a loosening of the cell cluster and increased cell migration[5]. Since many non-surgical tumor treatment modalities such as ionizing radiation, photodynamic therapy or hyperthermia (HT) massively induce ROS, the question arises as to whether therapeutically-induced oxidative stress may affect cell-cell- or cell-matrix-interactions in tumors thus leading to changes in the metastatic behavior. If ROS play a role in this process, the question also arises of whether oxygen radical scavenging by α-tocopherol may affect the stability of the tumor cell cluster after therapeutic ROS-induction *in vivo*.

2. MATERIAL AND METHODS

2.1. Animals and tumors

All studies were performed using the rat DS-sarcoma. Following s.c. injection of DS-ascites cells (0.4 ml; approx. 10^4 cells/μl) into the dorsum of the hind foot of male Sprague Dawley rats (Charles River Deutschland, Sulzfeld, Germany; body weight 150-210 g), experimental tumors grew as flat, spherical segments and replaced the subcutis and corium completely. Experiments were performed when tumors reached a target volume of 0.5 - 2.0 mL (mean 1.20±0.06 mL), approx. 7-12 days after inoculation of tumor cells. Studies were approved by the regional ethics committee and conducted according to UKCCCR guidelines[6] and the German Law for Animal Protection.

2.2. ROS induction and α-tocopherol treatment

The combination of 44°C-HT for 60 min, xanthine oxidase administration and inspiratory hyperoxia has previously been shown to cause massive ROS formation[7]. In the present study, the same treatment protocol was used. HT was performed by heating the tumor in a saline bath (9 g/l NaCl) set at 44.3°C. Animals were anesthetized with pentobarbital (40 mg/kg, i.p., Narcoren™, Merial, Hallbergmoos, Germany) and placed on a board in a ventral position above the saline bath in which the tumor-bearing legs were immersed. Fifteen minutes prior to HT, animals received an i.v. injection of 15 U/kg body weight of xanthine oxidase (XO). During HT, animals spontaneously breathed pure oxygen using a face mask. Control animals were also anesthetized, but not treated with HT, XO or oxygen breathing.

An additional group of animals received α-tocopherol (vitamin E emulsion, Fresenius Kabi, Bad Homburg, Germany) prior to ROS-generating therapy. Two doses of tocopherol were injected i.v. at a dose of 10 mg/kg 48 and 24 h prior to HT leading to an intratumoral tocopherol concentration of 0.288±0.035 μmol/g protein on the day of the ROS-generating treatment[8].

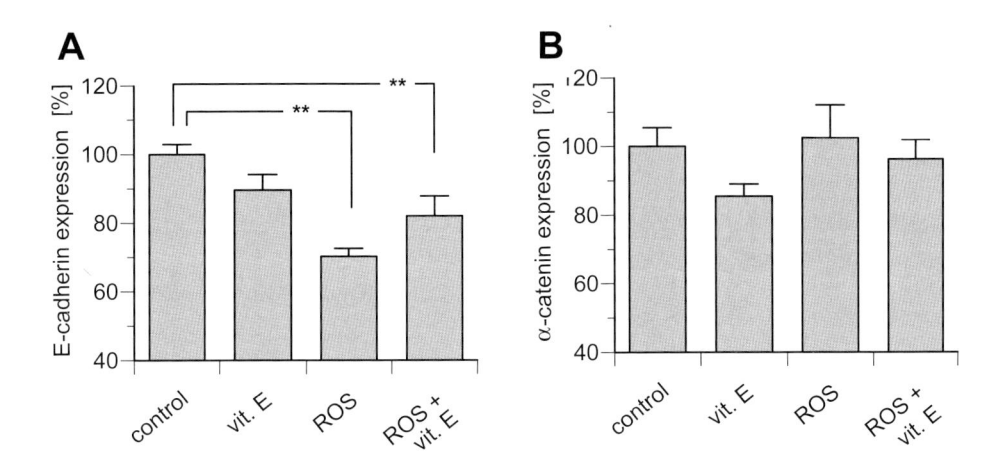

Figure 1. Relative expression of (A) E-cadherin and (B) α-catenin in DS-sarcomas 24 h after α-tocopherol application (*vit. E*), after ROS-inducing hyperthermia treatment (*ROS*), or a combination of both treatments (*ROS + vit. E*). Values are expressed as means±SEM of at least 11 tumors. (**) $p<0.01$. The differences between *ROS* and *ROS + vit. E* were not statistically significant.

In total, 4 experimental groups were studied. Animals underwent ROS-inducing HT with or without additional tocopherol application. In the second series, animals received sham treatment, again with or without vitamin E.

2.3. *In vivo*-expression of adhesion molecules

Adhesion molecule expression was determined by flow cytometry after staining for extra- or intracellular antigens. The expression of the following proteins was analyzed: E-cadherin (antibody Clone G-10; all antibodies supplied by Santa Cruz Biotechnology, Santa Cruz, CA, USA), N-cadherin (clone D-4), α-catenin (clone H-297), integrin αv (clone N-19), integrin β3 (clone F-11), integrin β5 (clone E-19). In addition, the expression of the integrin dimer αvβ3 (clone 23C6) was analyzed. Tumors were excised 24 h after treatment and mechanically disaggregated to produce a single cell suspension. After incubation of cells with the fluorochrome-labeled antibodies, fluorescence intensity was determined using a flow cytometer (Epics, Coulter, Hialeah FL, USA). The values measured in each tumor were normalized with respect to those from the control group.

2.4. Statistical analysis

Results are expressed as means±standard error of the mean (SEM). Differences between the groups were assessed using the two-tailed Wilcoxon test for unpaired samples. The significance level was set at $\alpha=5\%$.

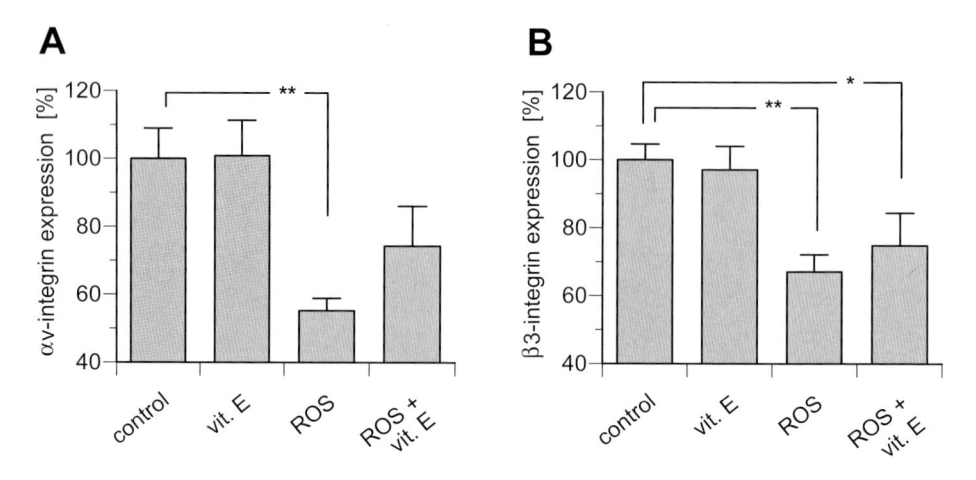

Figure 2. Relative (A) αv- and (B) β3-integrin expression in DS-sarcomas 24 h after α-tocopherol application (*vit. E*), after ROS-inducing hyperthermia treatment (*ROS*), or a combination of both treatments (*ROS + vit. E*). Values are expressed as means±SEM of at least 11 tumors. (*) $p<0.05$, (**) $p<0.01$. The differences between *ROS* and *ROS + vit. E* were not statistically significant.

3. RESULTS

Therapeutic ROS induction showed a pronounced impact on various adhesion molecules. Expression of these molecules was first measured 24 h after treatment since it was expected that changes in protein expression would require several hours to occur. E-cadherin expression was significantly reduced by approx. 30% by HT treatment (Fig. 1A). This effect could be partially reversed by α-tocopherol pre-treatment leading to a higher expression in combination with HT. However, this latter finding was not statistically significant but may indicate that the reduced expression was the result of oxidative stress. The intracellular α-catenin was almost unaffected by ROS-induction or vitamin E application (Fig. 1B).

The surface expression of the αv- and the β3-integrin was reduced by 45% and 33% respectively, following therapeutic ROS-induction (Fig. 2). In tumors pre-treated with α-tocopherol, the decrease was still detectable. Vitamin E, however, seems to reduce the impact of oxidative stress. In tumors treated with both HT and tocopherol, expression of these integrins was slightly (however, not statistically significantly) increased than with ROS-treatment alone (Fig. 2).

Integrins form heterodimers in order to anchor cells at the matrix proteins. Further experiments were undertaken to clarify whether the integrins were being affected independently or dependently of one another. The results show that ROS-inducing HT also down-regulated αvβ3-dimer expression (Fig. 3). Again, α-tocopherol slightly (but not at a statistically significant level) increased the expression indicating that ROS may play a causal role in this process (Fig. 3).

N-cadherin and β5-integrin expression was independent of all treatments used in the cell line investigated (data not shown).

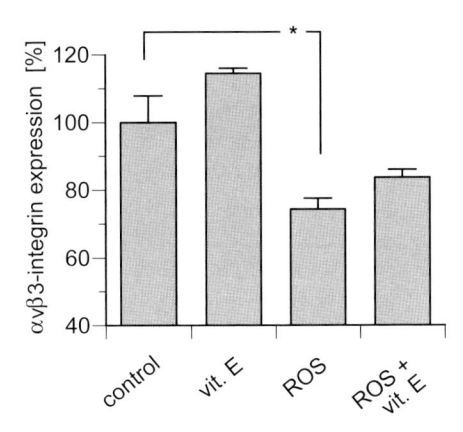

Figure 3. Relative expression of the αvβ3-integrin-dimer in DS-sarcomas 24 h after α-tocopherol application (*vit. E*), after ROS-inducing hyperthermia treatment (*ROS*), or a combination of both treatments (*ROS + vit. E*). Values are expressed as means±SEM of at least 4 tumors. (*) p<0.05. The difference between *ROS* and *ROS + vit. E* was not statistically significant.

4. DISCUSSION

Cell-cell contact relies on various cadherins which are connected to the cytoskeleton by intracellular catenins. Down-regulation of cadherins and catenins or mutations of the respective genes can cause increased shedding of tumor cells into the blood stream and, thus, a higher metastatic spread[2,3]. In the present study, E-cadherin expression was significantly reduced by oxidative stress (Fig. 1A). In animals pretreated with α-tocopherol, this reduction in E-cadherin expression following ROS treatment was at least partially reversed compared to animals not pretreated with α-tocopherol but still remained below the level in control animals. Other *in vitro* studies showed no or only a marginal reduction in cellular E-cadherin content upon oxidative stress[9,10]. These studies did, however, show that ROS led to a translocation of E-cadherin from the outer cell membrane to intracellular compartments leading to a reduction in tumor cell adhesion[9,10]. Since in the present study E-cadherin expression was measured only at the outer cell membrane, the findings seen in Fig. 1A are still in accordance with the previous results.

Adhesion of cells to the extracellular matrix mediated by integrins plays an important role in the mechanical stability of tumor tissue. Tumor cells shed into the blood stream often show a reduced integrin expression[11]. Integrins are also important for adhesion and migration of circulating tumor cells in new host tissue. Tumors with reduced expression of αv-integrin or in which αv-integrin was blocked show reduced metastatic spread[12-14].

In the present study, ROS-inducing HT led to a significant reduction in αvβ3-integrin (Figs. 2 & 3). Upon application of α-tocopherol there was a tendency (which was not statistically significant) that this reduction was at least partially reversible. Due to the different effects of these integrins in the formation of metastases, a prediction of whether this leads to an increased metastatic behavior (due to loosening of cells) or a reduced formation of metastases (due to decreased adhesion of circulating tumor cells in new host tissue) is not possible. In the study of Mori et al.[9], long-term exposure of cells to H_2O_2 *in vitro* led to increased expression of the integrins α2, α6 and β3. The differences to the findings of the present study concerning β3-integrin remain unclear. The duration of oxidative stress may play a role since in our study it was only applied for 1 h, whereas in the study of Mori et al., cells underwent oxidative stress for 48-96 h.

In conclusion, the results of this study show that the surface expression of several adhesion molecules (cadherins, integrins) is affected by a therapeutic ROS generation. However, the net effect of these changes on metastatic behavior cannot be predicted since down-regulation of integrins can also lead to a reduced adhesion of circulating tumor cells in new host tissues. Therefore, ROS *per se* may not have a purely pro- or anti-metastatic impact.

5. ACKNOWLEDGEMENT

This study was supported by the Deutsche Krebshilfe (grant 10-1798-Fr 1).

6. REFERENCES

1. J.D. Hood and D.A. Cheresh. Role of integrins in cell invasion and migration. *Nat. Rev. Cancer* **2**, 91-100 (2002).
2. I.R. Beavon. The E-cadherin-catenin complex in tumour metastasis: structure, function and regulation. *Eur. J. Cancer* **36**, 1607-1620 (2000).
3. S. Hirohashi and Y. Kanai. Cell adhesion system and human cancer morphogenesis. *Cancer Sci.* **94**, 575-581 (2003).
4. R.A. Cairns, R. Khokha, and R.P. Hill. Molecular mechanisms of tumor invasion and metastasis: an integrated view. *Curr. Mol. Med.* **3**, 659-671 (2003).
5. W.S. Wu. The signaling mechanism of ROS in tumor progression. *Cancer Metastasis Rev.* **25**, 695-705 (2006).
6. P. Workman, P. Twentyman, F. Balkwill, A. Balmain, D.J. Chaplin, J.A. Double, et al. United Kingdom Co-ordinating Committee on Cancer Research (UKCCCR) Guidelines for the Welfare of Animals in Experimental Neoplasia (2nd edit.). *Br. J. Cancer* **77**, 1-10 (1998).
7. J. Frank, D.K. Kelleher, A. Pompella, O. Thews, H.K. Biesalski, and P. Vaupel. Enhancement of oxidative cell injury and antitumor effects of localized 44°C hyperthermia upon combination with respiratory hyperoxia and xanthine oxidase. *Cancer Res.* **58**, 2693-2698 (1998).
8. O. Thews, C. Lambert, D.K. Kelleher, H.K. Biesalski, P. Vaupel, and J. Frank. Possible protective effects of α-tocopherol on enhanced induction of reactive oxygen species by 2-methoxyestradiol in tumors. *Adv. Exp. Med. Biol.* **566**, 349-355 (2005).
9. K. Mori, M. Shibanuma, and K. Nose. Invasive potential induced under long-term oxidative stress in mammary epithelial cells. *Cancer Res.* **64**, 7464-7472 (2004).
10. R.K. Rao, S. Basuroy, V.U. Rao, K.J. Karnaky Jr, and A. Gupta. Tyrosine phosphorylation and dissociation of occludin-ZO-1 and E-cadherin-β-catenin complexes from the cytoskeleton by oxidative stress. *Biochem. J.* **368**, 471-481 (2002).
11. M. Bockhorn, S. Roberge, C. Sousa, R.K. Jain, and L.L. Munn. Differential gene expression in metastasizing cells shed from kidney tumors. *Cancer Res.* **64**, 2469-2473 (2004).
12. A. Enns, T. Korb, K. Schlüter, P. Gassmann, H.U. Spiegel, N. Senninger, F. Mitjans, and J. Haier. αvβ5-integrins mediate early steps of metastasis formation. *Eur. J. Cancer* **41**, 1065-1072 (2005).
13. J.F. Harms, D.R. Welch, R.S. Samant, L.A. Shevde, M.E. Miele, G.R. Babu et al. A small molecule antagonist of the αvβ3 integrin suppresses MDA-MB-435 skeletal metastasis. *Clin. Exp. Metastasis* **21**, 119-128 (2004).
14. D.G. McNeel, J. Eickhoff, F.T. Lee, D.M. King, D. Alberti, J.P. Thomas et al. Phase I trial of a monoclonal antibody specific for αvβ3 integrin (MEDI-522) in patients with advanced malignancies, including an assessment of effect on tumor perfusion. *Clin. Cancer Res.* **11**, 7851-7860 (2005).

MINI SENSING CHIP FOR POINT-OF-CARE ACUTE MYOCARDIAL INFARCTION DIAGNOSIS UTILIZING MICRO-ELECTRO-MECHANICAL SYSTEM AND NANO-TECHNOLOGY

Jianting Wang[a], Bin Hong[a], Junhai Kai[b], Jungyoup Han[b], Zhiwei Zou[b], Chong H. Ahn[b], and Kyung A. Kang[a]

Abstract: A rapid and accurate diagnosis of acute myocardial infarction (AMI) is crucial for saving lives. For this purpose, we have been developing a rapid, automatic, point-of-care, biosensing system for simultaneous four cardiac marker quantification. This system performs a fluorophore mediated immuno-sensing on optical fibers. To improve the sensitivity of the sensor, novel nanoparticle reagents enhancing fluorescence were implemented. Micro-electro-mechanical system (MEMS) technology was applied in the sensing chip development and automatic sensing operation was implemented to ensure a reliable and user-friendly assay. The resulting system is a point-of-care, automatic four cardiac marker sensing system with a 2 x 2.5cm sensing chip. An assay requires a 200 μL plasma sample and 15-minute assay time.

1. INTRODUCTION

Acute myocardial infarction (AMI) is the world's leading cause of morbidity and mortality. According to the latest data by the American Heart Association (AHA),[1] in 2004, 7.9 million cases of myocardial infarction were reported in the US, including 865,000 new and recurrent cases and 157,600 deaths. AMI is also a disease with a high rate of misdiagnosis due to the low sensitivity of the current diagnostic tools used in the emergency room.[2,3] A better method for AMI diagnosis is to detect the elevation of cardiac markers in blood plasma.[4] Currently, the cardiac markers are usually quantified

---

[a] Jianting Wang, Bin Hong, and Kyung A. Kang, Department of Chemical Engineering, University of Louisville, Louisville, KY 40292. [b] Junhai Kai, Jungyoup Han, Zhiwei Zou, and Chong H. Ahn, Department of Electrical and Computer Engineering and Computer Science, University of Cincinnati, Cincinnati, OH 45221.

P. Liss et al. (eds.), *Oxygen Transport to Tissue XXX*, DOI 10.1007/978-0-387-85998-9_16,
© Springer Science+Business Media, LLC 2009

in a central laboratory, and it often takes hours from when it is ordered till the results are received. Therefore, a rapid and accurate point-of-care (POC) system is urgently needed.

Our fiber-optic, fluorophore-mediated, immuno-sensing system has demonstrated accurate, sensitive, rapid, and reliable simultaneous quantification of various biomarkers.[5,6] For the cardiac marker sensing, we have selected the following four markers: B-type natriuretic peptide (BNP) and C-reactive protein (CRP) are crucial markers for the diagnosis of congestive heart failure and acute coronary syndromes.[7,8] Myoglobin (MG) and cardiac troponin I (cTnI) are important markers for early diagnosis of a heart attack.[9] Based on the assessment of the clinical results, we have previously determined the clinically important sensing range of the marker[10,11,12] to be 26~260 pM for BNP; 30~300 pM for cTnI; 4~40 nM for MG and 5.6~56 nM for CRP. Hong et al. have successfully quantified the above four cardiac markers with four 3 cm sensors[10].

Our effort has been focused on both the sensitivity enhancement and sensor size reduction to develop a POC, AMI diagnostic system. Reduction of sensor size can usually reduce the sample and reagents volume (i.e., assay cost reduction) and improve portability. Smaller sensor size may, however, lower the sensitivity due to the less sensing surface. For BNP and cTnI, their concentrations in plasma at an early disease stage are only tens of pico molar level, requiring extremely high sensitivity. Since our sensors are mediated by fluorophore, enhancing the fluorescence can improve the sensitivity. Some nanometal particles have high-density surface plasmon polariton fields. Lone pair electrons of a fluorophore, often participating in the fluorescence self-quenching, can be transferred to the plasmon field, resulting in artificial fluorescence enhancement[13,14]. Some biocompatible solvents were also found to enhance the fluorescence significantly, possibly via effective dipole coupling between the fluorophore and the solvent molecule, resulting in an increase in the ennergy gap between the excited and the ground states[13,14]. Our research group has implemented nanomatal particles and sensor compatible organic solvents to the sensitivity[13,14]. Among nanometal particles and solvents, 5 nm-nanogold particles (5nm-NGP) dispersed in 1-butanol was found to be a very effective enhancer. MEMS technique was also incorporated for our micro sensing chip development and for a reliable and automated sensor operation.

This paper reports the results of the further reduction in the sensing chip size and its performance in quantifying four cardiac markers in plasma samples.

2. MATERIALS, INSTRUMENTS AND METHODS

Sensor preparation and assay protocol

For the cardiac markers and their respective monoclonal antibodies, BNP was purchased from Bachem (Torrance, CA) and monoclonal IgG against human BNP was from Strategic Biosolutions (Newark, DE). cTnI, MG, and CRP, and their antibodies were obtained from Fitzgerald Industries (Concord, MA). Plasma samples with cardiac markers were prepared by adding a known amount of cardiac markers to the human plasma.[10]

Four cardiac marker biosensors were constructed, following the protocol established by Tang et al.[12] The quartz fiber (Research International, Inc., Monroe, WA) was immobilized with streptavidin (Sigma/Aldrich, St. Louis, MO) and the monoclonal antibody (1°Mab) against the respective marker is conjugated with biotin

(Sigma/Aldrich) and immobilized on the surface of fiber *via* streptavidin-biotin bond and the sensor is encased in a micro-sensing chamber. The fluorophore Alexa Fluor® 647 (AF647; max. excitation/emission wavelengths, 649/666 nm) was from Invitrogen (Carlsbad, CA). The second monoclonal antibody (2°Mab) was conjugated with AF647 following the manufacture's instruction.

The cardiac marker assay proceeded as follows: a sample was introduced into the sensing chamber and incubated for 3 min at a rate of 1.2 cm/sec. Once the target marker binds specifically to the 1°Mab on the sensor surface, the fiber was washed with phosphate buffered saline with 0.1% Tween 20 (PBST, pH 7.4) for 1 minute to remove the unbound molecules in the sample. Then the fluorescence was measured (baseline). The fluorometer (Analyte 2000™, Research International, Monroe, WA) has four sensing ports. AF647 tagged 2°Mab, at a concentration of 10 µg/ml, was then applied to the chamber and incubated for 4 minutes to react with the bound analyte. The sensor was washed with PBST for 1 minute and another fluorescence reading was taken. The signal difference between this signal and the baseline subtracted by the negative control is the signal intensity of the sample.

For the fluorescence enhancing, 5nm-NGP coated with tannic acid was purchased from Ted Pella (Redding, CA). 1-butanol was purchased from Sigma/Aldrich. For the sensing with a nanogold particle reagent (NGPR), for the baseline, the NGPR was applied before the sample incubation. For the signal of the sample, it was applied after the second antibody was reacted with the analyte on the sensor surface.

In this paper, the enhancement is defined as the fluorescence signal increase due to the NGPR divided by the fluorescence from the sensor without NGPR.

Microfluidic system and automated sensing operation

The sensing chip and microchannel network of the sensing unit were microfabricated as described by Sohn et al.[15] and Hong et al[16]. The system includes a polycarbonate mother board with imbedded network of the microchannels; a four-channel, plastic sensing module; a micro-solenoid pump (12 v, 50 µL per stroke, 2 W); seven micro-solenoid valves (12 v, 280 mW, Lee Co.; Westbrook, CT); a data acquisition (DAQ) card (USB-6008, 8 inputs, 12 bits, 10000 samples/s, multifunctional I/O, National Instruments; Austin, TX); and a drive circuit with a power plug, a power switch, and a power LED.

3. RESULTS AND DISCUSSION

Reduction of BNP sensor size

As previously stated, here, we have attempted to reduce the sensor size from 3 to 1.5 cm. BNP sensing was tested first since BNP has the lowest sensing range among the four cardiac markers (see Introduction).

Figure 1 shows the performance of the BNP sensors at (a) 3 and (b) 1.5 cm, with and without NGPR. In terms of the signal intensity enhancement by NGPR, the signals were enhanced by 4 and 3 times for 3 cm and 1.5 cm sensor, respectively. For 1.5 cm the signal without NGPR was very low in the entire sensing range. Although the signal for the 1.5 cm sensor with NGPR was approximately only 40% that of the 3 cm sensor, the correlation coefficient was 0.96 (0.95 for 3 cm sensor), and the signal to noise ratio (S/N) was 4.1 (3.8 for 3 cm sensor), showing that the performance of the 1.5 cm sensor was still satisfactory, but with less sample and reagent.

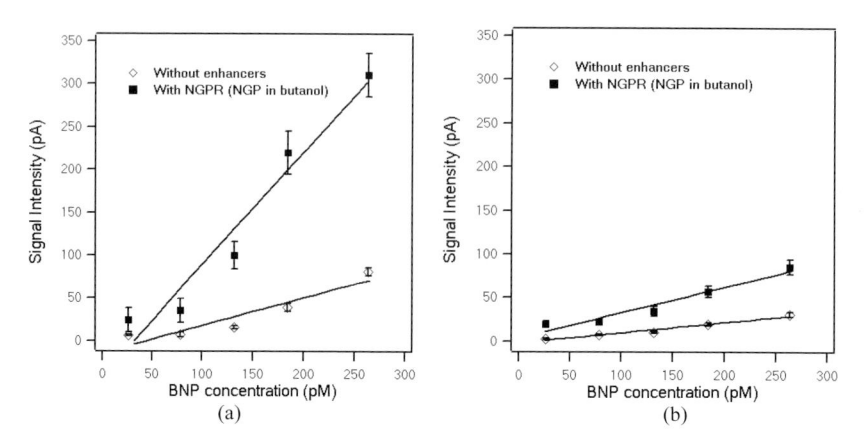

Figure 1. Performance of BNP sensor at sensor sizes of (a) 3 cm and (b) 1.5 cm with and without NGPR.

Four cardiac marker sensing by the mini-sensor

Since the sensing performance of the 1.5 cm BNP sensor was satisfactory, 1.5 cm was then tested for all four cardiac markers in a micro sensing chip. The size of the chip for four cardiac marker sensing was reduced from 4 x 4 cm (for 3 cm sensors) to 2 x 2.5 cm (for 1.5 cm sensors). A schematic diagram of the chip is shown in figure 2. As previously shown in Hong et al.[10], on the inner surface of the microchannel, bumps/baffles were microfabricated to create local turbulence, facilitating the analyte transport to the sensor surface better. The mini sensing chip requires a sample size of only 200 μl, and therefore, only 0.5 ml of a blood sample is needed for an assay.

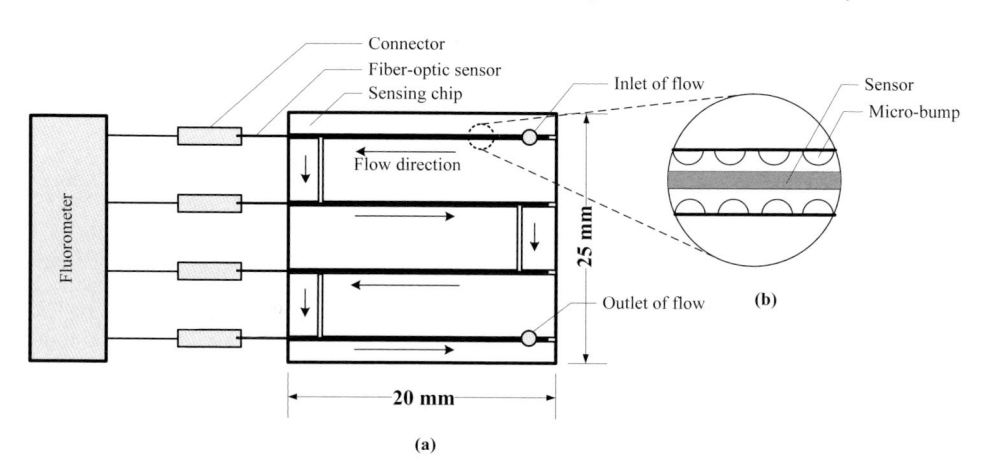

Figure 2. (a) Schematic diagram of mini-sensing chip and (b) Enlarged schematic diagram of a sensing chamber

For this study, the sample was the mixture of four cardiac markers in their sensing ranges in the human plasma (Fig. 3). For BNP and cTnI, the signal range without using NGPR was only 3~30 pA. With NGPR, the signal was enhanced by 3 and 4 times. The S/N ratios were 4.1 and 4.5, and the correlation coefficients were 0.96 and 0.97, respectively. For MG and CRP, the signal ranges without enhancer were as high as 4~105 pA and 90~400 pA, respectively. Nevertheless, since the four sensors need to be done simultaneously with a sample, NGPR was applied also to these two sensors. The signals were enhanced by 5 and 2 times and S/N were 5.2 and 6.7, respectively.

All four 1.5 cm sensors showed linear relationships between the marker concentration and the signal intensity, with correlation coefficients above 0.95.

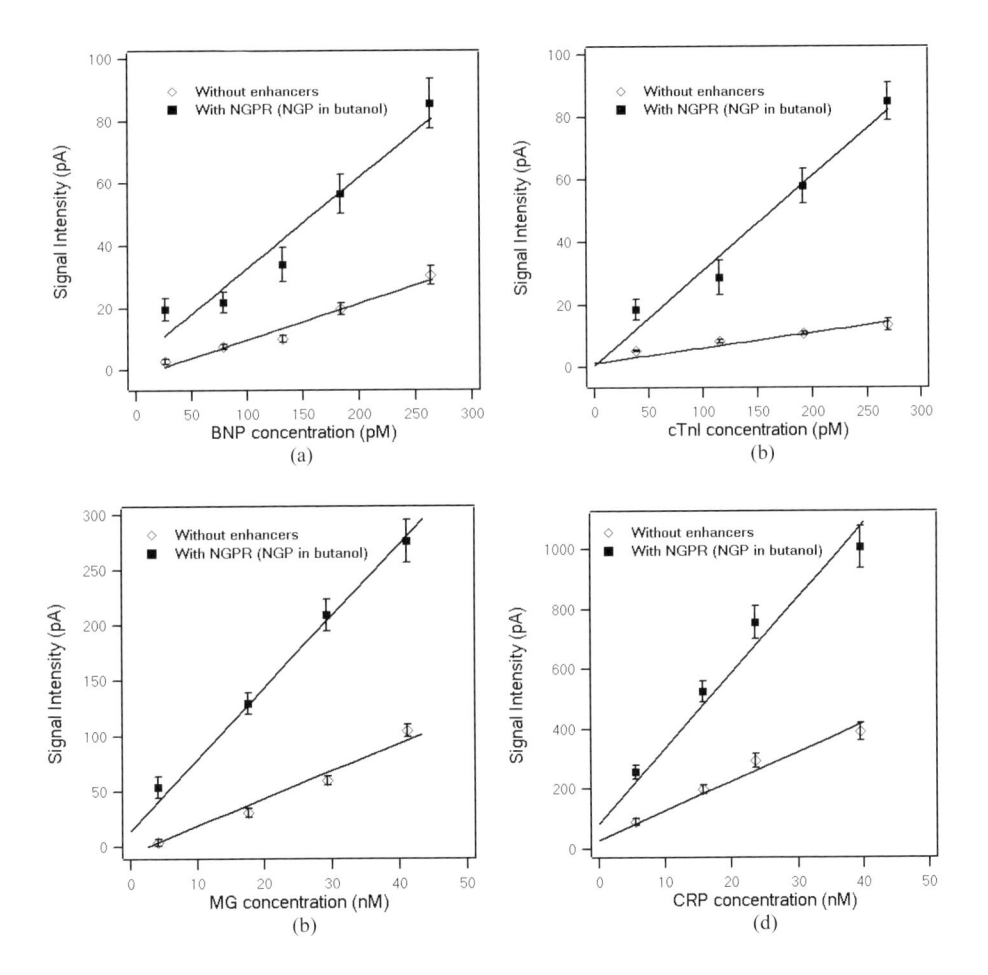

Figure 3. Sensing performance of 1.5 cm (a) BNP, (b) cTnI, (c) MG, and (d) CRP sensors with and without NGPR

Futuristic portable all-in-one device

In the near future, our current sensing system[16] of fluorometer, microfluidic flow control unit, display monitor and input keyboard may be put into a highly portable, all-in-one device at a size of 30 x 20 x 15 cm (Fig. 4). With disposable sensing chip and sample and reagent containers, a four cardiac marker sensing may be performed in the emergency room or even in an emergency-medical-service vehicle within 15 min.

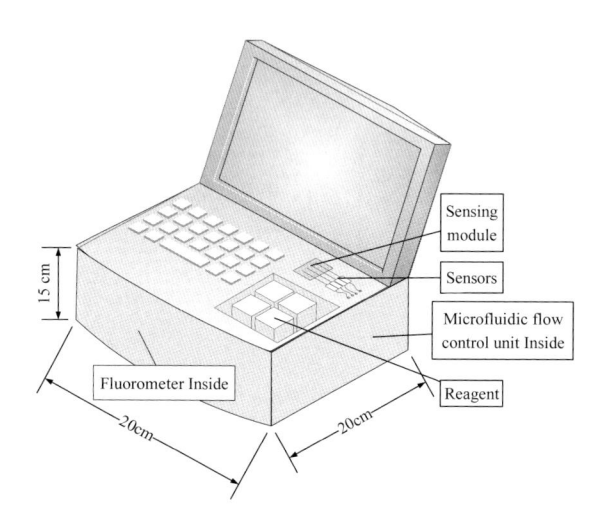

Figure 4. Schematic design diagram of all-in-one four cardiac marker sensing device.

4. CONCLUSIONS

Our MEMS based sensing device with mini-sensing chips (2 x 2.5 cm) and novel nano metal reagent can accurately quantify four cardiac markers simultaneously at clinically significant concentration range within 15 min, requiring a plasma sample volume of 200 µl. The entire sensing system may be converted to be a highly portable, all-in-one, POC sensing device.

5. ACKNOWLEDGEMENTS

The authors acknowledge the financial support from Kentucky Science and Engineering Foundation (KSEF-148-502-03-55) for fluorescence enhancement studies and National Science Foundation (BES-0330075) for cardiac marker biosensing.

6. REFERENCES

1. American Heart Association (AHA), Heart disease and stroke statistics, 2007 update

2. A. Wu, F. S. Apple, W. B. Gibler, R. L. Jesse, M. M. Warshaw, R. Valdes. National Academy of Clinical Biochemistry Standards of Laboratory Practice: recommendations for the use of cardiac markers in coronary artery diseases. *Clin Chem* **45**:1104–21 (1999).

3. J. H. Pope, T. P. Aufderheide, R. Ruthazer, Missed diagnoses of acute cardiac ischemia in the emergency department. *N Engl J Med* **342**: 1163–70 (2000).

4. R. Bernard, E. Corday, H. Eliasch, Nomenclature and criteria for diagnosis of ischemic heart disease. Report of the Joint International Society and Federation of Cardiology/World Health Organization task force on standardization of clinical nomenclature. *Circulation* **59**:607–9 (1979).

5. J. O. Spiker, K. A. Kang, W. Drohan, and D. F. Bruley, Protein C detection via fluorophore mediated immuno-optical biosensor. *Advances in Experimental Medicine and Biology* **428**, 621-627. (1999)

6. L. Tang and K. A. Kang. Preliminary study of simultaneous multi-anticoagulant deficiency diagnosis by a fiber optic multi-analyte biosensor. Proceedings of the 31st IOSTT Annual Meeting, Aug. 16-20, 2003, Rochester, NY.

7. A. S. Maisel, P. Krishnaswamy, H. C. Herrmann, and P. A. McCullough, Rapid measurement of B-type natriuretic peptide in the emergency diagnosis of heart failure, *New Engl. J. Med.* **347**, 161-167 (2002).

8. M. S. Sabatine , D. A. Morrow, C. P. Cannon, and E. Braunwald, Multimarker approach to risk stratification in non-ST elevation acute coronary syndromes: Simultaneous assessment of troponin I, c-reactive protein, and b-type natriuretic peptide, *Circ.* **105**, 1760-1763 (2002).

9. F. S. Apple, R. H. Christenson, R. Valdes, A. J. Andriak, K. Mascotti, and A. H.B. Wu, Simultaneous rapid measurement of whole blood myoglobin, creatine kinase MB, and cardiac troponin I by the triage cardiac panel for detection of myocardial infarction, *Clin. Chem.* **45**(2), 199-205 (1999).

10. L. Tang, Multi-analyte, fiber-optic immuno-biosensing system for rapid disease diagnosis: model systems for anticoagulants and cardiac markers. Dissertation. Chemical Engineering, University of Louisville, Louisville, KY. (2005)

11. L. Tang and K. A. Kang, Preliminary study of simultaneous multi-anticoagulant deficiency diagnosis by a fiber optic multi-analyte biosensor. *Advances in Experimental Medicine and Biology* **566**:303-309 (2006).

12. L. Tang, Y. J. Ren, B. Hong, and K. A. Kang, A fluorophore-mediated, fiber-optic, multi-analyte, immuno-sensing system for rapid diagnosis and prognosis of cardiovascular diseases, *J. Biomed. Optics* **11**, 021011 (2006).

13. B. Hong and K. A. Kang, Biocompatible, nanogold-particle fluorescence enhancer for fluorophore mediated, optical immunosensor, *Biosens. Bioelectron.* **21**(7), 1333-1338 (2006).

14. K. A. Kang and B. Hong, Biocompatible nano-metal particle fluorescence enhancers, *Crit. Rev. Eukar. Gene Expres.* **16**(1), 45-60 (2006).

15. Y. Sohn, J. H. Kai, C. H. Ahn, Protein array patterning on Cyclic Olefin Copolymer (COC) for disposable protein chip, *Sensor Lett.* **2**, 171-174 (2005).

16. B. Hong, J. H. Kai, Y. J. Ren, J. Y. Han, Z. W. Zou, C. H. Ahn, and K. A. Kang, Higly sensitive, rapid, reliable and automatic cardiovascular disease diagnosis with nanoparticle fluorescence enhancer and MEMS, Proceeding of the 34rd ISOTT annual meeting, August 12-17, Louisville, KY, USA, (in press)

A CHEMICAL BIOSYNTHESIS DESIGN FOR AN ANTIATHEROSCLEROSIS DRUG BY ACYCLIC TOCOPHEROL INTERMEDIATE ANALOGUE BASED ON "ISOPRENOMICS"

Yoshihiro Uto, Daisuke Koyama, Mamoru Otsuki, Naoki Otomo, Tadashi Shirai, Chiaki Abe, Eiji Nakata, Hideko Nagasawa, and Hitoshi Hori[*]

Abstract: Phytyl quinols, namely acyclic tocopherols, are key intermediates of tocopherol biosynthesis, but their biological activities remain unclear. We therefore investigated the structure-activity relationship of phytyl quinols to apply a chemical biosynthesis design for an antiatherosclerosis drug based on isoprenomics. We have achieved the biosynthesis-oriented design and synthesis of α- (TX-2254) and β- (TX-2247) phytyl quinol as an unnatural intermediate, other γ- (TX-2242) and δ- (TX-2231) phytyl quinol as a natural one. Geometry optimization and Molecular orbital (MO) calculation of TX-2254 showed a unique right-angle structure; however, MO energy of TX-2254 and d-α-tocopherol were very similar. Radical reactivity of TX-2231 was equal to dl-α-tocopherol, whereas TX-2254, TX-2247, and TX-2231 showed lower reactivity than dl-α-tocopherol. All four phytyl quinols showed almost the same moderate inhibitory activity against low-density lipoprotein (LDL) oxidation instead of their different degree of *C*-methylation with character different from tocopherols. In vivo toxicities of phytyl quinols against chick embryo chorioallantoic membrane (CAM) vasculature were hardly observed. We proposed phytyl quinols were possible antioxidants in plants and animals, like vitamin E.

1. INTRODUCTION

Tocopherols are frequently found in the chloroplast and constitute members of the

[*] Yoshihiro Uto, Daisuke Koyama, Mamoru Otsuki, Naoki Otomo, Tadashi Shirai, Chiaki Abe, Eiji Nakata, Hitoshi Hori, Department of Life System, Institute of Technology and Science, Graduate School, The University of Tokushima, Minamijosanjimacho-2, Tokushima, 770-8506 Japan. Hideko Nagasawa, Laboratory of Pharmaceutical Chemistry, Gifu Pharmaceutical University, Mitahorahigashi-5, Gifu, 502-8585 Japan.

P. Liss et al. (eds.), *Oxygen Transport to Tissue XXX*, DOI 10.1007/978-0-387-85998-9_17,   **109**

vitamin E group. They act as a potent lipid peroxidation inhibitor, for example, in the endothelial cell membrane in blood vessels. Phytyl quinols, namely acyclic tocopherols, are key intermediates of tocopherol biosynthesis and tocopherol cyclase substrate,[1] but their biological activities remain unclear.

Low-density lipoprotein (LDL) is a major cholesterol carrier in the blood, and it is suggested that the oxidative modification of LDL has an important role in the development of atherosclerosis.[2, 3] Therefore, the development of effective antioxidants are desirable to prevent the oxidation of LDL.

In this paper, we discuss the design and synthesis of phytyl quinols and their structure-activity relationship in terms of their inhibitory activity of LDL oxidation as new antiatherogenic antioxidants based on isoprenomics (medicinal chemistry of isoprenoid involved in structural analysis, biosynthesis, biological function, and chemotype).[4, 5]

2. MATERIALS AND METHODS

2.1. Chemicals

Synthesis of phytyl quinols and phytyl quinone was done in our laboratory and the details are currently under preparation for submission. Methyl cellulose, dl-α-Tocopherol, 2-thiobarbituric acid (TBA), and 1,1-diphenyl-2-picrylhydrazyl (DPPH) were purchased from Wako Pure Chemical Industries Ltd. (Osaka, Japan). Dimethylsulfoxide (DMSO), 1-Butanol, and 2-(N-morpholino)ethanesulfonic acid (MES) were obtained from Sigma Chemical Company (St. Louis, USA).

2.2. Molecular Orbital (MO) Calculation

The geometry optimization and highest occupied MO (HOMO) and lowest unoccupied MO (LUMO) calculation of TX-2254 and d-α-tocopherol by *ab initio* MO methods was demonstrated using the restricted Hartree-Fock (RHF) method at the 6-31G* level. All reported calculations were carried out with the GAUSSIAN 03 program package (Gaussian, Inc., USA).

2.3. Preparation of Human LDL

A LDL (density, 1.006-1.063 g/mL) was fractionated from human plasma by ultracentrifugation using the CS120 ultracentrifuge equipped with a RP80AT rotor (Hitachi Koki Co., Ltd), dialyzed at 4°C against 3 changes of phosphate-buffered saline (PBS; pH 7.35). The LDL solution was flushed with N_2, stored in the dark at 4°C, and used within 5 days from the time of the preparation. Protein was measured by the Bradford method using bovine serum albumin as standard.

2.4. Free Radical Scavenging Activity

Free radical scavenging activity was determined by using DPPH at 517 nm according to the method of Blois[6] with some modifications. To 3.0 mL of 100 μM DPPH ethanol solution (60%), containing 40 mM MES (pH 5.5), was added 3 µL of different concentrations of antioxidants tested. The mean effective concentration of antioxidants required for a 50% decrease of DPPH (EC_{50}) after 30 min was estimated by linear regression analyses at three individual measurements.

2.5. Antioxidative Activity for LDL Oxidation

The effects of antioxidants (10 µM) on kinetics of lipid oxidation of human LDL (50 µg protein/mL) were evaluated by spectrophotometric monitoring, at 234 nm, of conjugated diene lipid hydroperoxide formation,[7] during copper-induced oxidation (5 µM $CuSO_4$, 10 mM PBS) using a Hitachi U-3300 spectrophotometer at 37°C with a temperature controller SPR-10. The duration of the lag phase, that is the lag time, was calculated by extrapolating the propagation phase. Lipid peroxides were measured as thiobarbituric acid reactive substances (TBARS) by the method of Yagi.[8] The concentration (IC_{50}) leading to 50% decrease of the amount of TBARS was estimated by linear regression analyses at three individual measurements.

2.6. In Vivo Vascular Toxicity

Fertilized chicken eggs were purchased from Ishii Company (Tokushima, Japan). The onset-day of the incubation was called day 1. At day 10, an air cell region and a part of chorioallantoic membrane (CAM) blood vessel region (2×2 cm) on the egg shell were cut off with a grinder. A shell membrane was peeled off from the CAM and the shell membrane was removed. Opened shell window was sealed with opsite and then the egg was incubated at 37.6°C in a humidified atmosphere until day 13. At day 13, a silicon ring was placed on the CAM blood vessel and various concentration of the antioxidant solution including 1% methyl cellulose and 5% DMSO was dropped into the silicon ring. Opened shell window was sealed again with tegaderm and the egg was incubated at 37.6°C in a humidified atmosphere for 24 hours. Antiangiogenic toxicity of the antioxidant was judged by the ratio to the total number of egg (10 eggs) of the inhibited one (e.g. digestion or disappearance of CAM blood vessel in the silicon ring).

3. RESULTS AND DISCUSSION

3.1. Ex Vivo Synthesis of Phytyl Quinols and Quinones

We synthesized phytyl quinols and phytyl quinones according to our established regioselective *C*-prenylation reaction.[3] Chemical structures of four phytyl quinols (TX-2254, TX-2247, TX-2242, TX-2231), four phytyl quinones (TX-2229, TX-2232, TX-2233, TX-2230), and d-α-tocopherol are shown in Figure 1.

3.2. Optimized Geometry and Molecular Orbital of Phytyl Quinols and Quinones

We have obtained a right-angle structure and localization of LUMO at phytyl double-bond of TX-2254 by *ab initio* geometry optimization and molecular orbital calculation, although the HOMO and LUMO energy of TX-2254 and d-α-tocopherol were very similar.

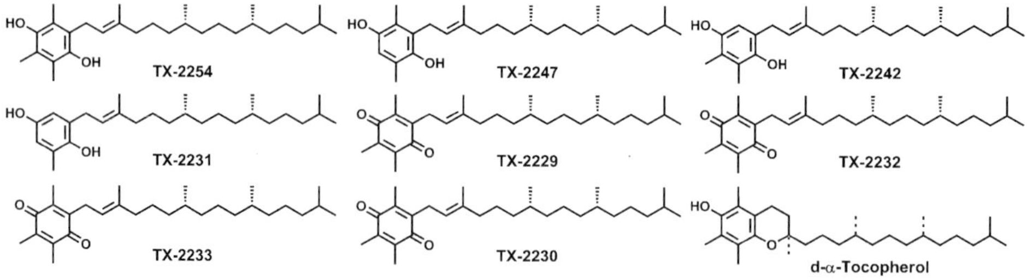

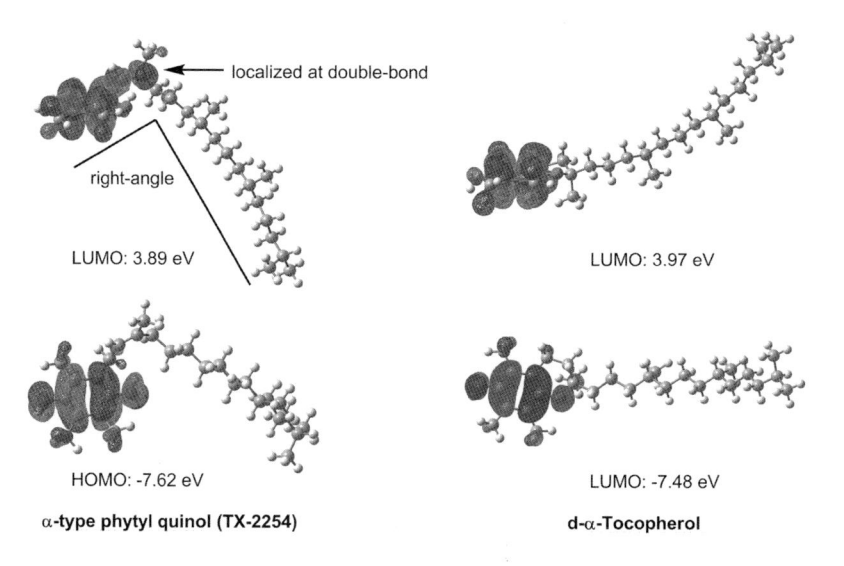

Figure 1. Chemical structures of phytyl quinols (TX-2254, TX-2247, TX-2242, TX-2231), phytyl quinones (TX-2229, TX-2232, TX-2233, TX-2230), and d-α-tocopherol.

Figure 2. Optimized geometry and HOMO and LUMO energies of TX-2254 (left) and d-α-tocopherol (right).

3.3. Antioxidative Properties of Phytyl Quinols and Quinones

We determined the EC_{50} value to evaluate the reactive potency with DPPH free radical in aqueous solution as shown in Table 1. DPPH radical reactivity of TX-2231 (EC_{50} = 20.9 μM) was equal to dl-α-tocopherol (EC_{50} = 21.4 μM), whereas TX-2254 (EC_{50} = 27.9 μM), TX-2247 (EC_{50} = 26.5 μM), and TX-2231 (EC_{50} = 24.3 μM) showed lower reactivity than dl-α-tocopherol. On the other hand, phytyl quinones showed no DPPH radical reactivity.

The inhibitory activities of phytyl quinols against LDL conjugated-diene formation (lag time) and TBARS formation (IC_{50}) are shown in Table 1. With regard to the lag time, phytyl quinols showed almost the same moderate antioxidative activities (lag time = 77.2

~ 84.4) with less inhibitory potency than that of dl-α-tocopherol (lag time = 100.8 min). Besides, phytyl quinones showed no inhibitory activities, similar to the results of the DPPH radical. The inhibitory activities of phytyl quinols against LDL TBARS formation were similar to the results of lag time.

Table 1. Antioxidative property of phytyl quinols and phytyl quinones.

Antioxidant	DPPH EC_{50} (µM)	LDL lag time (min)	LDL TBARS IC_{50} (µM)
TX-2254	27.9 ± 0.9	84.4	115.8 ± 5.8
TX-2247	26.5 ± 3.4	82.6	111.9 ± 2.3
TX-2242	24.3 ± 0.4	81.5	148.0 ± 10.1
TX-2231	20.9 ± 3.1	77.2	97.0 ± 3.4
TX-2229	> 200	58.2	> 1000
TX-2232	> 200	59.8	> 1000
TX-2233	> 200	50.2	> 1000
TX-2230	> 200	44.8	> 1000
dl-α-tocopherol	21.4 ± 0.9	100.8	66.5 ± 13.1

3.4. In Vivo Effect of Embryonic Chick Vascular Toxicity

According to figure 3, in vivo vascular toxicities of phytyl quinols were hardly observed except for TX-2247 (0.1 µg) and dl-α-tocopherol (0.1 and 10 µg).

4. DISCUSSION

We designed four phytyl quinols (including two biosynthetically unknown precursors) for LDL antioxidant based on isoprenomics. Phytyl quinols were similar moderate antioxidants instead of their different degree of C-methylation with a character different from tocopherols. Moreover, phytyl quinols having C_{20}-phytyl side-chain were less desirable LDL antioxidants than tocopherol having C_{16}-alkyl side-chain, because of more tightly-binding and less flexibility in the LDL membrane lipid.[5] Chick embryos are frequently used as an alternative experimental animal for angiogenic investigations.[9] We tried to use the chick embryo CAM in order to evaluate the vascular toxicities of phytyl quinols with prooxidant property. However, the vascular toxicity of phytyl quinols was very low as same as dl-α-tocopherol. We therefore proposed that phytyl quinols may be possible antioxidants in plants and animals, like vitamin E.

5. ACKNOWLEDGMENTS

This work was supported by the 2006 Research Project of the Faculty and School of Engineering, The University of Tokushima, Japan.

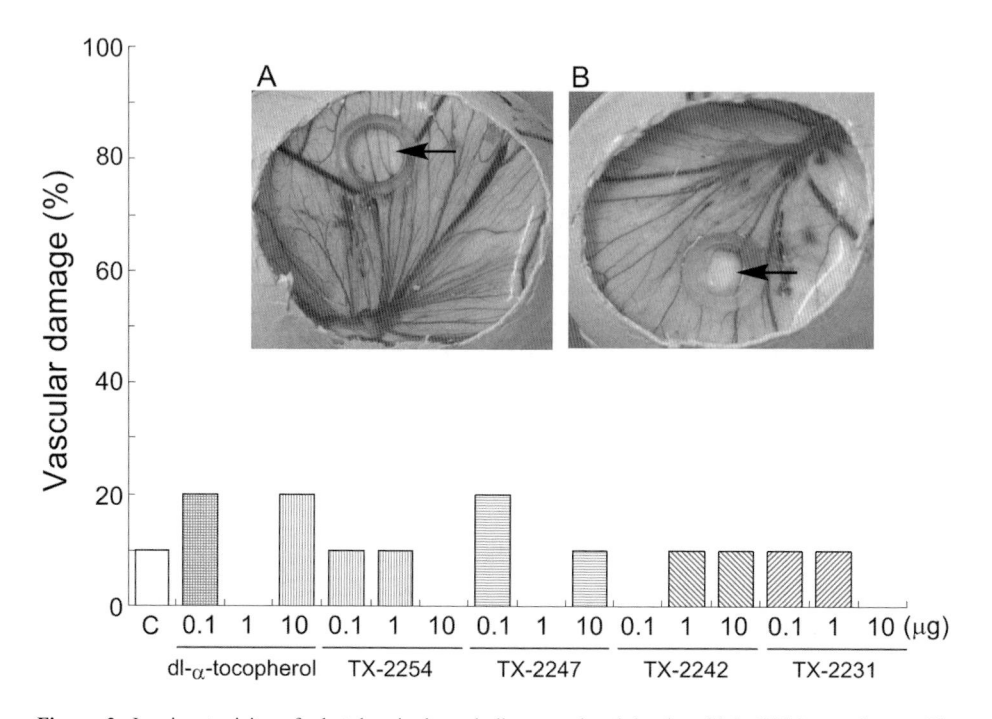

Figure 3. In vivo toxicity of phytyl quinols and dl-α-tocopherol in the chick CAM vasculature. These photograph (A, B) shows CAM vasculature without damage (A) or with damage (B) when 10 μg of TX-2247 was treated.

6. REFERENCES

1. A. Stocker, H. Fretz, H. Frick, A. Ruttimann, and W. D. Woggon, The substrate specificity of tocopherol cyclase, *Bioorg. Med. Chem.* **4**(7), 1129-1134 (1996).
2. D. Steinberg, Role of oxidized LDL and antioxidant in atherosclerosis, *Adv. Exp. Med. Biol.* **369**, 39-48 (1995).
3. J. L. Witztum, and D. Steinberg, Role of oxidized low density lipoprotein in atherogenesis, *J. Clin. Invest.* **88**(6), 1785-1792 (1991).
4. Y. Uto, S. Ae, A. Hotta, J. Terao, H. Nagasawa, and H. Hori, Artepillin C isoprenomics: design and synthesis of artepillin C analogues as antiatherogenic antioxidants, *Adv. Exp. Med. Biol.* **578**, 113-118 (2006).
5. Y. Uto, S. Ae, D. Koyama, M. Sakakibara, N. Otomo, M. Otsuki, H. Nagasawa, K. L. Kirk, and H. Hori, Artepillin C isoprenomics: design and synthesis of artepillin C isoprene analogues as lipid peroxidation inhibitor having low mitochondrial toxicity, *Bioorg. Med. Chem.* **14**(16), 5721-5728 (2006).
6. M. S. Blois, Antioxidant determinations by the use of a stable free radical, *Nature* **181**, 1199 (1958).
7. W. A. Pryor, and L. Castel, Chemical methods for the detection of lipid hydroperoxides, *Methods Enzymol.* **105**, 293-299 (1984).
8. K. Yagi, Simple assay for the level of total lipid peroxides in serum or plasma, *Methods Mol. Biol.* **108**, 101-106 (1998).
9. R. Auerbach, L. Kubai, D. Knighton, and J. Folkman, A simple procedure for the long-term cultivation of chicken embryos, *Dev. Biol.* **41**(2), 391-394 (1974).

ZYMOGEN PROTEIN C CONCENTRATE FOR SAFER HETEROZYGOTE SURGERY, "I AM A GUINEA PIG!"

Duane Frederick Bruley[*]

Abstract: Of great concern in medical surgery is the formation of blood clots, such as, deep vein thrombosis (DVT) with possible emboli reaching the lungs, thus diminishing or shutting down the oxygen transport to tissue processes. Our research has focused on producing an inexpensive protein C product *via* IMAC that could be used to bring blood levels to normal. Our work has emphasized safety for hereditary and acquired protein C deficient patients in all aspects of life. There are presently two expensive protein C products on the market. One is Xigris (activated protein C) produced by Eli Lilly & Co., and the second is Ceprotin, the zymogen product which is being manufactured by Baxter BioPharmaceuticals. I will be a guinea pig for my own research hypothesis in that I will have total hip replacement surgery using Ceprotin to help prevent blood clotting due to a protein C deficiency. It is difficult for surgeons and hematologists to agree on a safe procedure for protein C deficient patients undergoing major surgery. The hematologist is concerned about the potential for DVT and lung emboli, while the surgeon is also concerned about internal bleeding and infection after the surgery. The established approach would involve high doses of low molecular weight heparin (LMWH) for protein C deficient patients which would be unacceptable to the surgeon. My hypothesis has recommended administration of zymogen protein C to simulate a normal blood condition where the standard protocol could be used. The validity of this hypothesis is soon to be tested.

1. BACKGROUND

The control of blood hemostasis is critical for tissue health, including its oxygenation. When there is an imbalance between coagulants and anticoagulants in the plasma, at least two major pathological conditions can develop. First, if there is a deficiency of coagulants, such as Factor VII or IX, the patient becomes a hemophiliac

---

*Synthesizer, Inc., Ellicott City, MD, USA; UMBC, Baltimore, MD, USA

P. Liss et al. (eds.), *Oxygen Transport to Tissue XXX*, DOI 10.1007/978-0-387-85998-9_18,
© Springer Science+Business Media, LLC 2009

with excessive bleeding in the joints. Second, if there is a deficiency of an anticoagulant, such as protein C, a hyper-coagulable situation develops and typically the patient suffers a deep vein thrombosis (DVT).

This paper deals with the hyper-coagulable case and the resulting tissue oxygen deprivation due to blood clotting (agglutination). My first work with Dr. Melvin H. Knisely in 1962 began because he convinced me of the necessity of a good oxygen supply to the tissue for normal functioning, and he showed that many pathological states lead to blood agglutination and therefore poor oxygen transport to the tissue. Our goal was to use experimental and theoretical approaches (combined medical science and bioengineering) to better understand conditions that cause oxygen deficiency in tissue, the resulting damage, and in particular how to accommodate, prevent and cure the phenomena.[1]

My intention then and now is to investigate the safest and most effective therapy for the prevention and treatment of blood clotting in order to maintain normal oxygen transport to the tissue. Even though there are numerous anticoagulants/antithrombotics available today, my work has focused on protein C.[2] Protein C is the pivotal anticoagulant in the human coagulation cascade. It is a natural anticoagulant/anti-thrombotic/anti-inflammatory/anti-apoptotic[3] with little or no bleeding side effects when the zymogen is administered. All other anticoagulants, such as heparin, low molecular weight heparin, coumadin, and others being used clinically today, can lead to serious bleeding complications.

Protein C could be the ultimate anticoagulant for special medical conditions such as protein C deficiency. The zymogen is present in human blood plasma at an average concentration of 4 µg/mL with an *in vivo* half-life of about 6 hours. When the protein C level, *via* hereditary or acquired deficiency, is lowered (heterozygote or homozygote), serious problems can occur. When clots form it is possible for emboli to break away from the clot and enter the bloodstream. This can result in catastrophic stroke, heart attack, and pulmonary embolism. Studies performed in the 1980s[4] reported that thromboembolism was responsible for over 500,000 deaths each year, for which many could have resulted from protein C deficiency. The accepted treatment currently prescribes coumadin and heparin. These therapies can have serious side effects illustrated as excessive clotting and bleeding which can result in amputation and death. Zymogen protein C has no known side effects, primarily because it acts locally, therefore the therapeutic use of zymogen protein C appears to be very promising.

Human blood plasma is a natural source material for protein C production. Because blood plasma contains trace amounts of protein C and there are large amounts of other contaminants, including other clotting factors, immuno-affinity chromatography, because of its high specificity, is currently used to separate and purify protein C. This approach is very expensive.

Immobilized metal affinity chromatography (IMAC) is a newly developing separation technology that utilizes protein metal binding properties to separate them. This technology is an intermediate between high affinity, high specificity immuno-affinity chromatography and wider-spectrum, low specificity absorption such as ion exchange chromatography. IMAC is inexpensive and operates under mild conditions so the proteins do not denature. Also, there is little immunogenic contamination when compared to immuno-affinity chromatography.

Transgenic animal milk might be another important source material for the production of protein C and other blood factors. Present bioprocessing only recovers about 24% of the recombinant protein C expressed into the milk. IMAC is also being investigated as an alternate technology to improve the efficiency and recovery of the product. Our continued research is to investigate a variety of technologies and processes for the high volume, low cost production of protein C. In addition, we are exploring new administration and detection technologies and finally to "molecular engineer" the protein C molecule to increase its *in vivo* half-life.

The physiological importance of protein C can be illustrated by the fact that both the homozygous and double heterozygous forms of congenital protein C deficiency can result in life-threatening neonatal thrombosis and purpura fulminans.[5] Also, heterozygous patients are at risk of pathology that can result in amputation or death with no treatment without potential side effects presently available.

Protein C was first discovered by Seegers, and the new protein was named autothrombin IIa.[6] In 1976, the new protein was shown to be identical to the Vitamin K-dependent protein C that had been isolated by Stenflo.[6] The importance of protein C spread in 1981, when Griffin suggested a relationship between a hereditary protein C deficiency and the development of thrombo-embolytic disease.[6] Since Griffin's suggestion of the relationship, there has been extensive research and data collection concerning protein C and the anticoagulant pathway.

Protein C is a vitamin K-dependent serine protease that is produced in the liver, and circulates in the blood primarily as a two-chain inactive zymogen until it is activated by proteolytic cleavage. Protein C has the unique ability to act as both an anticoagulant and an antithrombotic.[7]

Human blood contains an entire array of proteins which prevent, form, and dissolve blood clots. If the delicate balance between the coagulation and anticoagulation of the blood is disturbed, then dangerous life-threatening results can occur.

Blood clots form whenever there is bleeding due to injury or disease. This causes the vessels at the injury site to constrict, known as vasoconstriction. Vasoconstriction will then slow the flow of blood to the damaged area Platelets then perform several important functions. First they adhere to each other and the cell wall. Second, they release clotting substances that aid in the formation of the clot. Finally, they release proteins that cause the clot to contract, and grow firm.[8]

The proteins involved form a complex cascade. When injury occurs, thromboplastin is released.[8] Thromboplastin then cleaves prothrombin to produce thrombin. Thrombin in turn cleaves fibrinogen to form fibrin. Fibrin is comprised of dense interlacing threads that entrap blood cells and platelets. The fibrin then shrinks and forms an insoluble fibrin mass.

Once the clot has served its purpose, then it must be dissolved. This is accomplished by the fibrinolytic cascade. This prevents permanent clotting and occlusion of the blood vessel. Normally, the fibronolytic mechanism remains localized to the site of clot formation. However, the mechanism may become overactive, resulting in hemorrhage or under active resulting in excessive clotting.[8]

The role of protein C as an anticoagulant in the coagulation cascade begins with the generation of thrombin. Thrombin then binds to thrombomodulin, which is a transmembrane cell surface binding protein. Thrombomodulin forms a high affinity complex with thrombin.[9] Thrombin itself can activate protein C, but the formation of the thrombin-thrombomodulin complex accelerates the activation of protein C as much as 20,000-fold.[9]

Activated protein C then complexes with Protein S, a nonenzymatic vitamin K-dependent cofactor. Protein S promotes the binding of activated protein C to membrane surfaces, and it is suggested that activated protein C and protein S form a complex on the surface of negatively charged phospholipids vesicles, endothelial cells and platelets.[10]

These proteins are then able to catalyze the proteolytic inactivation of factors Va and VIIIa in the presence of Ca^{2+} ions.[11] The inactivated factor Va then loses the ability to interact effectively with factor Xa or the substrated prothrombin. This in turn prevents the production of thrombin and fibrin.

The antithrombotic nature of activated protein C stimulates fibronolysis by neutralizing the inhibitor of tissue plasminogen activator (TPA). This stimulation is also enhanced by protein S.[12] It has also been shown that activated protein C causes the release of TPA from the surface of endothelial cells into blood. Finally, it has been shown that activated protein C inhibits an antifibrinolytic component released as a consequence of prothrombin activation.[13]

TPA is a fibrinolytic serine protease which has been found in many tissues and body fluids. TPA works by converting plasminogen to plasmin, which then dissolves fibrin. Single-chain TPA is a 68,000 g/mol peptide; two-chain TPA consists of 30,000 g/mol and 40,000 g/mol chains.[14]

Once protein C is activated, it circulates *in vivo* with a half-life of about 15 minutes.[15] Protein C is neutralized by either protein C inhibitor or alpha$_1$-antitrypsin. Both protein C inhibitor and alpha$_1$-antitrypsin form a complex with protein C *via* acylation of the active site. Protein C can also be inhibited by alpha$_2$-macroglobulin and alpha$_2$-antiplasmin.[16] Protein C is a complex glycoprotein that undergoes extensive post-translational modifications. It has a molecular weight of about 61,600 g/mol when fully glycosylated. The molecular weight without carbohydrate has been calculated to be 47,456 g/mol.[17]

Mature protein C is a two-chain disulfide linked hetrodimer consisting of a 155 amino acid 21 kD light chain and a 262 amino acid 41 kD heavy chain. It is a mosaic protein consisting of five modules whose homologous copies are found in many different proteins.[18] Starting from the NH$_2$-terminus, the first domain is the vitamin D-dependent, gamma-carboxyglutamic acid rich, or Gla domain. Then follows a short aromatic helical stack of hydrophobic residues.[13] The following two cystein-rich modules are epidermal growth factor-like, or EGF homologs.[19] These first four domains comprise the light chain. Finally, the serine protease module comprises the heavy chain ending in the COOH-terminus.[18]

Protein C folds into a tightly packed globular protein that forms two beta-barrel domains containing six anti-parallel beta strands each.[16] Several surface alpha helices surround the beta sheets, along with the two binding site loop regions. The first 3-D molecular model of the serine protease domain of protein C was based upon the coordinates derived from the homologous proteins thrombin, chymotrypsin, chymotrypsinogen, trypsin and trypsinogen.[16]

The most recent molecular model for protein C comes from X-ray crystallography data. A Gla domainless peptide was produced and crystallized. The peptide was solved at a 2.8 angstrom resolution. It was demonstrated that the heavy chain serine protease domain is very homologous to trypsin, but contains a large insertion loop near the active site.[11]

Protein C is held together by twelve disulfide bridges and by hydrogen bond interactions. The heavy and light chain is attached by a disulfide bridge. The inner protein core consists mainly of hydrophobic residues which further enhance its stability.[16] Protein C has a high denaturation temperature of 62°C.[18]

Protein C is naturally produced in the liver. The nucleotide sequence for human protein C has been determined, and the gene spans approximately 11 kb of DNA. The gene contains eight exons from 25 to 885 nucleotides, and seven introns from 92 to 2668 nucleotides.[17] Both the genomic and intronless cDNA coding sequences have been isolated for use in recombinant experiments from human liver libraries.[20] Once the protein has been transcribed in a liver cell, it is subjected to five types of post-translational modification in the endoplasmic reticulum and in the Golgi apparatus. These post-translation modifications include gamma carboxylation, beta hydroxylation, N-linked carboxylation, disulfide bond formation, and multiple proteolytic cleavages.[21]

2. POTENTIAL CLINICAL USES

My hypothesis for over 25 years has been that zymogen protein C should be used for heterozygote protein C deficient patients for many medical indications and clinical procedures.[22] For instance a hereditary or acquired protein C deficient patient faces potential abnormal complications during major surgery. The possibility of deep vein thrombosis (DVT) and thus emboli traveling to the lungs can lead to the patient's death. In my own case, I am to undergo total left hip replacement that was scheduled to take place before the Uppsala ISOTT meeting but has been delayed. Since I am heterozygote (50% level) protein C deficient the procedure is scheduled to include zymogen protein C to bring my blood PC levels to normal or slightly above. The protein C product is to be obtained from Baxter BioPharmaceuticals *via* the American Red Cross. The surgery is to be scheduled at Johns Hopkins Medical Complex. However, complications in getting the product have resulted in rescheduling the operation for a later date.

Starting in the early 1980s my research has focused on developing an inexpensive protein C zymogen product. Initiating an effort *via* the National Science Foundation through Virginia Polytechnical State University and the American Red Cross[2] we developed a product in conjunction with Baxter USA that was then dropped by the American Red Cross. A small company in Austria (Immuno AG) produced a similar product using the same technology, immuno-affinity chromatography. This product was approved for use in Europe in ~2002, after which Immuno AG was acquired by Baxter BioPharmaceutical. The product, CEPROTIN, has now been FDA approved for use in the United States (March 30, 2007). In both Europe and the United States the product has *been approved for, "Severe Congenital Protein C Deficiency" in children. Severe generally refers to Homozygotes. We have been studying immobilized metal affinity chromatography (IMAC)[23] for the last fifteen years in an attempt to reduce the cost of protein C zymogen product for medical prophylactic use for homozygote and heterozygote Protein C deficiency as well as for other special applications such as major surgery, pregnancy, etc..

An attempt was made to acquire Ceprotin for my hip surgery prior to this presentation and writing but we have not been successful in gaining use of the product. A complication has arisen that patients have to be established as congenitally deficient *via* genotyping of the patient. A problem that faces its use is there are only a two labs in the world that have the procedure to genotype congenital protein C deficiency. At present, there is one in Germany and one in England. It is my intention to follow through with the necessary requirements to be a guinea pig for the use of zymogen protein C concentrate for safer heterozygote surgery, in aging adults (I am presently seventy four years old).*

---

*My left hip surgery was completed successfully using Ceprotin which, hopefully will be reported at the ISOTT meeting in Japan, August 2008. A paper with the details will then be submitted for an ISOTT publication.

3. RECOMMENDATIONS AND CONCLUSIONS

Several other conditions and blood factors can lead to complications during surgery. For example, deficiencies in protein S or antithrombin III, and the Factor V Leiden Syndrome can also lead to potential abnormal clotting situations. It has been my recommendation that all children should be tested for these deficiencies at about age twelve years old to allow appropriate lifestyle adjustments that will help prevent blood clotting complications. In addition adults who are facing major surgery would be wise to have their medical team alert to abnormal levels of these important blood factors and other medical abnormalities such as Factor V Leiden. This would allow the medical team to attempt supplementing the necessary blood factors and establish protocols that would be safer for these problematic conditions. At present all medical surgeries to my knowledge have a standard procedure that is used for surgery and then if a complication occurs, the medical team attempts to prevent a disaster. I have stated clearly in many publications that a major difference for anticoagulant deficiencies is that, in most cases, it is a HIDDEN pathology, unlike hemophilia, where deficiencies in blood factors lead to obvious bleeding conditions.[22]

As this hypothesis unfolds, future papers will be written regarding the surgical and hematological strategies to be used and the ensuing results. Hopefully the results of my hip replacement surgery will be positive!

4. REFERENCES

1. H.I. Bicher., Bruley, D. F., and M. H. Knisely, "Anti-Adhesive Drugs and Tissue Oxygenation," edited by D. F. Bruley and H. I. Bicher, Advances in Experimental Medicine and Biology, Plenum, Press, Vol. 37B657-667, 1973.
2. D.F.Bruley. and W. N. Drohan, "Protein C and Related Anticoagulants," Advances in Applied Biotechnology Series, Volume 11, Gulf Publishing Company (Portfolio Publishing Company), 1990.
3. G.R. Sharma., Gerlitz, G., Berg, D.T., Cramer, M.S., Jakubowski, J.A., Galbreath, E.J., Heuer, J.G., and B.W.Grinnell, "Activated Protein C Modulates Chemokine Response, and Tissue Injuiry in Expereimental Sepsis",Advances in Experimental Medicine and Biology, Vol. 614, pp.83-91,Springer,2008.
4. D.F.Bruley, Hereditary protein C deficiency, ENCH 444 Class Handout, 1996.
5. P.C.Comp., "The clinical potential of protein C and activated protein C, advances in applied biotechnology series: protein C and related anticoagulants," edited by D. F. Bruley and W. Drohan (Portfolio Publ. Co., Texas, 1990), vol. 11.
6. D.F.Bruley, Hereditary protein C deficiency, ENCH 444 Class Handout, 1996.
7. R.Drews., et al., Proteolytic maturation of protein C upon engineering the mouse mammary gland to express furin, Proc. Natl. Acad. Sci. USA 92, 10462 (1995).
8. D.B.McClure., et al., Post-translational processing events in the secretion pathway of human protein C, a complex vitamin K-dependent antithrombotic factor, J. Bio. Chem. 267(27), 19710 (1992).
9. C.T.Esmon., The roles of protein C and thrombomodulin in the regulation of blood coagulation, J. Bio. Chem. 264 (9), 4743 (1989).
10. C.M.Henkens., et al., Plasma levels of protein S., protein C, and factor X; Effects of sex, hormonal state and age, Thrombosis and Haemostasis 74 (1995).
11. T.Mather., et al., The 2.8 Å crystal structure of Gla-domainless activated protein C, EMBO J. 15(24), 6822 (1996)
12. C.T.Esmon., The regulation of natural anticoagulant pathways, Science 235, 1348 (1997).

13. A.R.Rezaie and C. T. Esman, Tryptophan 231 and 234 in protein C report the Ca^{2+}-dependent conformational change required for activation by the thrombin-thrombomodulin complex, *Biochemistry* **34**, 12221 (1995).
14. Sigma Chemical Company (1997): http://www.sigmaaldrich.com.
15. C.T.Esmon., Regulation of coagulation: The nature of the problem, in: *Protein C and Related Coagulants*, edited by D. F. Bruley and W. N. Drohan (Gulf, Houston, 1990), p. 3.
16. C.I.Fisher., *et al.*, Models of the serine domain of the human antithrombotic plasma factor (1994).
17. D.C.Foster., *et al.*, The nucleotide sequence of the gene for human protein C, *Proc. Natl. Acad. Sci. USA* **82**, 4673 (1985).
18. L.V.Medved., *et al.*, Thermal stability and domain-domain interactions in natural and recombinant protein C, *J. Bio. Chem.* **270**(23), 13659 (1995).
19. C.L.Orthner., *et al.*, Conformational changes in an epitope localized to the NH_2-terminal region of protein C, *J. Bio. Chem.* **2647**(31), 18781 (1989).
20. B.W.Grinnell., *et al.*, Trans-activated expression of fully gamma carboxylated recombinant human protein C, an antithrombotic factor, *Bio/Technology* **5**, 1189 (1987).
21. S.C.B.Yan., *et al.*, Post-translational modifications of proteins; some problems left to solve, *TIBS* **14**, 264 (1989).
22. D.F.Bruley., Protein C – The Ultimate Anticoagulant/Antithrombotic?, in: *Anticoagulant, Antithrombotic, and Thrombolytic Thearapeutics II*, edited by C. A. Thibeault and L. M. Savage (IBC Library Series, Southborough, MA, 1998).
23. H.Wu and D. F. Bruley, Chelator, metal ion and buffer studies for protein C separation, *Comparative Biochemistry and Physiology Part A* **132**, 213 (2002).

STEADY STATE REDOX LEVELS IN CYTOCHROME OXIDASE: RELEVANCE FOR IN VIVO NEAR INFRARED SPECTROSCOPY (NIRS)

Chris E. Cooper[1], Martyn A. Sharpe[2], Maria G. Mason[1] and Peter Nicholls[1]

Abstract: In the visible/NIR (600 – 900 nm) three different redox centres are potentially detectable *in vivo* in mitochondrial cytochrome *c* oxidase: haem *a* (605nm), the binuclear haem a_3/Cu$_B$ centre (655 nm) and Cu$_A$ (830 nm). In this paper we report changes in the steady state reduction of these centres following increases in the rate of electron entry into the purified enzyme complex under conditions of saturating oxygen tension. As turnover is increased all three centres becomes progressively reduced. Analysis of the steady states indicated that all three centres remained in apparent equilibrium with cytochrome *c* throughout the titration. The calculated redox potentials of Cu$_A$ (+224 mV) and haem *a* (+267 mV) were consistent with previous equilibrium data. The 655 nm band was also found to be oxygen and flux sensitive. It may be a useful additional *in vivo* detectable chromophore. However, it titrated with an apparent redox potential of +230mV, far from its equilibrium value (+400 mV). The implications of these results for the interpretation of non invasive measurements of mitochondrial function are discussed.

1. INTRODUCTION

There has been renewed interest recently in using optical techniques to study the mitochondrial cytochrome oxidase (CCO) redox state *in vivo*, both to address basic science[1, 2] and clinical[3, 4] questions. In most cases changes in redox state have been interpreted as due to changes in intracellular pO$_2$[4]. However, there are a number of other parameters that can change the redox state of CCO. CCO reacts with ferrocytochrome *c*

[1] Department of Biological Sciences, University of Essex, Wivenhoe Park, Colchester Essex, CO4 3SQ, UK
ccooper@essex.ac.uk
[2] Biochemistry and Molecular Biology Department, Michigan State University, East Lansing, MI 48824-1319, USA

P. Liss et al. (eds.), *Oxygen Transport to Tissue XXX*, DOI 10.1007/978-0-387-85998-9_19,   **123**
© Springer Science+Business Media, LLC 2009

and oxygen to produce water, ferricytochrome c and a proton motive force ($\Delta \bar{\mu}_{H+}$) across the inner mitochondrial membrane. In general flux through CCO can be modelled as a Michaelis-Menton function of its substrates (oxygen, ferrocytochrome c), modulated by product inhibition (from ferricytochrome c and $\Delta \bar{\mu}_{H+}$). Thus, in addition to oxygen, changes in the redox state of cytochrome c and Δp will contribute to any observed *in vivo* changes in redox state. Our "grand aim" is to develop a model of CCO function to aid the interpretation of *in vivo* mitochondrial spectroscopy. The interpretation of mitochondrial $\Delta \bar{\mu}_{H+}$ effects *in vivo* is complicated by the fact that $\Delta \bar{\mu}_{H+}$ can act at three parts of the chain (complexes I, II and III), making it impossible to *a priori* predict the effects at one site[5]. Similarly it is a non trivial, and controversial, process to measure redox states accurately at low oxygen tension[6, 7]. Therefore the initial, more tractable, aim of our project outlined here is to understand the relationship of CCO redox centres to changes in the cytochrome c redox state at saturating oxygen and zero $\Delta \bar{\mu}_{H+}$.

2.   METHODS

Bovine heart CCO was purified according to the method of Kuboyama *et al.*[8] with Tween-80 substituting for Emasol. Maximum turnover (pH 7.4) was approximately 380 $e^- aa_3^{-1} s^{-1}$. In a cuvette 10 µM CCO was incubated at 30 °C. with 10 µM horse heart cytochrome c and 100 nM bovine liver catalase, in 1 ml of 25 mM K^+-HEPES, 0.1% lauryl maltoside, pH 7.8. Sodium ascorbate (40 mM) was added to the cuvette, to initiate turnover of the enzyme; the solution was then allowed to go anaerobic. Following anaerobiosis hydrogen peroxide (H_2O_2) was added to the cuvette (final concentration 0.5 - 2 mM). The catalase rapidly decomposed the H_2O_2 into oxygen in the incubation medium, thus re-initiating enzyme turnover. Following the second anaerobiosis, the cuvette was re-oxygenated with H_2O_2 and a second steady state was produced. No difference was observed between the two steady states.

Absorbance changes between 520 and 800 nm were monitored using a Beckman DU 7400 diode array spectrophotometer. Redox changes of cytochrome c were monitored at 550-540 nm. Redox changes of haem a were monitored at 605-630 nm, assuming a 14% contribution at this wave pair of haem a_3 in the reduced enzyme and no contribution of haem a_3 in the steady state[9]. Absorbance changes in the near infrared were recorded using a Aminco DW2 spectrophotometer under identical assay conditions; the Cu_A redox state was measured using the 780-830 nm wavepair. 100% oxidation of all chromophores was asssumed in the absence of added reductants and 100% reduction was assumed upon anaerobiosis. Repeating the experiment at increasing concentrations of the reductant TMPD (N,N,N',N' -tetramethyl-p-phenylenediamine) resulted in an increase in the cytochrome c redox state and a consequent increase in enzyme turnover. The latter was measured by using the time taken for the sample to go anaerobic (independent measurements with an oxygen electrode demonstrated that under these conditions the rate of oxygen consumption is essentially linear with respect to the complete time of the assay, due to the low, submicromolar, K_m of the enzyme for oxygen under these conditions).

3. RESULTS

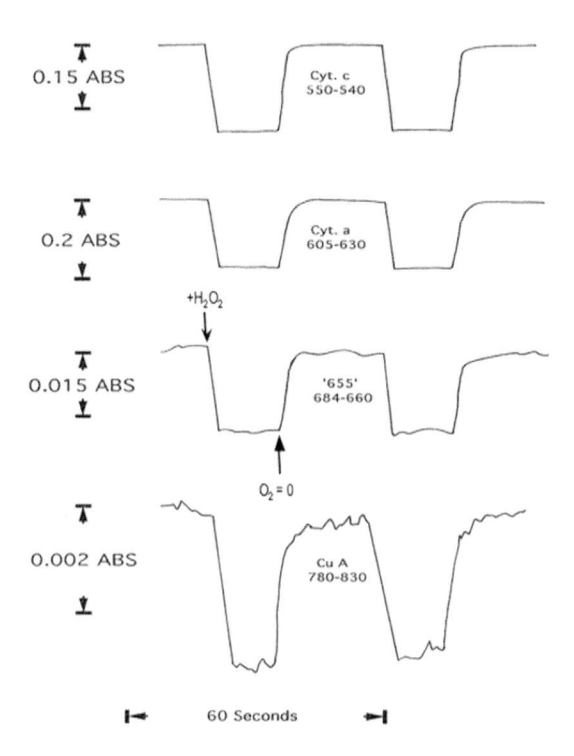

Figure 1. Absorbance changes during repeated cycles of oxygenation/deoxygenation (illustrative additions of peroxide and effects of anaerobiosis indicated)

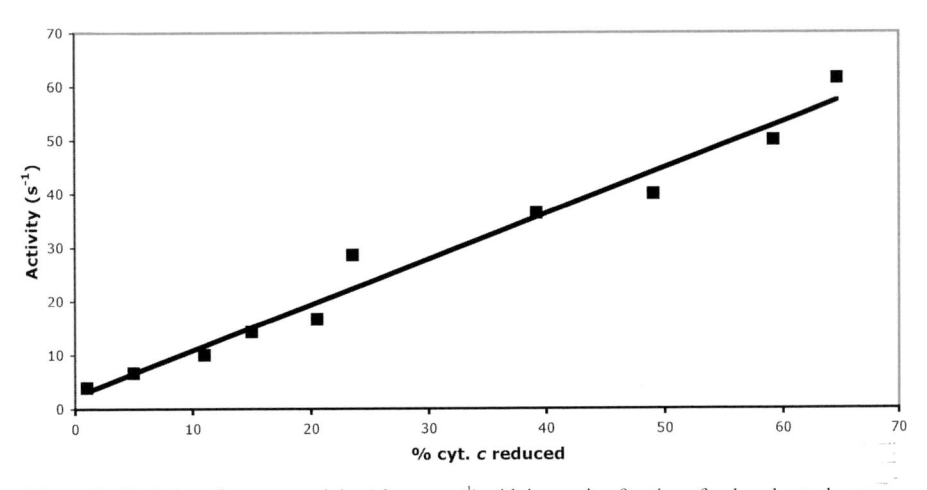

Figure 2. Variation of enzyme activity (electrons s^{-1}) with increasing fraction of reduced cytochrome c

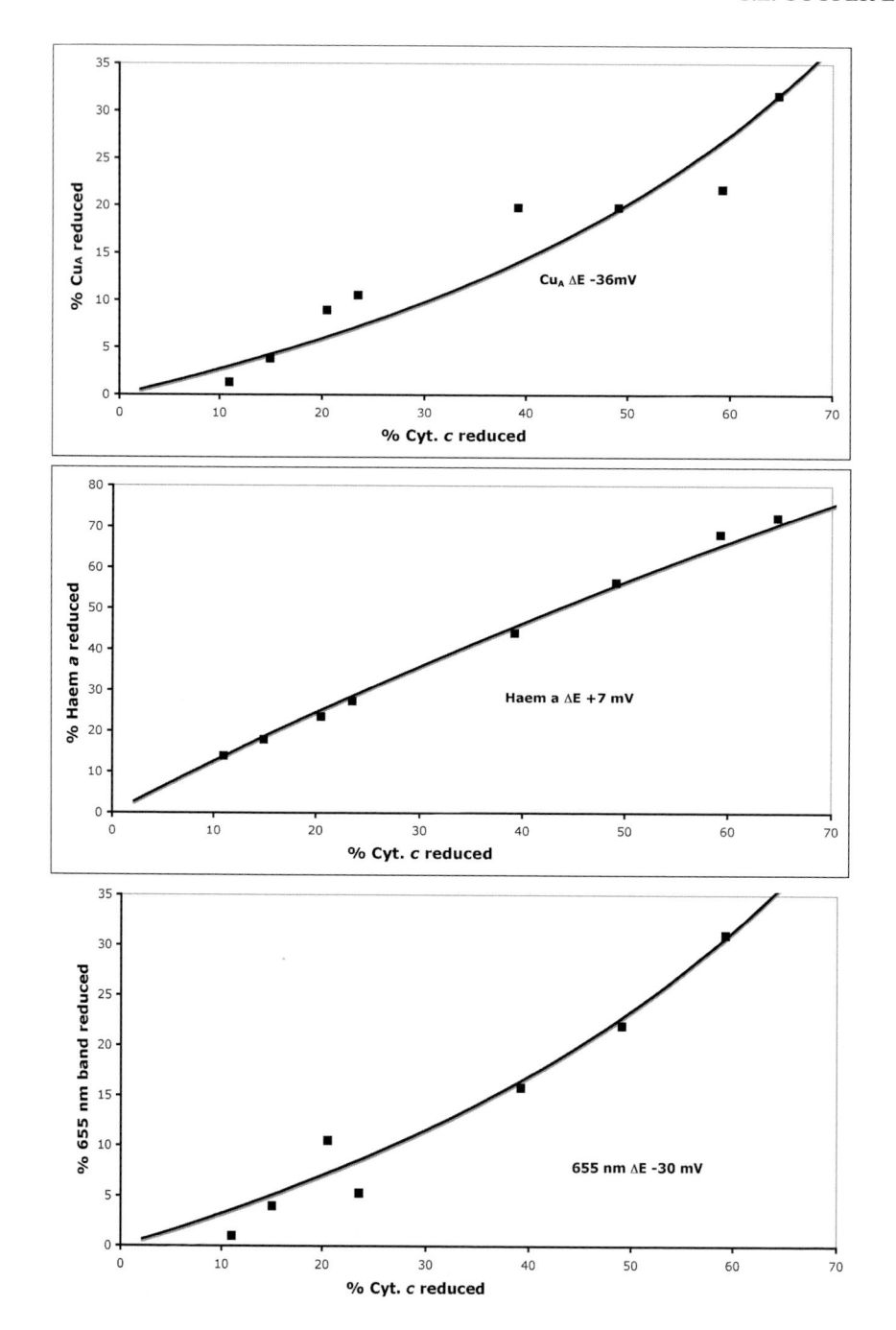

Figure 3. Relationship between cytochrome oxidase and cytochrome c redox states during turnover. (data fitted to Nernst equation, relative redox potential with respect to cytochrome c indicated)

Figure 1 illustrates the changes in redox states during repeated cycles of respiration driven deoxygenation and peroxide/catalase–driven reoxygenation of the system. All centres respond essentially simultaneously, becoming partially reduced in the presence of the oxygen and fully reduced when the oxygen is used up. It is possible to set-up different steady states at different concentrations of reductant (TMPD). Increasing the TMPD concentration increases the steady state reduction of cytochrome c and hence the rate of enzyme turnover, which is linearly dependent on the cytochrome c redox state (Figure 2). It is possible to compare the steady state of cytochrome c to that of the optical chromophores in CCO. By fitting to the Nernst equation it can be seen whether an individual redox centre behaves as if it were in redox equilibrium with cytochrome c and, if so, what the relative redox potentials are (illustrated for the single electron transition between cytochrome c and Cu_A in Equation 1).

$$E_{Cu_A} - E_{cytc} = \frac{RT}{F}\left(\left(\ln\frac{cytc^{3+}}{cytc^{2+}}\right) - \left(\ln\frac{Cu_A^{2+}}{Cu_A^+}\right)\right) \qquad \text{Equation 1}$$

Figure 3 indicates that Cu_A, haem a and the 655 nm band all behave as if in apparent redox equilibrium with cytochrome c. Assuming a redox potential for cytochrome c in free solution of + 260 mV, this results in apparent potentials for the other centres of: Cu_A (+ 224 mV), haem a (+ 267 mV) and 655 nm band (+ 230mV).

4. DISCUSSION

These findings indicate that a variety of redox centres within mitochondrial CCO (Cu_A, haem a, haem a_3/Cu_B) are sensitive to changes in the concentration of the substrates ferrocytochrome c and oxygen. This is not surprising[10] although the sensitivity to substrate delivery reported here is rarely discussed when interpreting *in vivo* data. Although all signals appear in an apparent redox equilibrium with cytochrome c, the values for these "steady state" redox potentials vary.

In the case of Cu_A the potential measured (+230 mV) falls within the range of values (230 - 265 mV) calculated from true equilibrium experiments performed in the absence of oxygen or in the presence of an inhibitor of oxygen consumption[11-14]. For haem a the situation is more problematic. The finding that the haem a potential is slightly more positive than cytochrome c is consistent with other steady state[15] or pseudo steady-state experiments[16,17]. However, when haem a_3 is allowed to become reduced in an equilibrium titration there is a strong redox interaction between haem a and a_3[14] such that haem a no longer titrates as an n = 1 electron donor; its apparent redox potential can be as high as 360 mV. It is therefore difficult to make quantitative comparisons between the steady state and true equilibrium data for haem a. The 655 nm band is more problematic, both in spectral assignment and redox potential. It probably a charge transfer band of ferric high-spin haem a_3, which is modulated by the redox state of Cu_B. In equilibrium titrations it titrates at a potential of 400 mV. Clearly this is far removed from our value of +230 mV. Perhaps not surprisingly the signal that is part of the oxygen reduction site titrates far from equilibrium in a steady state when oxygen is being consumed.

What implications does this have for *in vivo* detection of CCO? Clearly these three centres that can all detect flux changes in the enzyme. However, the interpretation of changes in the haem a 605 and haem a_3 655 nm band is problematic, due to their complex redox and oxygen interactions. Furthermore these two centres are buried inside the enzyme and their response to $\Delta \mu_{H+}$ is unpredictable. In contrast the Cu_A centre is likely to remain in equilibrium with cytochrome c under most conceivable conditions. It is physically close to the cytochrome c binding site, sensing a similar $\Delta \bar{\mu}_{H\pm}$[12]. Although more work is needed, particularly at low $[O_2]$ and in the presence of $\Delta \mu_{H+}$, it seems likely that Cu_A will prove to be a robust marker of the cytochrome c oxidation state *in vivo*.

5. ACKNOWLEDGEMENTS

We would like to thank BBSRC (D0609821/1) and EPSRC (EP/F006551/1) for financial support.

6. REFERENCES

1. M. M. Tisdall, I. Tachtsidis, T. S. Leung, C. E. Elwell, and M. Smith, Near-infrared spectroscopic quantification of changes in the concentration of oxidized cytochrome c oxidase in the healthy human brain during hypoxemia, *J. Biomed. Opt.* **12**, 024002 (2007).
2. G. De Visscher, R. Springett, D. T. Delpy, J. Van Reempts, M. Borgers, and K. van Rossem, Nitric oxide does not inhibit cerebral cytochrome oxidase in vivo or in the reactive hyperemic phase after brief anoxia in the adult rat, *J. Cereb. Blood Flow. Metab.* **22**, 515-519. (2002).
3. I. Tachtsidis, M. Tisdall, T. S. Leung, C. E. Cooper, D. T. Delpy, M. Smith, and C. E. Elwell, Investigation of in vivo measurement of cerebral cytochrome-c-oxidase redox changes using near-infrared spectroscopy in patients with orthostatic hypotension, *Physiol. Meas.* **28**, 199-211 (2007).
4. Y. Kakihana, T. Kuniyoshi, S. Isowaki, K. Tobo, E. Nagata, N. Okayama, K. Kitahara, T. Moriyama, T. Omae, M. Kawakami, Y. Kanmura, and M. Tamura, Relationship between redox behavior of brain cytochrome oxidase and neurological prognosis, *Adv. Exp. Med. Biol.* **530**, 413-419 (2003).
5. M. Banaji, A generic model of electron transport in mitochondria, *J. Theor. Biol.* **243**, 501-516 (2006).
6. N. Oshino, T. Sugano, R. Oshino, and B. Chance, Mitochondrial function under hypoxic conditions: the steady states of cytochrome a+a₃ and their relation to mitochondrial energy state, *Biochim. Biophys. Acta* **368**, 298-310 (1974).
7. D. F. Wilson, M. Erecinska, C. Drown, and I. A. Silver, The oxygen dependence of cellular energy metabolism, *Arch. Biochem. Biophys.* **195**, 485-493 (1979).
8. M. Kuboyama, F. C. Yong, and T. E. King, Studies on cytochrome oxidase VIII. Preparation and some properties of cardiac cytochrome oxidase., *J. Biol. Chem.* **247**, 6375-6383 (1972).
9. P. Nicholls, The steady state behaviour of cytochrome c oxidase in proteoliposomes, *FEBS Lett.* **327**, 194-198 (1993).
10. C. E. Cooper, S. J. Matcher, J. S. Wyatt, M. Cope, G. C. Brown, E. M. Nemoto, and D. T. Delpy, Near infrared spectroscopy of the brain: relevance to cytochrome oxidase bioenergetics, *Biochem. Soc. Trans.* **22**, 974-980 (1994).
11. A. J. Moody and P. R. Rich, The effect of pH on redox titrations of haem a in cyanide-liganded cytochrome-c oxidase: experimental and modelling studies, *Biochim. Biophys. Acta* **1015**, 205-215 (1990).
12. P. R. Rich, I. C. West, and P. Mitchell, The location of Cu_A in mammalian cytochrome c oxidase, *FEBS Lett.* **233**, 25-30 (1988).
13. D. F. Wilson, M. Erecinska, and P. L. Dutton, Thermodynamic relationships in mitochondrial oxidative phosphorylation, *Annu. Rev. Biophys. Bioeng.* **3**, 203-230 (1974).
14. E. A. Gorbikova, K. Vuorilehto, M. Wikstrom, and M. I. Verkhovsky, Redox titration of all electron carriers of cytochrome c oxidase by Fourier transform infrared spectroscopy, *Biochemistry* **45**, 5641-5649 (2006).
15. P.-E. Thörnström, P. Brzezinski, P.-O. Fredriksson, and B. G. Malmström, Cytochrome c Oxidase as an Electron-Transort-Driven Proton Pump: pH Dependence of the Reduction Levels of the Redox Centres during Turnover, *Biochemistry* **27**, 5441-5447 (1988).
16. M. Brunori, A. Colosimo, G. Rainoni, M. T. Wilson, and E. Antonini, Functional intermediates of cytochrome oxidase. Role of "pulsed" oxidase in the pre-steady state and steady state reactions of the beef enzyme, *J. Biol. Chem.* **254**, 10769-10775 (1979).
17. I. Belevich, D. A. Bloch, N. Belevich, M. Wikstrom, and M. I. Verkhovsky, Exploring the proton pump mechanism of cytochrome c oxidase in real time, *Proc. Natl. Acad. Sci. U S A* **104**, 2685-2690 (2007).

BICUCULLINE-INDUCED SEIZURES: A CHALLENGE FOR OPTICAL AND BIOCHEMICAL MODELING OF THE CYTOCHROME OXIDASE CU_A NIRS SIGNAL

Chris E. Cooper[1], Mark Cope[2], Clare E. Elwell[2] and David T. Delpy[2]

Abstract: The effect of seizures on brain blood flow and metabolism has been extensively studied. However, few studies have focused on mitochondria. We used near infrared spectroscopy (NIRS) to study hemoglobin and cytochrome oxidase changes during seizures, induced by the GABA antagonist bicuculline, in the adult rat. A broadband spectroscopy system was used with the optodes placed across the rat head. We focused on the initial seizures post-bicuculline addition during which oxyhemoglobin (HbO_2) increased, deoxyhemoglobin (HHb) decreased and total hemoglobin (Hbtot) increased. The NIRS signal associated with the oxidised Cu_A centre of mitochondrial cytochrome c oxidase (oxCCO) decreased. At the highest bicuculline doses (0.25 mg/animal) the maximum values recorded were: $\Delta HbO_2 = +19 \pm 7$ μM; $\Delta HHb = -12 \pm 4$ μM; $\Delta Hbtot = +7 \pm 4$ μM, $\Delta oxCCO = -1.7 \pm 0.3$ μM. These results are broadly in line with other NIRS studies. However, previous measurements of NADH fluorescence indicate *oxidation* of the mitochondrial redox chain under these conditions. The changes induced by bicuculline provide an interesting challenge to the physics and biochemistry of using NIRS to study mitochondrial redox states in vivo and we explore the possible spectroscopic and/or biochemical meaning of these apparent anomalies.

1. INTRODUCTION

Since 1977[1], near infrared spectroscopy (NIRS) has been extensively used to study redox state changes in mitochondrial cytochrome c oxidase (CCO). In the wavelength range 700 – 900 nm these are dominated by a broad band centred at 830 nm and associated with the oxidised form of the binuclear Cu_A redox centre[2]. This centre is the initial electron acceptor from the substrate, reduced cytochrome c. Electron transfer

[1] Department of Biological Sciences, University of Essex, Wivenhoe Park, Colchester Essex, CO4 3SQ, UK
ccooper@essex.ac.uk
[2] Department of Medical Physics and Bioengineering, Malet Place Engineering Building, University College London, Gower Street, London WC1E 6BT, UK

P. Liss et al. (eds.), *Oxygen Transport to Tissue XXX*, DOI 10.1007/978-0-387-85998-9_20,
© Springer Science+Business Media, LLC 2009

to/from cytochrome c and Cu_A is fast compared to enzyme turnover[3] and it is likely that changes in the Cu_A redox state reflect those in cytochrome c and *vice versa*.

There has been considerable controversy about the ability of NIRS techniques to detect changes in the redox state of CCO *in vivo* (oxCCO) in the presence of the generally larger concentration changes of hemoglobin. We have reviewed this field previously[2] and suggested the use of mitochondrial inhibitors (e.g. cyanide) to test algorithms claiming to be sensitive to changes in oxCCO[4]. However, it would also be useful to have reproducible tests that reported on activations of mitochondrial metabolism. We have looked at the increased mitochondrial oxygen consumption induced by the mitochondrial uncoupler, dinitrophenol[5]. In this paper we look at an alternative, more physiological, method to cause increases in cerebral mitochondrial metabolism, the use of the GABA antagonist, bicuculline; bicuculline is interesting in this regard as it causes changes in both cerebral biochemistry/physiology[6] (increases in cerebral glucose utilisation, blood flow, nitric oxide production, oxygen consumption and ATP turnover) and cerebral optical properties (due to seizures and increases in blood volume). In this study we report on the changes in oxCCO NIRS in the adult rat brain following the initial addition of bicuculline and comment on the implications for our ability to detect, and importantly *interpret*, these changes in oxCCO.

2. METHODS

All the work described here was carried out in accordance with UK regulations for the use of animals in research. Male Wistar rats (n = 6) weighing 300 to 500 g were anaesthetised with urethane (ethyl carbonate 36% w/v solution, 0.5 ml/100 g body weight intra-peritoneal). Tracheostomy was performed and a femoral artery and vein cannulated. The skull was then exposed by an incision along the top of the head lengthwise, the scalp tissues reflected and the temporal muscles removed. Cautery sealed the exposed tissue to prevent blood loss. The head was immobilised in a stereotactic holder and a source and detector optical fibre bundle were positioned either side of the rat's skull in contact with the parietal bones. A clear gel was used between the optical fibres and the skull to improve optical coupling. This enabled transillumination of approximately 1.4 cm width across the rat's brain. Light was delivered to the source fibre from a broadband white light source and a cooled CCD detector was used to detect the transmitted NIR light between 700 and 900 nm[7]. Electroencephalographic (EEG) electrodes to monitor cerebral function were placed in burr holes made in the skull. In some studies the animal was paralysed with curare (intra-venously 0.2 ml of 1.5 mg/ml solution, additional doses given throughout when required). Anaesthesia was maintained throughout by giving 0.2 ml of 24% urethane intra-venously every 2 to 3 hours. Body temperature was maintained at $37 \pm 1°C$ via a heated bed. Arterial blood pressure was monitored continuously by a transducer (Elcomatic model 750, U.K.) attached to the arterial catheter. The attenuation changes measured by the broadband spectroscopy system were converted into changes in HbO_2, HHb and oxCCO using an algorithm that allowed for a non-linear conversion of attenuation to extinction coefficient at each wavelength and at each time point[8,9]. Multi-linear regression was then used to convert the modified data to changes in $[HbO_2]$, $[HHb]$ and $[oxCCO]$. The wavelength range used was between 700 and 900 nm.

3. RESULTS

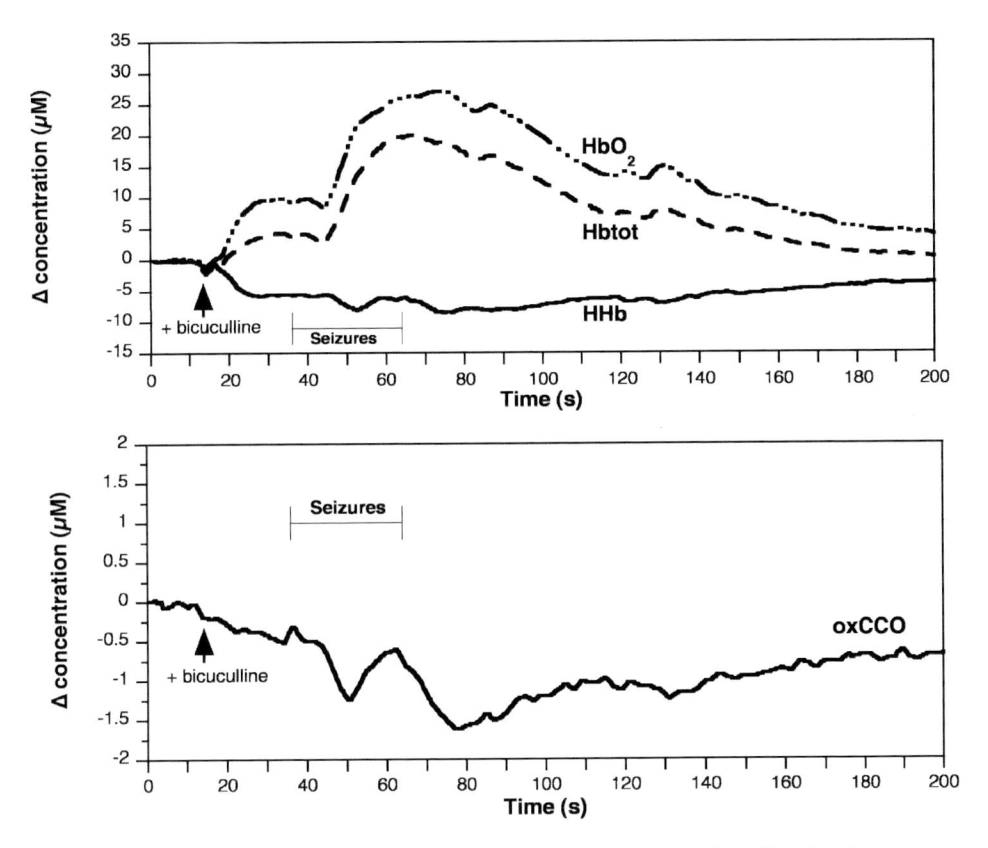

Figure 1. Typical concentration changes following the addition of 0.25 mg bicuculline in a single rat

Figure 1 illustrates the changes observed in one animal following a single high dose of bicuculline. The typical experiment illustrated consisted of 115 spectra measured at 2 second intervals during which time the animal was breathing 21% oxygen; 12 seconds after the start of the collection the animal was given a bolus injection of bicuculline which caused an immediate large increase in arterial blood pressure from 95 to 140 mmHg. A large increase in EEG activity, accompanied by general seizures, followed some 36 seconds after the injection and lasted for 28 seconds. [HbO$_2$] increased and [HHb] fell immediately after the injection. The increase in HbO$_2$ was larger than the fall in HHb, resulting in an increase in total Hb concentration (Hbtot). The degree of [oxCCO] decrease indicated a significant reduction of the CCO Cu$_A$ centre. After the seizures (36 – 64 s) there was a large secondary [Hbtot] increase. Figure 2 shows group data for the maximum chromophore changes following the addition of different concentrations of bicuculline. The pattern of changes seen in Figure 1 was consistent for all animals. Paralysis (via the muscle relaxant curare) had no significant effect on the changes observed at the highest bicuculline concentration used (Figure 3).

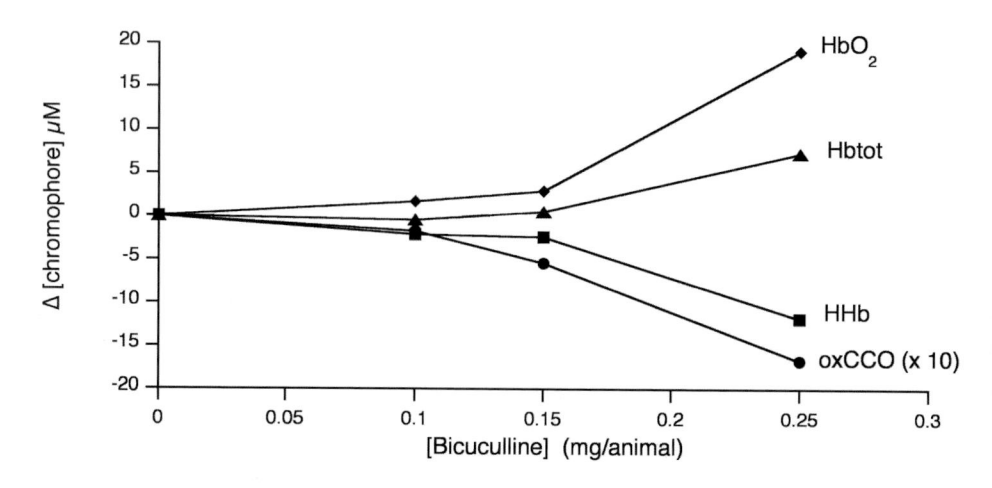

Figure 2. Dose response for maximum chromophore changes following bicuculline addition. Data are means from six animals (n = 1 at 0.1 mg, n=2 at 0.15 mg and n=3 at 0.25 mg). Note that oxCCO concentration changes displayed have been multiplied by a factor of 10.

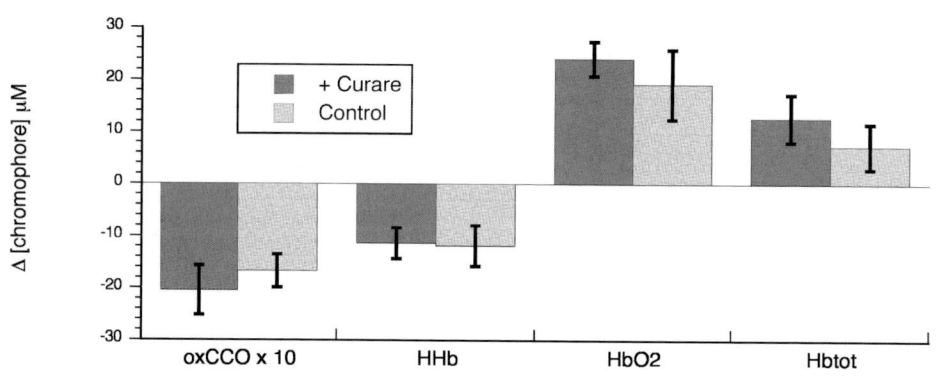

Figure 3. Effect of curare on chromophore concentration changes (0.25 mg bicuculline / animal). Mean ± SD (n = 3). No significant differences were observed in the presence of curare.

4. DISCUSSION

The addition of bicuculline poses a number of challenges to the use of NIRS to measure the cytochrome oxidase redox state *in vivo*. The physics challenge relates to the large optical changes induced in the brain, caused by both strong absorption and, possibly, scattering changes. We can rule out physical movement artefacts, given that the data are insensitive to the use of muscle relaxants. However, non-linear changes in the wavelength-dependence of optical pathlength are more difficult to correct for. A fundamental problem with converting optical density changes to relative chromophore concentrations is the wavelength and absorption coefficient dependence of the optical pathlength. This leads to a non-linear and variable relationship between optical pathlength, attenuation and wavelength. Strictly speaking the standard Beer-Lambert law therefore cannot be applied. Whilst this is likely to lead to only quantitative errors in the estimated changes in [HbO_2] and [HHb], it could lead to qualitative (i.e. sign) changes in the much smaller oxCCO signal. For example one possibility is that the strong absorbance changes due to changes in hemoglobin oxygenation and, especially, concentration alter the wavelength dependence of the optical pathlength such that there is "cross-talk" between the hemoglobin and oxCCO signals. In principle, our non-linear algorithm should correct for this situation and, in this study, the changes in [oxCCO] were generally not simply a function of the other chromophore concentrations. For example, in Figure 1, [Hbtot] rose during the initial 50 s and [oxCCO] fell. At 50s [Hbtot] rose again, but was accompanied by a rise in [oxCCO], ruling out an obvious cross-talk artefact[10]. The oxCCO change therefore seems to be a genuine "third" experimental parameter, not directly mirroring the hemoglobin chromophores.

A different approach can be taken to convert the measured attenuation changes to concentration changes. Measurements of the wavelength dependence of the optical pathlength[11] are used to modify the extinction coefficients of these chromophores in order to reflect the *in vivo* situation. Using no pre-processing, the attenuation changes can be fitted directly to these modified extinction coefficient changes of the chromophores of interest. One would expect such algorithms to perform less well if the wavelength dependence of pathlength altered during the experiment, as might happen for large changes in hemoglobin oxygenation or concentration. However, when one such algorithm (UCLn[12]) was used to analyse the data collected in this current study the resulting chromophore changes still showed a *decrease* in oxCCO despite an *increase* in cerebral HbO_2. The widely used commercial NIRS systems developed by Hamamatsu Photonics (the NIRO series) use this type of algorithm to convert the attenuation changes measured no more than four wavelengths (e.g. 775, 810, 850 and 910 nm). When we analysed our data using only these wavelengths we still saw a decrease in oxCCO accompanied by an increase in HbO_2. However if the data were analysed using chromophore extinction coefficients which had not been modified to account for the wavelength dependence of optical pathlength we saw an *increase* in oxCCO. This suggests that this bicuculline study provides a genuine challenge to the ability of NIRS to detect changes in oxCCO.

Although these data are not inconsistent with previous NIR studies using other seizure-inducing agents[13], they nevertheless provide a challenge to current interpretations of the oxCCO signal. These almost entirely focus on oxygen as a modulator. It is frequently assumed that a decrease in oxCCO must be due to a decrease in mitochondrial pO_2. Yet here we see a decrease in oxCCO when pO_2 is likely to, if anything, rise. However, given that the Cu_A centre is likely to be in redox equilibrium with cytochrome *c*, all the factors that affect the *c* redox state are likely to affect Cu_A as well[14,15]. These include the supply of electrons to the respiratory chain, as well as the magnitude of the mitochondrial proton motive force. An increase in substrate supply at the top of the chain (e.g. NADH) could decrease oxCCO even at high pO_2. The effect of ATP turnover (i.e. a drop in proton motive force) is less easy to predict and could, in theory, oxidise or reduce a mitochondrial centre[14,15]. Nitric oxide (a strong inhibitor of CCO[16]) would be expected to increase both NADH reduction as well as Cu_A reduction. Bicuculline would be expected to increase ATP turnover, local nitric oxide concentration and, possibly, substrate supply. However, *in vivo* NADH fluorescence data following seizures[17,18] consistently shows a decrease in [NADH]. This would argue against a nitric oxide or substrate supply driven reduction in oxCCO, and might suggest instead a "cross-over" point in mitochondria where NADH is oxidised and Cu_A reduced following a drop in membrane potential[15]. The data in this paper therefore makes strong predictions that can be tested experimentally. To our knowledge there have been no simultaneous measurements of changes in CCO Cu$_A$ and NADH *in vitro* or *in vivo*. A final interpretation of the optical changes following bicuculline addition must therefore await these measurements.

5. ACKNOWLEDGEMENTS

We would like to thank Susan Wray (now at University of Liverpool) for technical assistance and BBSRC (D0609821/1) and EPSRC (EP/F006551/1) for financial support.

6. REFERENCES

1.   F. F. Jöbsis, Non-invasive infrared monitoring of cerebral and myocardial oxygen sufficiency and circulatory parameters, *Science* **198**, 1264-1267 (1977).
2.   C. E. Cooper, S. J. Matcher, J. S. Wyatt, M. Cope, G. C. Brown, E. M. Nemoto, and D. T. Delpy, Near infrared spectroscopy of the brain: relevance to cytochrome oxidase bioenergetics, *Biochem. Soc. Trans.* **22**, 974-980 (1994).
3.   F. Malatesta, F. Nicoletti, V. Zickermann, B. Ludwig, and M. Brunori, Electron entry in a CuA mutant of cytochrome c oxidase from Paracoccus denitrificans. Conclusive evidence on the initial electron entry metal center, *FEBS Lett.* **434**, 322-324 (1998).
4.   C. E. Cooper, M. Cope, R. Springett, P. Amess, J. Penrice, L. Tyszczuk, S. Punwani, R. Ordidge, J. Wyatt, and D. T. Delpy, Use of mitochondrial inhibitors to demonstrate that cytochrome oxidase near-infrared spectroscopy can measure mitochondrial dysfunction non-invasively in the brain, *J. Cereb. Blood Flow Metab.* **19**, 27-38 (1999).
5.   R. Springett, J. Newman, D. T. Delpy, and M. Cope, Oxygen dependency of cerebral CuA redox state during increased oxygen consumption produced by infusion of a mitochondrial uncoupler in newborn piglets, *Adv. Exp. Med. Biol.* **471**, 181-188 (1999).
6.   B. S. Meldrum and B. Nilsson, Cerebral blood flow and metabolic rate early and late in prolonged epileptic seizures induced in rats by bicuculline, *Brain* **99**, 523-542 (1976).
7.   M. Cope, D. T. Delpy, J. S. Wyatt, S. C. Wray, and E. O. R. Reynolds, A CCD spectrometer to quantitate the concentration of chromophores in living tissue utilising the water absorption peak of water at 975nm, *Adv. Exp. Med. Biol.* **247**, 33-41 (1989).
8.   M. Cope. The application of near infrared spectroscopy to non-invasive monitoring of cerebral oxygenation in the newborn infant (PhD): University of London, 1991.
9.   M. Cope, P. van der Zee, M. Essenpreis, S. R. Arridge, and D. T. Delpy, Data analysis methods for near infrared spectroscopy of tissue: problems in determining the relative cytochrome *aa*3 concentration, *Proc. SPIE* **1431**, 251-262 (1991).
10. I. Tachtsidis, M. Tisdall, T. S. Leung, C. E. Cooper, D. T. Delpy, M. Smith, and C. E. Elwell, Investigation of in vivo measurement of cerebral cytochrome-c-oxidase redox changes using near-infrared spectroscopy in patients with orthostatic hypotension, *Physiol. Meas.* **28**, 199-211 (2007).
11. M. Essenpreis, C. E. Elwell, P. van der Zee, S. R. Arridge, and D. T. Delpy, Spectral dependance of temporal point spread functions in human tissues, *Applied Optics* **32**, 418-425 (1993).
12. S. J. Matcher, C. E. Elwell, C. E. Cooper, M. Cope, and D. T. Delpy, Performance Comparison of Several Published Tissue Near-Infrared Spectroscopy Algorithms, *Anal. Biochem.* **227**, 54-68 (1995).
13. M. Yanagida and M. Tamura, Changes in oxygenation states of rat brain tissues during glutamate-related epileptic seizures--near-infrared study, *Ad.v Exp. Med. Biol.* **345**, 579-586 (1994).
14. M. Banaji, A generic model of electron transport in mitochondria, *J. Theor. Biol.* **243**, 501-516 (2006).
15. B. Chance and G. R. Williams, Respiratory enzymes in oxidative phosphorylation. III. The steady state, *J. Biol. Chem.* **217**, 409-427 (1955).
16. C. E. Cooper, Nitric oxide and cytochrome oxidase: substrate, inhibitor or effector?, *Trends Biochem. Sci.* **27**, 33-39 (2002).
17. A. Mayevsky and G. G. Rogatsky, Mitochondrial function in vivo evaluated by NADH fluorescence: from animal models to human studies, *Am. J. Physiol. Cell Physiol.* **292**, C615-C640 (2007).
18. E. Lothman, J. Lamanna, G. Cordingley, M. Rosenthal, and G. Somjen, Responses of electrical potential, potassium levels, and oxidative metabolic activity of the cerebral neocortex of cats, *Brain Res.* **88**, 15-36 (1975).

A METHOD TO CALCULATE ARTERIAL AND VENOUS SATURATION FROM NEAR INFRARED SPECTROSCOPY (NIRS)

Jan Menssen, Willy Colier, Jeroen Hopman, Djien Liem, and Chris de Korte[*]

Abstract: For adequate development and functioning of the neonatal brain, sufficient oxygen (O_2) should be available. With a fast sampling ($f_s > 50$ Hz) continuous wave NIRS device, arterial (SaO_2) and venous (SvO_2) saturation can be measured using the physiological fluctuations in the oxyhemoglobin (O_2Hb) and total hemoglobin (tHb) concentrations due to heart action and respiration. Before using this technique in a neonatal setting, the method was verified on adult volunteers (n=7) by decreasing inspired oxygen down to an arterial saturation of 70% using a pulse oximeter as reference. NIRS optodes were placed on the left forehead; the pulse oximeter sensor was placed on the right forehead. The experiments were repeated with different optode spacings.

SaO_2 and SvO_2 were determined using the ratio between the O_2Hb and tHb value in the amplitude spectrum at the heart rate and respiration rate, respectively.

A good agreement between calculated SaO_2 and reference SaO_2 from pulse oximetry was found (bias range -3.5% to 5.2%, SD of the residuals 1.3% to 3.5%). Optode spacing of 15 mm yielded a negative bias compared to optode spacing of 45 mm. It was not always possible to calculate SvO_2 because the respiration peak could not always be detected.

1. INTRODUCTION

Oxygen is important for adequate development and functioning of the neonatal brain. Measuring the arterial saturation (SaO_2) and venous saturation (SvO_2) in the brain could provide us with information about the changes in oxygen consumption over time. SaO_2 merely reflects oxygen supply and delivery and SvO_2 offers insight into oxygen demand.[1]

[*] Jan Menssen, Jeroen Hopman, Djien Liem, and Chris de Korte, Dept. of Pediatrics, Radboud University Nijmegen Medical Centre, Nijmegen, The Netherlands. Willy Colier, Artinis Medical Systems, Zetten, The Netherlands. e-mail : j.menssen@cukz.umcn.nl

P. Liss et al. (eds.), *Oxygen Transport to Tissue XXX*, DOI 10.1007/978-0-387-85998-9_21,
© Springer Science+Business Media, LLC 2009

Our goal was to further explore the possibility of using a method that offers both arterial and venous saturation continuously using Near InfraRed Spectroscopy (NIRS) in a clinical setting. The intention is to apply this method to the neonatal brain to get insight in the oxygen consumption. The method should not be harmful for the patient which implies that methods changing the venous blood volume by tilting the head down or partial jugular occlusion are not suitable.

SaO_2 measurements, using NIRS have been previously discussed in the literature.[2-4] Also methods to offer SvO_2 based on respiration-induced oscillations in the venous blood volume have been published.[5-7] Wolf et al.[8] describes a method that calculates both SaO_2 and SvO_2, however in this study, it is not fully clear how the results were validated. SaO_2 validation studies in adults are done in the range from 90% to 100%. Critically ill newborns quite often have arterial saturation values far below this range.

Tissue heterogeneity may also influence saturation measurements with NIRS. Therefore, optode distance and application could affect the results. Little is known about the reproducibility of saturation measurements with NIRS. In this study, validation of the method was done by lowering SaO_2 to values of around 70%. Furthermore reproducibility was tested by repeated measurements and measurements at different optode distances were performed.

Before introducing our method in a neonatal clinical setting, the method was verified on adult volunteers.

2.   MATERIAL AND METHODS

In seven healthy volunteers (four male and three female) SaO_2 decreased by diminishing the inspired oxygen fraction (FiO_2). The volunteers were lying on a bed and were breathing through a face-mask while FiO_2 was steadily reduced until an arterial saturation of 70-75% was reached. Then FiO_2 was quickly increased to a level of 30% oxygen until a stable baseline condition was reached. Changes in FiO_2 were induced by means of a computer-controlled mass flow system (Bronckhorst High Tech BV, Veenendaal, The Netherlands); a standard pulse oximeter (Nellcor N200, Nellcor, Pleasanton CA, USA) was used as reference. The reflectance sensor was placed on the right forehead using adhesives. NIRS optodes were placed on the left side of the forehead with a special tool to make sure the same inter optode spacing was used in all volunteers. A one-channel continuous wave NIRS device (Oxymon, Artinis Medical Systems, Zetten, The Netherlands) with 3 lasers (770 nm, 848 nm and 901 nm) and one receiver was used.[9] Data were collected with a sampling frequency of 50 Hz and stored for off-line analysis. Data obtained from the pulse oximeter (heart rate and SaO_2) were acquired simultaneously in the NIRS device using the external analog digital converter board.

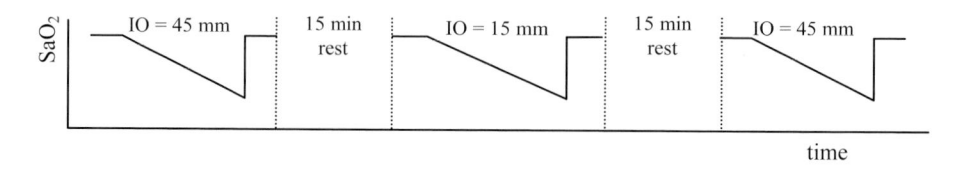

Figure 1. Schematic drawing of the measurement protocol.

For each volunteer the procedure was repeated three times. After each desaturation a 15 minutes rest period was included. The inter optode spacing (IO) at the first and last measurement was 45 mm and at the second measurement 15 mm.

Analysis was done using software written in Matlab (The Mathworks, Natick MA, USA). First the optical densities changes were converted to oxyhemoglobin (O_2Hb), de-oxyhemoglobin (HHb) and total hemoglobin (tHb) concentration changes using a modified Lambert-Beer algorithm.[10] The next step was to detect heartbeats from the optical density signal at 901 nm. For each detected heartbeat a window of 10.24 s (for SaO_2) or 20.48 s (for SvO_2) signal was used to calculate the amplitude spectrum. The reason for the smaller window size for SaO_2 is to obtain a faster response. A Fast Fourier Transform (FFT) was performed on these windows after the baseline was removed and the signals were detrended. From the FFT spectrum, SaO_2 and SvO_2 were determined as the amplitude ratio between O_2Hb and tHb at, respectively, the heart rate peak and respiration rate peak found in the spectrum. The amplitude was calculated as the mean value in a small window around the detected peak. O_2Hb and HHb were visually inspected to confirm they are in phase.

3. RESULTS

An example of a measurement is given in Figure 2. The upper panel shows the changes in hemoglobin concentrations from the NIRS device; the lower panel shows the heart rate and SaO_2 obtained from the pulse oximeter.

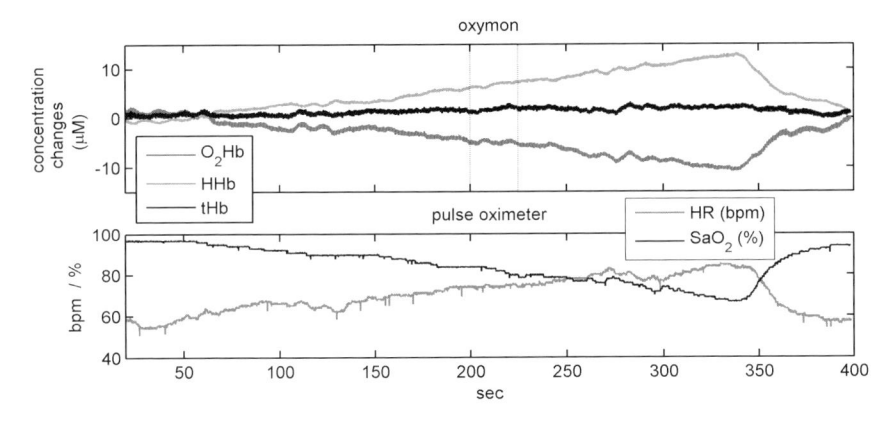

Figure 2. Example of a measurement. The upper panel contains the NIRS signals; the lower panel contains signals from the pulse oximeter. Cleary visible is the decrease in oxyhemoglobin concentration simultaneously with the decrease in arterial saturation while the total hemoglobin is almost constant

In Figure 3a, we have zoomed in to take a closer look at the pulsatile nature of the NIRS signals. Both the fast pulsatility due to the heart rate and the slow fluctuations due to breathing are clearly visible in the O_2Hb and tHb signal and in the corresponding FFT spectra from these signals. As expected at a high saturation, the pulsatility of the de-oxyhemoglobin (HHb) signal is small. This is illustrated in Figure 3b, where the amplitude spectrum is given.

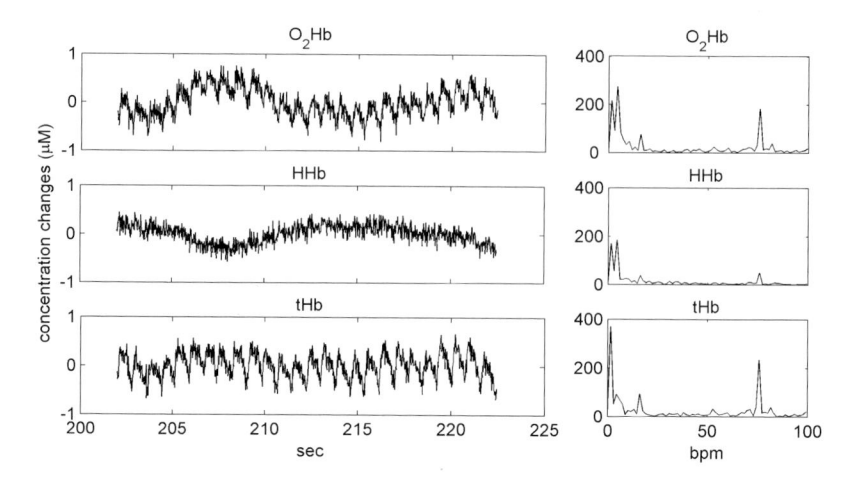

Figure 3. The left panels (a) contain 20.48 sec of concentration signals. The right panels (b) contain the amplitude spectrum (μM/Hz). Peaks of the heart rate (76 BPM) and the respiration rate (16 min^{-1}) are clearly visible.

3.1. Arterial Saturation

The comparison between SaO_2 measured with the pulse oximeter and SaO_2 calculated from the pulsatile NIRS traces is shown for one subject in Figure 4. Bland-Altman[11] analysis was used to calculate the bias and the standard deviation of the residuals. At high saturation values, the difference between SaO_2 obtained from the pulse oximeter and SaO_2 from NIRS is larger than at low saturation values.

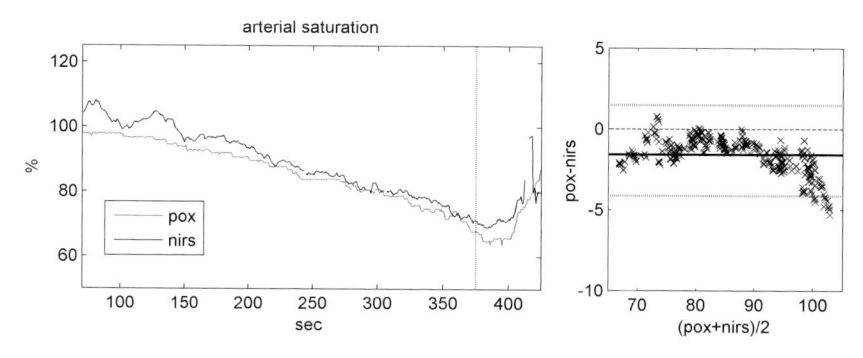

Figure 4. Comparison between arterial saturation measured with a pulse oximeter (pox) and NIRS. In the right panel a Bland-Altman plot is generated over the first 375 seconds (the fast recovery to normoxia is discarded). The mean difference (-1.32%) and 2 * standard deviation of the residuals (1.40%) lines are drawn.

As Table 1 shows, a good correlation between SaO_2 measured with a pulse oximeter and SaO_2 calculated from the NIRS signals exists. The mean difference (bias) varies from -3.4% to 5.2% and the error after correction for the bias (standard deviation of the residuals) is between 1.28% and 3.54%. There is a difference in bias between the measurements done with optode distance = 45 mm compared with measurements done with optode distance = 15 mm (paired t-test, p=0.03, 1st and 2nd measurement and p=0.02,

2nd and 3rd measurement). Also the standard deviation of the residuals (std$_{res}$) in the 3rd measurement is smaller than the 2nd measurement (p=0.04).

Table 1. Comparison between SaO$_2$ from pulse oximeter with NIRS.

subject	1st measurement (IO=45 mm)			2nd measurement (IO = 15 mm)			3rd measurement (IO = 45 mm)		
	r	bias (%)	std$_{res}$ (%)	r	bias (%)	std$_{res}$ (%)	r	bias (%)	std$_{res}$ (%)
1	0.98	-3.13	1.67	0.83	-3.51	2.09	0.89	0.63	1.99
2	0.85	2.13	2.15	0.89	-2.46	2.56	0.76	5.25	2.57
3	0.95	-3.42	2.29	0.92	-2.34	1.98	0.96	-0.54	1.28
4	0.98	1.90	2.81	0.92	-1.44	1.96	0.97	-1.32	1.40
5	0.91	4.04	3.54	0.94	-2.85	2.70	0.95	1.02	1.76
6	0.93	4.67	1.45	0.80	2.45	2.17	0.91	3.60	1.73
7	0.91	2.04	2.70	0.90	-0.69	1.88	0.94	1.96	1.96

3.2. Venous Saturation

In 5 of the 7 subjects respiration rate detection was sufficient to calculate venous saturation but not in all measurements. For the detection no relation with optode distance or respiration rate was found. Respiration rate varied from 10-15 min^{-1} to 40-50 min^{-1}.

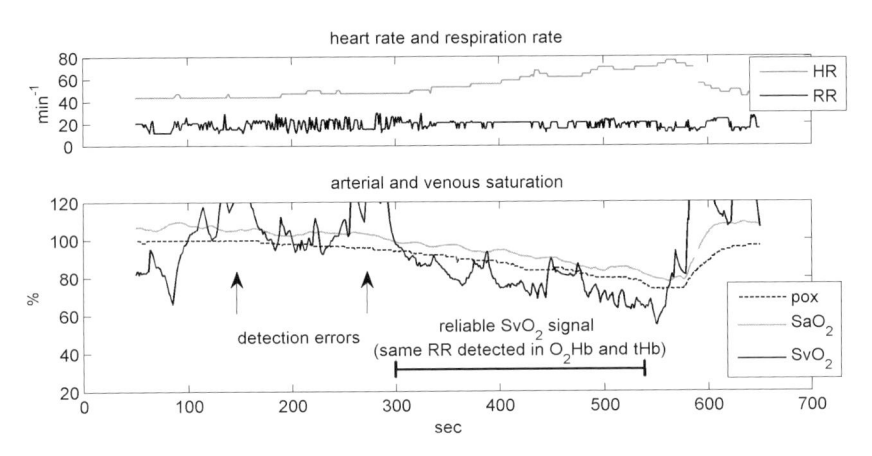

Figure 5. Example of venous saturation. SvO$_2$ is reliable when in both O$_2$Hb and tHb signal same respiration rate (RR) is detected and the value is smaller as SaO$_2$

An example is given in Figure 5. Although it was not always possible to detect the respiration rate and calculate the venous saturation, the signal part between 300 and 550 seconds looks reliable because the decrease in arterial saturation is followed by a similar decrease in venous saturation as expected at a stable metabolic rate. However, sometimes in the reliable period, artefacts in O$_2$Hb and tHb are present, which results in a calculated SvO$_2$ value that is higher than the arterial saturation.

4. DISCUSSION AND CONCLUSIONS

We have used a previously described non-invasive method to calculate SaO_2 and SvO_2 in the human brain using NIRS. There was a good agreement in the trends of the SaO_2 values measured with the pulse oximeter and the calculated values. The method was reproducible. Our method confirms "Method B" as described by Leung et al.[4] In accordance with this paper, we also found SaO_2 values > 100%, but we could not detect pulsations in HHb due to the poor signal-to-noise ratio. At small optode distance SaO_2 was overestimated, possibly due to changes in the venous volume induced by arterial pulsations as suggested by Leung. A relatively small volume is measured that is fixed between skin and skull and the effect can be relative strong. At large distance the measured volume lies mostly within the brain and the arterial volume can pulsate without influencing the venous compartment that is penetrated by the NIR light.

Calculation of the venous saturation based on respiration induced oscillations in the venous compartments of the brain is complicated, because they are not always detectable as Wolf et al.[8] also found. If oscillations are detected, O_2Hb and HHb are not always in phase indicating the model is not valid. Our out-of-phase results support the oscillating flow theory as described by Franceschini et al.[5] A combination of volume oscillations and flow oscillations can also occur. Additional information is needed in the model. Therefore we are evaluating a new multi parameter model in an animal study.

In this study, SvO_2 was compared with SaO_2. At a stable metabolic rate, a decrease in SaO_2 is followed by a similar decrease in SvO_2 as we found. Better validation is required and will be done in the mentioned study.

REFERENCES

1. K.M. White, Completing the hemodynamic picture : SvO_2. *Heart Lung. vol 14, pp 272-280 (1985).*
2. W.N.J.M. Colier, M.C. van der Sluijs, J.J.M. Menssen, R.J.F. Houston, and B. Oeseburg, Validation of an algorithm to detect arterial saturation from a near infrared spectrophotometric signal. *Presentation at SPIE BIOS '99 San Jose (1999).*
3. M.A. Franceschini, E. Gratton, and S. Fantini, Noninvasive optical method of measuring tissue and arterial saturation: an application to absolute pulse oximetry of the brain *Opt Lett. vol 24, pp 829-831 (1999)*
4. T.S. Leung, I. Tachtsidis, P. Velayuthan, C. Oliver, J.R. Henty, H. Jones, M. Smith, C.E. Elwell, and D.T. Delpy, Investigation of oxygen saturation derived from cardiac pulsations measured on de adult head using NIR spectroscopy *Adv Exp Med Biol. vol 578, pp 209-215 (2006)*
5. M.A. Franceschini, D.A. Boas, A. Zourabian, S.G. Diamond, S. Nadgir, D.W. Lin, J.B. Moore, and S. Fantini, Near-infrared spiroximetry: noninvasive measurements of venous saturation in piglets and human subjects, *J. Applied Physiol. vol 92, pp 372-384 (2002)*
6. T.S. Leung, M.M. Tisdall, I. Tachtsidis, M. Smith, D.T. Delpy, and C.E. Elwell, Cerebral tissue oxygen saturation calculated using low frequency haemoglobin oscillations measured by near infrared spectroscopy in adult ventilated patients. *Adv Exp Med Biol. 2008 (in press)*
7. D.W. Brown, D. Haensse, A. Bauschatz, and M. Wolf, NIRS Measurements of venous oxygen saturation. *Adv Exp Med Biol. vol 578, pp 251-256 (2006)*
8. M. Wolf, G. Duc, M. Keel, P. Niederer, K. von Siebenthal, and H.U. Bucher, Continuous noninvasive measurement of cerebral arterial and venous oxygen saturation at the bedside in mechanically ventilated neonates, *Critical Care Medicine. vol 25, pp 1579-1582 (1997)*
9. M.C. van der Sluijs, W.J.N.M. Colier, R.J.F. Houston, and B. Oeseburg, A new and highly sensitive continuous wave near infrared spectrophotometer with mutiple detectors, *SPIE 3194: 6372 (1997)*
10. Y. Wickramasinghe and P. Rolfe, Modified NIR coefficients taking into account the wavelength dependence of optical pathlength, in: *Newsletter EC Concerted Action Near Infrared Spectrophotometry and Imaging of Biological Tissue, Issue 2,* edited by J. Bancroft, Stoke on Trent (1993)
11. J.M. Bland and D.G. Altman, Statistical methods for assessing agreement between two methods of clinical measurement, *Lancet. vol 32, pp 307-310 (1986)*

DEVELOPMENT OF A DYNAMIC TEST PHANTOM FOR OPTICAL TOPOGRAPHY

Peck H. Koh, Clare E. Elwell, and David T. Delpy[*]

Abstract: Optical topography (OT) is a near infrared spectroscopy (NIRS) technique that provides spatial maps of haemodynamic and oxygenation changes. When developing, testing and calibrating OT systems it is often necessary to use tissue simulating phantoms that are capable of providing realistic changes in attenuation properties. We present a novel dynamic tissue phantom that enables spatially and temporally varying tissue properties to be reproduced in a controlled manner.

This new dynamic test phantom consists of a modified liquid crystal display (LCD) (enabling flexible and rapid changes in attenuation across different regions of the phantom) sandwiched between two layers of tissue simulating epoxy resin (providing static and homogeneous optical absorption and scattering). By activating different pixels in the liquid crystal display it is possible to produce highly localised and dynamic changes in attenuation which can be used to simulate the changes associated with the cerebral haemodynamic response to functional activation. The reproducibility of the dynamic phantom will be described with examples of its use with an OT system.

1. INTRODUCTION

The development and testing of most imaging systems requires the use of tissue phantoms that are capable of providing realistic changes in attenuation properties. These phantoms can be used to test design prototypes and evaluate instrument performance. Many useful optical phantoms have been developed in recent years with a wide range of attenuation properties that are characteristic of biological tissues. Optical phantoms for NIRS studies were first developed for breast imaging studies[1] and subsequent research into different types of optical imaging techniques has led to the generation of a variety of tissue phantoms. Many of these have focused on the design of regular-shaped objects with specific attenuation properties including the use of biological molecules such as haemoglobin and melanin as absorbing components[2] and the generation of hybrid

[*] Department of Medical Physics and Bioengineering, University College London, Gower Street, London WC1E 6BT, United Kingdom. pkoh@medphys.ucl.ac.uk

P. Liss et al. (eds.), *Oxygen Transport to Tissue XXX*, DOI 10.1007/978-0-387-85998-9_22,
© Springer Science+Business Media, LLC 2009

phantoms for multimodality imaging.[3] Classical phantoms generally have static properties i.e. absorption (μ_a) and reduced scattering coefficients (μ_s') matching those of the biological tissues[4] that are not spatially variable.

NIR topography systems are increasingly being used to monitor the haemodynamic response to functional activation via the resulting spatially and temporally varying changes in cortical attenuation. Simulation of this varying physiological signal cannot be carried out using the static phantom approach and there is a need for a dynamic phantom that can reproduce these optical changes spatially and temporally in a controlled manner.

Design of a Dynamic Phantom using Liquid Crystal Display (LCD)

One approach to producing a phantom which enables rapid and spatially varying changes in attenuation is to use the existing design of a multilayer resin phantom and to insert an additional layer of material within it, the optical properties of which can be altered easily. An LCD seems to fit the requirements for such an inserted layer since 1) it relies on external light to provide the image and does not emit light itself and 2) it has a layer of light-polarising liquid molecules whose orientation is electrically controlled which allows its attenuation properties to be changed rapidly and with high spatial specificity. This combination of static layers in which the optical properties are fixed (representing skin/skull and white matter) and a dynamic layer in which the attenuation properties can vary (representing cortical tissue) provides a reasonable simulation of the type of physiological signal changes that the OT systems are required to monitor.

Figure 1a shows the configuration of the LCD dynamic phantom. The phantom incorporates a top layer of epoxy resin with a realistic thickness (to represent the attenuation due to extracerebral layers). The light then passes through the LCD display. In the current study the size of LCD unit used allowed six laser sources and three detectors to be configured to produce measurements over seven channels with a fixed optode spacing of 30 mm (Figure 1b). The lower block phantom material (representing the brain white matter) is large enough to ensure light is scattered back to the surface and light loss to the boundary is small.

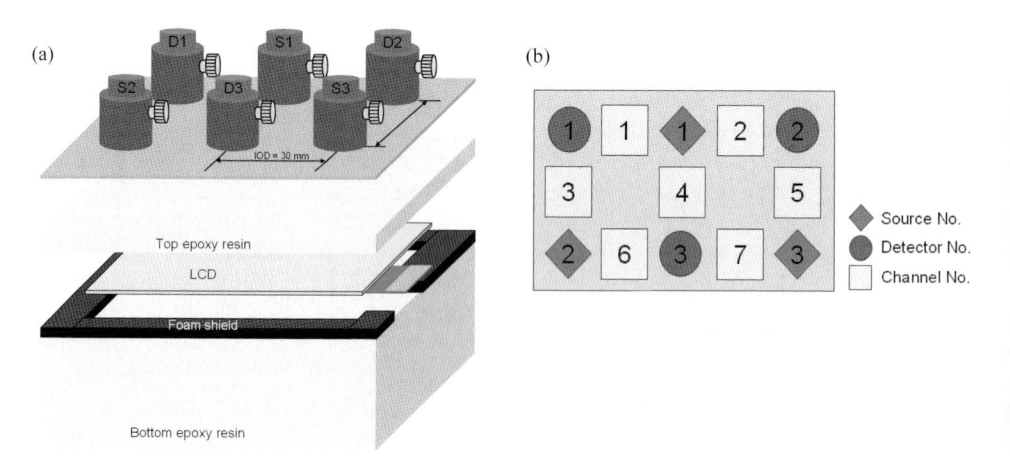

Figure 1. (a) Schematic of the dynamic phantom and (b) optode arrangement used in the current study.

2.   METHODS

A QVGA (resolution: 320 x 240 pixels) transflective mode LCD (Nan Ya Plastics Corporation) was used. The 2.2 mm-thick passive display (active area: 76 x 57 mm) uses a Chip-on-Flex configuration. Each active pixel in the LCD has an area of 5.64×10^{-3} mm^2. An external controller (CB-GT380, Amulet Technologies Limited) was selected to drive and control the display. A HTML script describing the sequence of the animation is programmed into the controller. All optical measurements were made using a ETG-100 OT system (Hitachi Medical Corporation, Japan).[5] The results described are from the light intensity measurements of the 780 nm-wavelength, expressed in arbitrary units.

The recipe used to construct the solid phantom has previously been published.[4, 6] The matrix material was constructed using epoxy resin (MY 753 Aeropia Chemical Supplies). A concentrated dye solution (Pro Jet 900NP Zeneca Ltd.) as absorber and a Superwhite polyester pigment (Alec Tiranti Ltd.) as the scatterer were mixed with the solution. The resulting phantom has a μ_a of 0.01 mm^{-1} and μ_s' of 1.0 mm^{-1} (at 800 nm) and the mean cosine of scatter is about 0.5.[4] The area of the epoxy resin block was 110 x 90 mm^2 with the base block having a thickness of 30 mm and the top block 5 mm. The absorption and scattering coefficients of the static phantom block were measured using a time-resolved optical system (MONSTIR).[7] The test result showed that the variation between expected and measured coefficients was less than 5 %.

3.   RESULTS

In OT it is often assumed that the measured intensity correlates with the size of the activated attenuating *region*. However this will only operate over a limited range of sizes and shapes since in a scattering medium light can take various routes between the source and detector including paths which do not pass through the activated region. Initial studies were therefore conducted to investigate (i) the size of attenuating region that can be detected by OT, (ii) the effect of different region geometries and (iii) "crosstalk" between different optodes with spatially varying attenuation regions.

Effect of Attenuation Region Size

A circular attenuation region was programmed on the display area positioned centrally between source 1 and detector 3 (i.e. at the position of channel 4). The diameter of the region was varied between 1 and 27 mm. To determine the *contrast* (i.e. intensity difference between two different conditions) the detected intensities were normalised to an "all-dark display" *baseline* condition where all the pixels were activated.

Figure 2 shows the normalised intensity as a function of region diameter. The results show a generally linear relationship between detected intensity and attenuator size down to a diameter of 5 mm. Below 5 mm, the contrast, which we have arbitrarily defined as a change of less than 1 %, approached the system noise level.

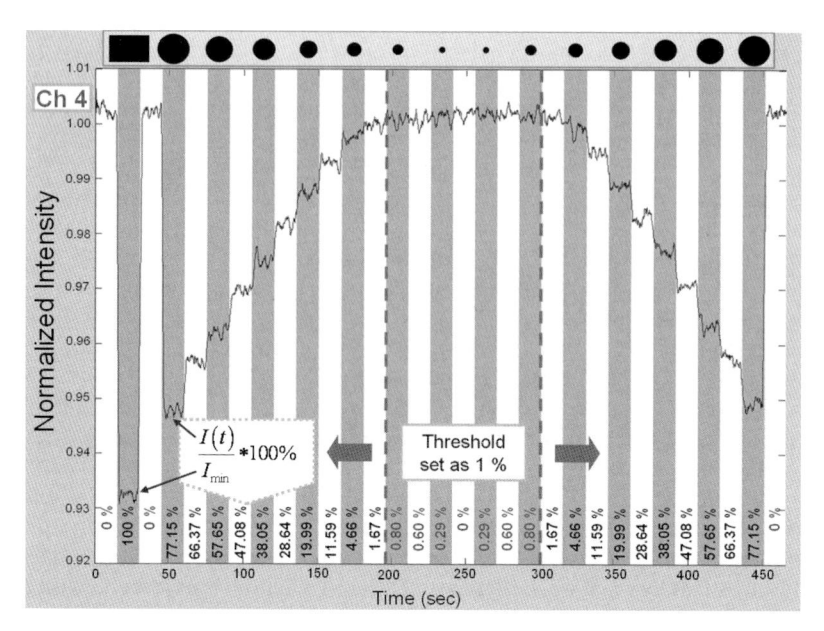

Figure 2. Attenuation time course for varying attenuator sizes (indicated at top of figure). The alternate grey and white areas indicate the period for each size variation and the normalised intensity changes (indicated at bottom of figure) were calculated with reference to a baseline condition where the whole LCD was dark.

Effect of Attenuation Region Shape

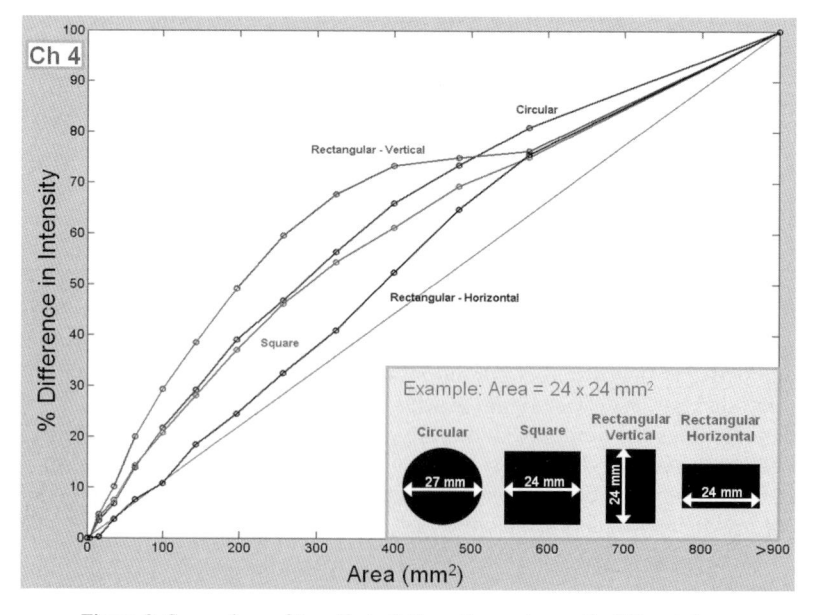

Figure 3. Comparison of the effect of attenuation regions with different shapes.

The LCD was programmed to produce a range of attenuation regions with equal areas but different shapes, including squares, circles, rectangles with fixed width (vertical) and rectangles with fixed height (horizontal). The regions were centred on channel 4 and their areas were varied between 1 and 576 mm^2. The intensity changes were subsequently normalised to the previously described baseline condition. Figure 3 shows a plot of the normalised intensity changes (expressed in percentage difference) for each of the attenuator shapes as a function of attenuator area. The result shows a generally linear trend between intensity and region sizes, but identifies a significant difference between the two rectangular shapes with same areas. The variation is due to the orientation of the optode pair relative to the region. The results suggest that along with size variation, differences in shape and orientation of the activated region can affect the OT signal.

Effect of Attenuator Position on Crosstalk between Channels

A circular attenuation region with a diameter of 27 mm was programmed on the LCD to move horizontally in sequence along the midline from channels 1-2, 3-5 and 6-7. The amount of "crosstalk" was defined as a residual change detected in the neighbouring channels. Figure 4 shows the intensity changes on all 7 channels as the 27 mm diameter attenuator was moved across the phantom. 10 episodes of crosstalk were identified (highlighted as circles) during this test. Subsequently, the attenuator region size was gradually reduced until the crosstalk effect was considered to be negligible (less than 1 % change in intensity). This occurred when the diameter of the attenuator region was less than 21.5 mm.

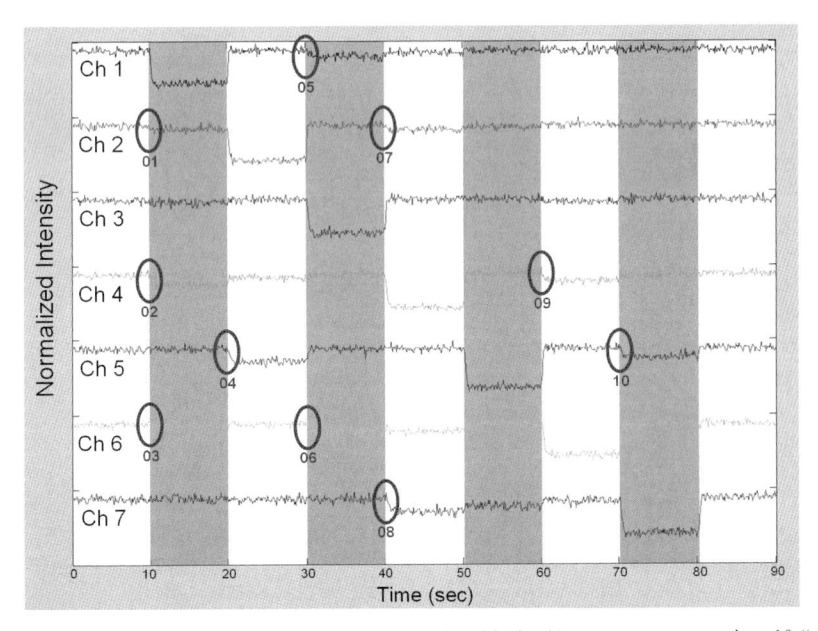

Figure 4. Normalised intensity changes for all 7 channels with the 27 mm attenuator region. 10 "crosstalk" effects are circled.

4. DISCUSSION

A new dynamic phantom using a modified LCD has been developed as a tool for evaluating and optimising OT instruments. By placing the LCD between two blocks of static phantom materials which provide well-characterised optical properties, a more realistic simulation of the spatially and temporally varying attenuation changes seen during cortical functional activation can be achieved. The effects of attenuator size, shape and orientation on measured intensity have been described. In addition, evidence of "crosstalk" between different optode pairs has been identified. This initial assessment suggests that, for the phantom geometry described in this study, the OT system was able to provide reasonable spatial differentiation with various attenuator shapes and sizes.

There are several issues which were not addressed in this study; the attenuation effects associated with the LCD particularly the loss of light due to the polarizers and effect of the clear glass on the LCD have not been quantified. While the contribution due to the difference in refractive indices between the epoxy resin (1.56 at 800 nm) and the LCD glass (~1.5) is considered to be negligible, coupling between the LCD and static phantom layer has not been investigated. The current setup can also only simulate attenuation changes in two dimensions, when in reality light attenuation due to physiological changes in tissue occurs in three dimensions.

In most cases optical topography is used to measure the spatial *changes* in attenuation arising from changes in chromophore concentration and as such *absolute* quantification of the attenuation properties is not required. Since the dynamic phantom is also intended to be used as a calibration tool for system, the actual photon path (which in turn affects the sampling area) will not affect the measurement as long as the variation remain fairly consistent throughout the measurement volume.

5. ACKNOWLEDGEMENTS

This work was funded with grants from EPSRC/MRC, (GR/N14248/01) and in collaboration with Hitachi Advanced Research Laboratory, Japan.

6. REFERENCES

1.   J.Linford, S.Shalev, J.Bews, R.Brown, H.Schipper, Development of a tissue-equivalent phantom for diaphanography, *Medical Physics*, **13**(6): 869-875 (1986).
2.   A.M.De Grand, S.J.Lomnes, D.S.Lee, M.Pietrzykowski, S.Ohnishi, T.G.Morgan, A.Gogbashian, R.G.Laurence, J.V.Frangioni, Tissue-like phantoms for near-infrared fluorescence imaging system assessment and the training of surgeons, *Journal of Biomedical Optics*, **11**(1): 014007 (2006).
3.   W.D.D'Souza, E.L.Madsen, O.Unal, K.K.Vigen, G.R.Frank, B.R.Thomadsen, Tissue mimicking materials for a multi-imaging modality prostate phantom, *Medical Physics*, **28**(4): 688-700 (2001).
4.   M.Firbank and D.T.Delpy, A design for a stable and reproducible phantom for use in near infra-red imaging and spectroscopy, *Physics in Medicine and Biology*, **38**: 847-853 (1993).
5.   Y.Yamashita, A.Maki, H.Koizumi, Measurement system for noninvasive dynamic optical topography, *Journal of Biomedical Optics*, **4**(4): 414-417 (1999).
6.   M.Firbank, M.Oda, D.T.Delpy, An improved design for a stable and reproducible phantom material for use in near-infrared spectroscopy and imaging, *Physics in Medicine and Biology*, **40**: 955-961 (1995).
7.   F.E.W.Schmidt, M.E.Fry, E.Hillman, J.C.Hebden, D.T.Delpy, A 32-channel time-resolved instrument for medical optical tomography, *Review of Scientific Instruments*, **71**(1): 256-265 (2000).

NUMERICAL SIMULATION OF OXYGEN TRANSPORT IN CEREBRAL TISSUE

Toshihiro Kondo, Kazunori Oyama, Hidefumi Komatsu, and Toshihiko Sugiura*

Abstract: The physiological mechanism of coupling between neuronal activity, metabolism and cerebral blood flow (CBF) is not clarified enough. In this study, the authors have examined activity-dependent changes in oxygen partial pressure (pO_2) and CBF response in an arteriole by 2-dimensional numerical simulation with a mathematical model of O_2 transport from the arteriole to its surrounding tissue including an adjusting function of CBF. In the steady state of O_2 consumption, an area in the tissue where O_2 is supplied from the arteriole becomes smaller as O_2 consumption rate of the tissue increases, which is accompanied by increase of CBF. Therefore decrease of the O_2-supplied area gradually becomes stagnant. Unsteady responses of the local pO_2 and CBF were also examined. The response of pO_2 in the upstream area of the arteriole is monophasic increment corresponding to CBF response, whereas the response in the middle area is biphasic response showing an initial decrease followed by a positive peak. In the downstream area, advective flow holds decrement of the pO_2. Delay in CBF response to neuronal activity has also been found.

1. INTRODUCTION

NIRS (near-infrared spectroscopy) has received attention as a new method of brain function. It can measure neuronal activity-dependent changes in concentration of oxy-Hb and deoxy-Hb and has revealed fast O_2 consumption closely related to neuronal response in capillaries.[1] In recent NIRS data analysis, it is a general idea that increase of oxy-Hb corresponds to brain activation. However, importance of early increase of deoxy-Hb before the increase of oxy-Hb has been shown by experiments using animals[2-4] and humans.[5] Thus, the meaning of signals measured by NIRS still have unclear points. One

* *Department of Mechanical Engineering, Keio University, 3-14-1, Hiyoshi, Kouhoku-ku, Yokohama, Japan.
E-mail:sugiura@mech.keio.ac.jp

P. Liss et al. (eds.), *Oxygen Transport to Tissue XXX*, DOI 10.1007/978-0-387-85998-9_23,

of the reasons is that the physiological mechanism of coupling between neuronal activity, metabolism and CBF is not clarified enough.

By numerical simulation, the correlation between O_2 consumption and CBF in steady state was examined with Krogh's cylindrical model.[6,7] Steady-state O_2 transport from an arteriole and capillaries to tissue was simulated by using the structure of microvascular networks[8]. However, changes in O_2 consumption are usually accompanied by a transient change in CBF due to its adjusting function, which was also simulated, under the assumption that O_2 concentrations in the tissue are uniform in space [9].

In this study, as a first step to examine hemoglobin dynamics, considering pO_2 spatial distributions in the tissue, the authors have simulated transient O_2 transport from an arteriole to its surrounding tissue with an adjusting function of CBF. We discuss neuronal activity-dependent unsteady responses of local pO_2 and CBF caused by unsteady O_2 consumption.

2. MODELING AND FORMULATION

2.1. Analytical model

Figure 1 shows our analytical model for O_2 transport from an arteriole to its surrounding tissue. Here, for simplicity, we use 2-Dmodel in order to observe reacting qualitative change of pO_2 in space and in time. We assume pO_2 distributions in the tissue as well as in the arteriole. The length in the axial direction is L, the arteriole radius is r, and the distance between the center of the arteriole and the far end of the tissue in the radial direction is R_0.

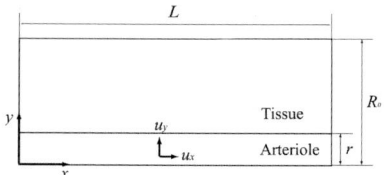

Figure 1. 2-D analytical model for O_2 transport

2.2. Governing equations

O_2 in blood vessels is transported by advective flow and diffusion. On the other hand, O_2 in tissue is transported only by diffusion and O_2 consumption occurs only in tissue. Governing equations for O_2 partial pressure $p(x,y,t)$ in the vessel and the tissue are shown below.

In the vessel

$$\alpha_1 \frac{\partial p}{\partial t} = \alpha_1 k_1 \left(\frac{\partial^2 p}{\partial x^2} + \frac{\partial^2 p}{\partial y^2} \right) - \alpha_1 \left(u_x \frac{\partial p}{\partial x} + u_y \frac{\partial p}{\partial y} \right) \tag{1}$$

In the tissue

$$\alpha_2 \frac{\partial p}{\partial t} = \alpha_2 k_2 \left(\frac{\partial^2 p}{\partial x^2} + \frac{\partial^2 p}{\partial y^2} \right) - f \tag{2}$$

where $f(x,y,t)$ denotes the O_2 consumption rate. $u_x(x,y,t)$ and $u_y(x,y,t)$ are the components of CBF velocity. k_1 and k_2 are diffusivity coefficients of O_2 in blood and in the tissue, respectively. α_1 and α_2 are solubility coefficients for O_2 in blood and in the tissue, respectively. It can be assumed that flux of O_2 from the vessel equals that into the tissue as the boundary condition at the vessel wall. O_2 flux at the tissue edges is zero. At the inlet of the vessel, O_2 concentration c_0 ($=\alpha_1 p_0$) keeps being injected and at outlet O_2 does not outflow by diffusion. The following equations show these boundary conditions.

In the vessel

$$c_0 u_x = \alpha_1 p(0,y,t) u_x - k_1 \alpha_1 \frac{\partial p(0,y,t)}{\partial x} \tag{3}$$

$$\alpha_1 \frac{\partial p(L,y,t)}{\partial x} = 0 \tag{4}$$

$$\frac{\partial p(x,0,t)}{\partial y} = 0 \tag{5}$$

In the tissue

$$\frac{\partial p(x,y,t)}{\partial x} = 0 \quad \left(at \quad x = 0, L \right) \tag{6}$$

$$\frac{\partial p(x,y,t)}{\partial y} = 0 \quad \left(at \quad y = R_0 \right) \tag{7}$$

Since blood is an incompressible liquid, its flow distribution in a vessel can be determined with the Navier-Stokes equations and the equation of continuity. Furthermore, the Reynolds number in an arteriole is so small that Hagen-Poiseuille's flow can hold. By assuming the velocity in the radial direction equals zero, the velocity of CBF in the arteriole can be expressed as below.

$$u_x(0, y) = \frac{\Delta P}{4\eta L} \left(r^2 - y^2 \right) \tag{8}$$

2.3. Adjusting function of CBF

CBF increases to compensate for an increase of energy demand caused by regional neural activity, which is called an adjusting function of CBF. To increase CBF, the radius of a blood vessel or the velocity of CBF has to be increased. In brain microcirculation, increase of CBF is caused by a vasodilating action of a vascular smooth muscle which covers an arteriole, and CO_2 is a dominant factor in this mechanism. To be concrete, decrease in pO_2 and increase in CO_2 partial pressure (pCO_2), thus caused by glucose metabolism in cell, causes increase in pH in the vessel wall and CBF increases. Changes in pO_2 equals changes in pCO_2, because O_2 consumption and CO_2 generation arise at the same time. In this study, the authors assume that pO_2 is a factor of vasodilating action and decrease in pO_2 at the vessel wall is proportional to increase in the vessel radius r, as follows.

$$r(p_w)=\begin{cases} r_{max} & (p_w \le p_l) \\ \dfrac{p_s - p_w}{p_s - p_l}(r_{max} - r_c) + r_c & (p_l < p_w \le p_s) \\ r_c & (p_s < p_w) \end{cases} \tag{9}$$

where p_w denotes the minimum value of pO_2 at the vessel wall. p_s and p_l are the threshold and the critical value of pO_2, respectively. r_c is the vessel radius at rest, and r_{max} is the maximum of the vessel radius.

The vessel radius is assumed as a function of pO_2 and hence time. This assumption leads to a moving boundary problem, which is tough to solve. In this study, instead of changing the vessel radius, we adjust the maximum velocity of the blood flow with keeping the vessel radius constant, so that it is equal to the maximum velocity for the changed radius.

2.4. Nondimensionalization

Here the steady vessel radius r_c is chosen as the characteristic length, the diffusion time in the vessel r_c^2/k_1 is chosen as the characteristic time, and c_0/α_1 is chosen as the characteristic pressure of O_2. Equations (1)-(2) can be transformed into nondimensional governing Eqs. (10)-(11).

In the vessel
$$\frac{\partial p^*}{\partial t^*} = \left(\frac{\partial^2 p^*}{\partial x^{*2}} + \frac{\partial^2 p^*}{\partial y^{*2}}\right) - Pe(1 - y^{*2})r^{*2}\frac{\partial p^*}{\partial x^*} \quad \left(Pe = \frac{\Delta P r_c^3}{4\eta L k_1}\right) \tag{10}$$

In the tissue
$$\frac{\partial p^*}{\partial t^*} = k\left(\frac{\partial^2 p^*}{\partial x^{*2}} + \frac{\partial^2 p^*}{\partial y^{*2}}\right) - \alpha f^* \quad \left(k = k_2/k_1, \alpha = \alpha_1/\alpha_2, f^* = r_c^2 f/k_1 c_0\right) \tag{11}$$

where variables are nondimensionalised by $x^* = x/r_c$, $y^* = y/r_c$, $t^* = k_1 t/r_c^2$, and $p^* = \alpha_1 p/c_0$. Nondimensional parameters are Pe, k, α, f^* and r^*. Pe is the Peclet number, which is the ratio of effect of diffusion to that of advective flow, k is the ratio of diffusion coefficients, α is the ratio of solubility coefficients for O_2, f^* is the nondimensional oxygen consumption rate in tissue, and r^* is the nondimensional radius, given by r/r_c.

3. NUMERICAL SIMULATION

3.1. Response in steady state

Table 1 shows physiological parameters used in this study. The Peclet number (Pe) is 133. Figure 2 shows distribution of pO_2 in the tissue and the arteriole at rest. The Peclet number is so large that most of the incoming O_2 is carried downstream by advective flow and the pO_2 level in the downstream is kept high. An area in tissue where O_2 is supplied from the arteriole depends on the oxygen consumption rate. The responses of the radial length of this area R and CBF Q for different O_2 consumption rate f are shown in Figure 3. Here Q is calculated by Eq. (12), as follows.

$$Q(r) = \frac{\pi \Delta P}{8\eta L} r^4 \tag{12}$$

where ΔP is a difference of the pressure. The radial length of this area at rest R_c is 0.135mm and CBF at rest Q_c is $6.28 \times 10^{-3} mm^3/s$ with $f=f_c$. R decreases with increase of f, though the rate of this decrease in R is gradually reduced. The result suggests that increase of CBF caused by increase of O_2 consumption rate causes decrease of pO_2 to be wiped out.

3.2. Response in unsteady state

Transient responses have also been examined by simulating the case in which neuronal activities last 1 second in 80 percent of the tissue, which causes the O_2 consumption rate to be 10 times as high as that at rest. Though this rate is larger than the actual value, we used this rate in order to see what qualitative reactions happen in early time if the O_2 consumption is larger. Figure 4 shows changes in pO_2 at four different points on the central axis of the vessel ($y=0$), while Figure 5 shows change in CBF with the gray part indicating the period of the neuronal activities. The response of pO_2 in the upstream area of the arteriole is monophasic increment corresponding to CBF response. The response in the middle area is biphasic response showing an initial decrease followed by a positive peak. The response largely decreases in the downstream area. Advective flow negates O_2 diffusion effect in the upstream area. Delay in CBF response from the neuronal activities equals the time passing until the pO_2 at the vessel wall decreases by diffusion beyond the threshold P_s. It can be found that CBF keeps increasing even after the end of the neuronal activities, which can result from continuance of the hypoxic condition in the downstream area.

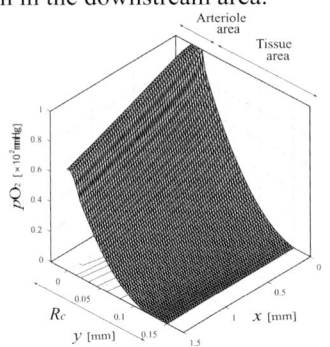

Figure 2. Distribution of pO_2 in tissue and arteriole at rest

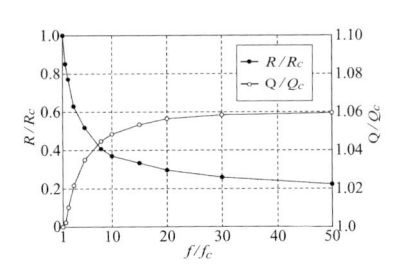

Figure 3. Radius of O_2-supplied area and steady blood flow as a function of O_2 consumption rate

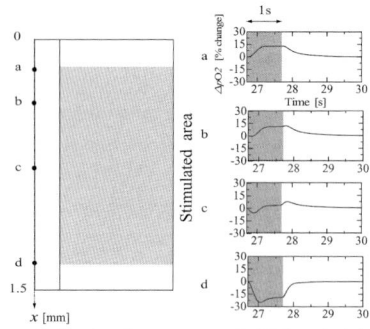

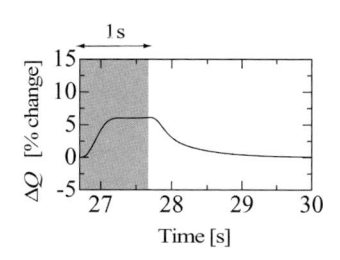

Figure 4. Regional responses of pO_2 in arteriole to 1 second stimulation ($f=10f_c$)

Figure 5. Response of CBF to stimulation ($f=10f_c$)

Table 1. Parameters employed in this model

Parameters		Values
r_c	resting arteriole radius	0.02mm
r_{max}	maximum of arteriole radius	0.022mm
R_0	outer radius of tissue	0.2mm
L	arteriole length	1.5mm
k_1	diffusivity of O_2 in blood	$1.5 \times 10^{-3} mm^2/s$
k_2	diffusivity of O_2 in tissue	$2.4 \times 10^{-3} mm^2/s$
α_1	solubility coefficient for O_2 in blood	$2.8 \times 10^{-3} mM/mmHg$
α_2	solubility coefficient for O_2 in tissue	$3.9 \times 10^{-3} mM/mmHg$
f_c	O_2 consumption rate at rest	$3.72 \times 10^{-2} mM/s$
p_0	pO_2 in inflow blood	90mmHg
p_s	threshold pO_2 in vessel reaction	50mmHg
p_l	critical pO_2 in vessel reaction	30mmHg
η	blood viscosity of arteriole	$0.04g/mm \cdot s$
ΔP	difference of the pressure	90mmHg

4. CONCLUSION

The authors have numerically simulated responses of pO_2 and CBF in steady and unsteady state of O_2 consumption with a mathematical model of O_2 transport from an arteriole to its surrounding tissue including an adjusting function of CBF.

In steady state, O_2 transport in the arteriole is highly affected by the effect of advective flow than that of diffusion and so pO_2 in the downstream area stays at a comparatively high level. An area in the tissue where O_2 is supplied from the arteriole decreases with increase of the O_2 consumption rate, while CBF increases at the same time and decrease of the O_2-supplied area is gradually lowered.

In unsteady state, pO_2 in the arteriole, which is affected by the O_2 consumption rate, depends on the location. It has also been observed that there is a delay on CBF response to neuronal activity. It is considered that that delay equals the time passing until the pO_2 at the vessel wall decreases by diffusion beyond the threshold of pO_2 in vessel reaction.

5.REFERENCES

1. T Kato, Principle and technique of NIRS-Imaging for human brain FORCE: fast-oxygen response in capillary event, International Congress Series, 1270, 85-90 (2004).
2. Fox PT, Raichle ME, Focal physiological uncoupling of cerebral blood flow and oxidative metabolism during somatosensory stimulation in human subjects, Proc Natl Acad Sci USA. 83 (4), 1140-4 (1986).
3. Buxton RB, Frank LR, A model for the coupling between cerebral blood flow and oxygen metabolism during neural stimulation, J Cereb Blood Flow Metab. 17 (1), 64-72 (1997).
4. Offenhauser N, Thomsen K, Caesar K, Lauritzen M, Activity-induced tissue oxygenation changes in rat cerebellar cortex: interplay of postsynaptic activation and blood flow, J Physiol. 565(pt1), 279-94 (2005).
5. T Akiyama, T Ohira, T Kawase and T,kato, TMS orientation for NIRS functional motor mapping, Brain Topogr, 19(1-2),1-9 (2006).
6. Hudetz AG, Mathematical model of oxygen transport in the cerebral cortex, Brain Res. 817 (1-2), 75-83 (1999).
7. Wang CH, Popel AS, Effect of red blood cell shape on oxygen transport in capillaries, Math Biosci. 116 (1), 89-110 (1993).
8. Tanishita, K, Masamoto, K, Negishi, T, Takizawa, N, Kobayashi, H, Oxygen transport in the microvessel network, In Organ Microcirculation, (ed, Ishii, H et al), Springer, pp. 13-20 (2004)
9. Y Zheng, J Martindale, D Jhonston, M Jones, J Berwick, and J Mayhew, A model of the hemodynamic responses and oxygen delivery to brain, NeuroImage 16, 617-637 (2002)

NON-INVASIVE ESTIMATION OF METABOLIC FLUX AND BLOOD FLOW IN WORKING MUSCLE: EFFECT OF BLOOD-TISSUE DISTRIBUTION

Nicola Lai[1,3], Gerald M. Saidel[1,3], Matthew Iorio[1], Marco E. Cabrera[1,2,3,*]

Abstract: Muscle oxygenation measurements by near infrared spectroscopy (NIRS) are frequently obtained in humans to make inferences about mechanisms of metabolic control of respiration in working skeletal muscle. However, these measurements have technical limitations that can mislead the evaluation of tissue processes. In particular, NIRS measurements of working muscle represent oxygenation of a mix of fibers with heterogeneous activation, perfusion and architecture. Specifically, the relative volume distribution of capillaries, small arteries, and venules may affect NIRS data. To determine the effect of spatial volume distribution of components of working muscle on oxygen utilization dynamics and blood flow changes, a mathematical model of oxygen transport and utilization was developed. The model includes blood volume distribution within skeletal muscle and accounts for convective, diffusive, and reactive processes of oxygen transport and metabolism in working muscle. Inputs to the model are arterial O_2 concentration, cardiac output and ATP demand. Model simulations were compared to exercise data from human subjects during a rest-to-work transition. Relationships between muscle oxygen consumption, blood flow, and the rate coefficient of capillary-tissue transport are analyzed. Blood volume distribution in muscle has noticeable effects on the optimal estimates of metabolic flux and blood flow in response to an exercise stimulus.

*Department of Biomedical Engineering[1] and Pediatrics[2], Center for Modeling Integrated Metabolism Systems and Rainbow Babies and Children's Hospital[3], Case Western Reserve University, Cleveland, OH 44106, USA

P. Liss et al. (eds.), *Oxygen Transport to Tissue XXX*, DOI 10.1007/978-0-387-85998-9_24,   **155**

1. INTRODUCTION

For analysis of regulation mechanisms of oxygen homeostasis during exercise, dynamic measurements of pulmonary oxygen uptake (VO_2) by indirect calorimetry and of muscle oxygenation by NIRS are performed simultaneously.[1,2,3] These measurements have intrinsic limitations that can cause inaccurate evaluation of underlying tissue processes.[4] Specifically, estimates of metabolic flux and blood flow in response to exercise depend on factors that are not precisely known such as the local microvascular volume distribution and the oxygenated hemoglobin and myoglobin concentrations. To evaluate the effects of extra-vascular volume and blood volume distribution during exercise, a mathematical model was applied to quantify how these factors affect oxygen transport and utilization.[5,6,7] The mathematical model presented here is used to examine the effect of local muscle blood-tissue volume distribution on the responses of muscle oxygen saturation during exercise. Together with NIRS data, optimal estimates can be obtained of maximal metabolic flux (V_{max}) and blood flow increase (ΔQ) during exercise.[7] This model accounts for spatial and temporal distribution of oxygen concentration in tissue and between arterioles, capillaries and venules. Model simulations of the dynamic response of muscle oxygenation are compared with NIRS measurements of StO_2 from the vastus lateralis muscle during bicycle exercise in human subjects.[6]

2. METHODS

Model Development. The metabolic response of skeletal muscle to an exercise stimulus can be described by transport and metabolic processes associated with Oxygen, ATP, and PCr.[6,7] The processes of oxygen transport and utilization in muscle can be modeled as spatially distributed in blood and tissue.[7] The total muscle volume, $V_{mus}=V_{bl}+V_{tis}$, consists of (artery, capillary and venous) blood volume (V_{bl}) and extra-vascular muscle cells of tissue (V_{tis}). Total oxygen concentration ($C^T_{O_2,x}$) in blood (x=b) and in tissue cells (x=c) within the muscle is the sum of the free ($C^F_{O_2,x}$) and bound oxygen ($C^B_{O_2,x}$) concentrations, which are related by local equilibrium.[7] Total oxygen concentrations in the capillary blood, and in muscle cells vary with time (t) and tissue location as indicated by the cumulative muscle volume (v) from the arterial input $v=0$ to the venous output $v=V_{mus}$:

$$\frac{\partial C^T_{O_2,b}}{\partial t} = -\frac{Q}{f_{cap}}\frac{\partial C^T_{O_2,b}}{\partial v} + D_b\frac{\partial^2 C^T_{O_2,b}}{\partial v^2} - \frac{PS}{f_{cap}}\left(C^F_{O_2,b} - C^F_{O_2,c}\right) \qquad 0 < v < V_{mus} \qquad (1)$$

The first term on the right side represents convective transport of oxygen in the direction of blood flow Q in which f_{cap} is the ratio of capillary blood volume to total muscle volume; the second term represents axial dispersion characterized by an effective dispersion coefficient D_b; the third term represents transport between capillary blood and extra-vascular tissue, which depends on the permeability-surface area, PS. The dynamic, spatial distribution of total oxygen concentration in muscle cells of extravascular tissue is

$$\frac{\partial C^T_{O_2,c}}{\partial t} = D_c\frac{\partial^2 C^T_{O_2,c}}{\partial v^2} + \frac{PS}{f_{tis}}\left(C^F_{O_2,b} - C^F_{O_2,c}\right) - \frac{uO_{2m}}{f_{tis}} \qquad 0 < v < V_{mus} \qquad (2)$$

where D_c is an effective dispersion coefficient in muscle tissue; f_{tis} is the ratio of extra-vascular muscle tissue volume to total muscle volume. The oxygen utilization rate per unit volume of total muscle volume is proportional to the oxidative phosphorylation flux, $uO_{2m}=f_{tis}\,\phi_{OxPhos}$.

The metabolic reaction processes that involve oxidative phosphorylation are associated with the concentration dynamics of ATP and PCr. These cellular concentrations depend implicitly on spatially distributed oxygen transport flux between blood and muscle cells. The dynamic mass balances of cellular concentrations are related to metabolic fluxes:

$$\frac{\partial C_{ATP}}{\partial t} = -\phi_{ATPase} + \beta\,\phi_{OxPhos} + \Delta\phi_{CK}\,; \qquad \frac{\partial C_{PCr}}{\partial t} = -\Delta\phi_{CK} \qquad (3,4)$$

where ϕ_{ATPase} is the ATP utilization flux, β is the stoichiometric coefficient that relates oxidative phosphorylation to ATP production, and $\Delta\phi_{CK}$ is the net forward flux of the creatine kinase reaction. The metabolic fluxes are functions of O_2, ADP, ATP, PCr, and Cr. However, the concentration pairs ATP-ADP and PCr-Cr are related by mass conservation of adenosine and creatine, respectively, whose total concentrations are constant: $C_A^T = C_{ADP} + C_{ATP}$ and $C_C^T = C_{Cr} + C_{PCr}$.

In response to a step increase in work rate from rest, the dynamic response of blood flow Q at exercise is assumed to be exponential:[7]

$$Q(t) = Q_0 + \Delta Q\left[1 - exp\left[-(t-t_0)/\tau_Q\right]\right] \qquad (t > t_0) \qquad (5)$$

where Q_0 is the steady-state flow before exercise, $\Delta Q = Q_{SS} - Q_0$ is the increase in blood flow between steady states, τ_Q is the time constant of muscle blood flow, and t_0 is the time at the onset of exercise. Also, associated with blood flow increase in response to exercise, there is an effective increase in the rate coefficient of capillary-tissue transport:

$$PS(t) = PS_0 + \Delta PS\left[1 - exp\left[-(Q-Q_0)/q_C\right]\right] \qquad (Q > Q_0) \qquad (6)$$

where PS_0 is the steady-state rate coefficient before exercise, ΔPS is the increase in the rate coefficient, Q_0 is the steady-state blood flow, and q_C is an arbitrary parameter.[6]

For comparison of oxygen responses to exercise from model simulations and experimental data, the simulated muscle oxygen saturation, StO_2, is intended to reflect the muscle volume distribution in the region of the NIRS measurement (Fig. 1a)

$$StO_2 = \frac{f_{bl}C_{HbO_2} + C_{MbO_2}f_{tis}}{C_{Hb}f_{bl} + C_{Mb}f_{tis}} \qquad (7)$$

where f_{bl}, f_{tis} are the volume fractions of blood and extravascular (cells and interstitial fluid) in muscle ($f_{bl}+f_{tis}=1$); C_{Hb}, C_{Mb} are the total concentrations of hemoglobin in blood and myoglobin in tissue. The total composite oxyhemoglobin in blood and oxymyoglobin concentrations in tissue is defined as:

$$C_{HbO_2} = C_{art}^B\,W_{art} + <C_{cap}^B>W_{cap} + C_{ven}^B\,W_{ven}, \quad C_{MbO_2} =<C_{tis}^B>W_{tis} \qquad (8,9)$$

where C_{art}^B and C_{ven}^B are oxyhemoglobin concentrations in red blood cells of arteries and veins; $<C_{cap}^B>$ and $<C_{tis}^B>$ are volume-averaged capillary and tissue concentrations. Since these concentrations are defined with respect to different domain volumes, volume weighting fractions ($W_{art}, W_{ven}, W_{cap}, W_{tis}$) are introduced. W_{tis} is the ratio of myocyte volume to total tissue volume. In domain X (art,cap,ven) the fraction W_X is the ratio of blood cell volume to total blood volume so that $W_{art}+W_{ven}+W_{cap}=Hct$. Furthermore, we can scale these ratios so that $\omega_{ar} + \omega_{cap} + \omega_{ven} =1$, where $\omega_X=W_X/Hct$.

Model simulation and parameter estimation. Numerical solution of the partial differential equations is based on the method of lines. The spatial derivatives are discretized so that the model consists of a set of ordinary differential-difference equations. Parameter values of ΔQ and V_{max} were estimated by comparing simulations of StO_2 evaluated by Eq. 7 with experimental measurement of StO_2 using a least-squares (Φ) objective function.[6] The data, obtained from an experiment with a normal human subject, are typical of responses measured by NIRS for a step response to moderate intensity exercise from a warm-up steady state.[5,6] The transport processes (Eqs. 1 and 2) depend on the muscle blood composition and distribution (f_{bl}, ω_{cap}) in which $f_{cap} = f_{bl}\omega_{cap}$ and ($f_{tis}=1-f_{bl}$). In these equations that relate to the whole muscle, the same parameter values for muscle composition (Table 1) were used for all simulations. NIRS measurements for StO_2 depend on blood composition and distribution parameters (f_{bl}, ω_{art}, ω_{cap}, ω_{ven}) might have a range of values associated with the working muscle (e.g., Table 1). Simulations with these parameter values affect the oxygen saturation model (Eq. 7). Values of all other model parameters were the same as those determined previously.[5,6,7]

Table 1. Blood composition and distribution in the whole working muscle volume ($V_m=10L$) and within a local (NIRS) region.[3] Note that $\omega_{art}=1-\omega_{cap}-\omega_{ven}$.

Muscle Region	$f_{bl}(\%)$	$\omega_{cap}(\%)$	$\omega_{ven}(\%)$
Whole muscle	7	15	75
Local (NIRS)	2-15	5-20	65-90

3. RESULTS

Muscle blood volume distribution and ΔPS affect optimal estimates of V_{max} and ΔQ from model simulations of StO_2 data (Fig. 1a). This was quantified by the minimum objective function (Φ), i.e., sum of square differences, which varied by less than 1% for the range of parameter values in Table 1. The StO_2 response depends on the relative contributions of oxyhemoglobin C_{HbO2} and oxymyoglobin C_{MbO2}. In these simulations, the absolute concentration of C_{HbO2} at steady-state warm up was 4 times greater than that of C_{MbO2}. From warm up to exercise steady states, the C_{HbO2} decreases 7 times more than C_{MbO2}. Blood composition and distribution (f_{bl}, ω_{cap}, ω_{ven}) in working muscle affected the optimal estimates of V_{max} (Fig. 1b) and ΔQ. When f_{bl} increased from 2% to the reference level of 7%, the estimated values of V_{max} decreased by 14% while ΔQ increased by 25%. When f_{bl} increased from 7 to 15% the estimated values of V_{max} decreased by 7% while ΔQ increased by 20%. Over the entire range of blood distribution parameters (ω_{ven}, ω_{ven}) in Table 1, estimates of V_{max} and ΔQ changed less than 4%.

4. DISCUSSION

Non-invasive, simultaneous measurements[1,2,3] of pulmonary oxygen uptake by indirect calorimetry and of muscle oxygenation by NIRS provide the database for analyzing regulation mechanisms of oxygen homeostasis during exercise. These measurements combined with biophysically based mathematical models can be used to estimate parameters that are essential in quantifying various factors that affect oxygen transport and utilization during exercise.[5,6,7]

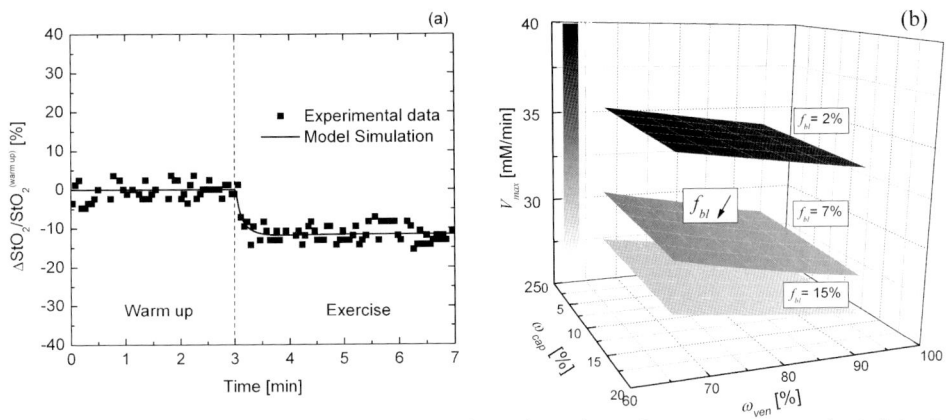

Figure 1. (a) Dynamic response of StO_2 to a step change in work rate from warm up to moderate intensity exercise. Simulations of StO_2 are obtained assuming the NIRS muscle region volume composition to be the same as that of the whole muscle; (b) Effect of blood volume composition (f_{bl}) and its distribution within the muscle region investigated by NIRS (ω_{cap}, ω_{ven}) on parameter estimation of V_{max} (ΔPS=600 L L^{-1} min^{-1}).

The optimal estimation of key muscle parameters requires comparison of model simulations to measured NIRS signals, which depend on blood volume fraction and its distribution in the muscle region. Several confounding aspects must be addressed to interpret the NIRS signal including (a) Hb absorbance within the blood (arterial, capillary, and venous) is a summed signal; (b) Oxygenation concentration in blood changes continuously with location with the capillary bed from arterial to venous blood. (c) The concentrations of bound oxygen (C_{HbO2}, C_{MbO2}) have different values and responses to exercise depending on the rates of oxygen transport (convection and diffusion) and utilization.

Analysis of several confounding aspects of the data requires a mathematical model that describes the spatial variation of oxygen transport in muscle in which (arterial, capillary, venous) blood and tissue are distinguished and the related dynamics of cellular metabolism. Furthermore, the model must be able to simulate the dynamic response to exercise, which requires incorporation of muscle blood flow dynamics, $Q(t)$, and the flow dependence of capillary-tissue O_2 transport, $PS(Q)$. Finally, the model must simulate the dependence of StO_2 on the relative concentrations of (C_{HbO2}, C_{MbO2}) and volume fraction of blood in muscle (f_{bl}). With a variety of model parameter values, simulations of StO_2 agree closely with data (e.g., Fig. 1a). To determine how well V_{max} and ΔQ can be estimated from such sparse data with many unknown parameters, a sensitivity analysis was performed in which the effect of blood tissue composition on these estimates was investigated. The responses of C_{HbO2} and C_{MbO2} to exercise depend on blood composition and distribution (f_{bl}, ω_{cap}, ω_{ven}). These affect StO_2 and estimation of V_{max} (Fig.1b) and ΔQ. Associated with an increased f_{bl}, (a) StO_2 responded faster, (b) estimated V_{max} was lower, and (c) estimated ΔQ was higher. Changes of f_{bl} in the physiological range (7-15%) had a small effect on the estimation of V_{max} (7%) and a larger effect on ΔQ (20%).

The accuracy of model predictions is limited by uncertainties of blood tissue composition, blood flow, and permeability-surface area dynamics. Thus, future studies are needed to improve the estimation of structural and functional parameters affecting oxygen transport and metabolism during exercise. These studies require advances in NIRS technology in combination with optimal experimental design and mathematical modeling.[8,9] More detailed analysis of NIRS signals could lead to better absolute values of C_{HbO2} and C_{MbO2}[10,11] and blood volume distribution in working muscle. Although the mathematical model applied in this study incorporated the effects of blood volume distribution, the next level of investigation requires a model that distinguishes active and inactive muscle as well as heterogeneities of blood flow and O_2 utilization.[12] These developments are important for quantitative understanding of pathological alterations of oxygen transport and metabolism in subjects with diabetes and vascular disease.[13,14]

5. ACKNOWLEDGEMENTS

Supported by a grant (P50 GM-66309) from the National Institute of General Medical Sciences (NIH).

6. REFERENCES

1. M. Ferrari, L. Mottola, and V. Quaresima, Principles, Techniques, and limitations of Near infrared Spectroscopy, *Can. J. Appl. Physiol.* **29**(4): 463-487 (2004)
2. B. Grassi, S. Pogliaghi, S. Rampichini, V. Quaresima, M. Ferrari, Muscle oxygenation and pulmonary gas exchange kinetics during cycling exercise on-transitions in humans, *J Appl Physiol* 95:149-58 (2003).
3. R. Boushel, H. Langberg, J. Olesen, J. Gonzales-Alonzo, J. Bülow, M. Kjær, Monitoring tissue oxygen availability with near infrared spectroscopy (NIRS) in health and disease, *Scand J Med Sci Sports*, 11:213-222 (2001).
4. Pittman, R.N., Oxygen supply to contracting skeletal muscle at the microcirculatory level: diffusion vs. convection, *Acta Physiol. Scand.* **168**, 593-602 (2000).
5. N. Lai, R.K. Dash, M.M. Nasca, G.M. Saidel, M.E. Cabrera, Relating pulmonary oxygen uptake to muscle oxygen consumption at exercise onset: in vivo and in silico studies, *Eur J Appl Physiol* 97: 380-394 (2006).
6. N. Lai, M. Camesasca, G.M. Saidel, R.K. Dash, M.E.Cabrera. Linking pulmonary oxygen uptake, muscle oxygen utilization and cellular metabolism during exercise, *Ann Biomed Eng* 35(6): 956-968 (2007).
7. N. Lai, G.M. Saidel, B. Grassi, L.B. Gladden, M.E. Cabrera, Model of oxygen transport and metabolism predicts effect of hyperoxia on canine muscle oxygen uptake dynamics, *J Appl Physiol*, 103:1366-1378 (2007).
8. S. Fantini, M.A Franceschini-Fantini, J.S. Maler, S.A. Walker, B. Barbieri, and E. Gratton, Frequency-domain multichannel optical detector for non-invasive tissue spectroscopy and oximetry, *Opt Eng* 34: 32-42, (1995).
9. P. Rolfe, In vivo near-infrared spectroscopy. *Ann. Rev. Biomed Eng* 2: 715-754, (2000).
10. D.M. Mancini, L. Bolinger, H. Li, K. Kendrick, B. Chance, and J.R. Wilson. Validation of near-infrared spectroscopy in humans, *J Appl Physiol* 77: 2740–2747 (1994).
11. T.K. Tran, N. Sailasuta, U. Kreutzer, R. Hurd, Y. Chung, P. Mole´, S. Kuno, and T. Jue, Comparative analysis of NMR and NIRS measurements of intracellular PO_2 in human skeletal muscle, *Am J Physiol Regul Integr Comp Physiol* 276: R1682–R1690 (1999).
12. H. Miura, K. McCully, S. Nioka, and B. Chance, Relationship between muscle architectural features and oxygenation status determined by near infrared device, *Eur J Appl Physiol* 91: 273-278 (2004).
13. T. Hamaoka, K. Mccully, V. Quaresima, K. Yamamoto and B. Chance, Near-infrared spectroscopy/imaging for monitoring muscle oxygenation and oxidative metabolism in healthy and diseased humans, *Journal of Biomedical Optics* 12(6), 062105, 2007.
14. M. Wolf, M. Ferrari and V. Quaresima, Progress of near-infrared spectroscopy and topography for brain and muscle clinical applications, *Journal of Biomedical Optics* 12(6), 062104, 2007.

OXYGEN EXTRACTION INDEX MEASURED BY NEAR INFRARED SPECTROSCOPY - A PARAMETER FOR MONITORING TISSUE OXYGENATION?

Oskar Baenziger, Matthias Keel, Hans-Ulrich Bucher, and Martin Wolf[*]

Abstract: The objective was to assess the ability of near infrared spectrophotometry (NIRS) to detect changes in tissue oxygenation due to alterations in oxygen delivery. Ten hemodynamically stable preterm neonates with a median gestational age of 27.9 weeks (range 25.1-31.2), a median birth weight of 840g (range 690-1310), and a postnatal age of 29 days (range 2-45) were included in this prospective trial. Tissue oxygenation of the lower leg was measured by NIRS and the oxygen extraction index (OEI) was calculated prior and after a transfusion of 10-20ml/kg body weight packed red blood cells. The OEI decreased from 0.31 (range 0.13-0.39) to 0.24 (range 0.12-0.36, p<0.005). This decrease correlated positively with the weight matched amount of packed red cell transfusion (r^2=0.40, p<0.05) and with the increase in hematocrit (r^2=0.58, p<0.005). The OEI obtained by a NIRS may allow to monitor changes in tissue oxygenation.

1. INTRODUCTION

Many critically ill patients have abnormal tissue perfusion and oxygenation, which may lead to multiple organ failure, a complication with high morbidity and mortality[1]. Despite constant advances in monitoring and treatment of critically ill neonates the parameters most widely used for hemodynamic monitoring (arterial blood pressure, heart rate, cardiac output, diuresis, arterial and venous blood pH and lactate) are rather insensitive for the assessment of abnormal tissue perfusion[2,3]. Changes in these parameters appear usually late, when tissue ischemia and hypoxia are already significant and systemic lactic acidosis is present. Moreover in neonates an invasive monitoring is difficult and potentially harmful. Therefore a noninvasive sensitive parameter for early detection of insufficient tissue perfusion is needed. Near infrared spectrophotometry (NIRS) is a widely accepted non-invasive method to assess hemodynamic state and

[*] Clinic for Neonatology, University Hospital Zurich email: martin.wolf@usz.ch

P. Liss et al. (eds.), *Oxygen Transport to Tissue XXX*, DOI 10.1007/978-0-387-85998-9_25,
© Springer Science+Business Media, LLC 2009

oxygenation of the neonatal brain[4-6]. Neonatal anemia is another condition that leads to a decreased tissue oxygenation due to a reduced oxygen transport capacity. The peripheral oxygenation in anemic preterm infants was studied by calculating a peripheral fractional oxygen extraction using NIRS and the venous occlusion technique[7]. A correlation of the peripheral fractional oxygen extraction with hemoglobin concentration was demonstrated. As the venous occlusion technique to measure peripheral venous saturation is a more complicated method that still needs postprocessing, a NIRS algorithm was developed that calculates tissue oxygenation (TO). This TO should theoretically correlate with the oxygenation of the tissue under the optodes. However, only little is known about the ability of this algorithm to detect changes in tissue oxygenation due to altered oxygen delivery in the newborn.

The aim of our study was to assess the ability of a new near infrared spectrophotometry algorithm to detect changes in tissue oxygenation due to alterations in organ oxygen delivery.

2.   METHODS

Patients: Ten clinically and hemodynamically stable preterm neonates receiving a red blood cell transfusion for clinical reasons were enrolled. None received inotropic support. All were on theophyllin for neonatal apneas and breathing spontaneously, four on nasal CPAP, of which three needed additional O_2 (FiO$_2$ 25-30%). During the measurement the infants were asleep and great care was given to avoid movement artifacts. Exclusion criteria: Congenital heart defects, patent ductus arteriosus, necroticing enterocolitis or venous catheters at the leg of the measurement.

NIRS method: The tissue oxygenation (TO) was calculated from the attenuation of near infrared light measured through the lower leg in transmission by the Cerebral Redox Monitor 2001 NIRS from Critikon, Newport, UK. The calculations are based on the algorithm and follow similar assumptions as previously described[6]. The output intensity (initial intensity) of the four laser diodes (776nm, 819nm, 871nm, 909nm) at the tip of the emitting fiber was measured. The drift of the temperature stabilized laser diodes was tested and was negligible. Thus the attenuation of the lower leg was determined with emitted intensity and intensity received after the light had penetrated the tissue (corresponding to the signals at detector 1 and detector 2[6]). Although the errors made in the assumptions concerning the differential pathlength factor, light scattering and coupling may considerably affect the absolute values of oxy- and deoxyhemoglobin (O_2Hb and HHb) concentration, numerical tests showed that the proportion of O_2Hb and HHb was virtually unaffected and thus TO=O_2Hb/(O_2Hb+HHb) can be measured. Heart rate, arterial oxygen saturation (SaO$_2$) and blood pressure were recorded simultaneously. The oxygen extraction index (OEI) was calculated OEI=(SaO$_2$-TO)/SaO$_2$.

Protocol: The probes were fixed to the lower leg by a cohesive elastic bandage with the gastrocnemius muscle between the emitter and detector. The interoptode distance was measured. The SaO$_2$ probe was placed distally on the same leg. The data recording (sample time 0.56s) was started 10 minutes prior to the red blood cell transfusion. Infants were transfused either at hematocrit<35% if on nasal CPAP or <30% if they had recurrent apneas or <20% for chronic asymptomatic anemia. The amount and rate (30ml/kg/hour) of transfused red blood cells was determined by the physician in charge according to the standard clinical transfusion protocol. The data recording was continued for 60 minutes after the end of the transfusion. A capillary hematocrit was measured 60 minutes after the end of the transfusion.

Test-Retest variability (TRV): To determine TRV of the OEI we performed 10 repeated measurements on the resting adductor pollicis muscle between the thumb and the index finger in 10 healthy adult subjects. The optodes were completely removed and replaced before restarting for each measurement. The interoptode distance was 1.8-2.5cm and similar to the one (1.5-2.1cm) in the 10 preterm infants.

Statistical methods: The TO was averaged over 10min, first prior to the transfusion and second 60minutes after the end of the transfusion. Data between the patients were analyzed by nonparametric tests (Wilcoxon); the results are expressed as median and range. The analyses were performed by the SPSS Version 7.5.1 (SPSS Inc., Chicago, IL)

The study was approved by the local ethics committee, and informed consent was obtained from the parents prior to the study.

3. RESULTS

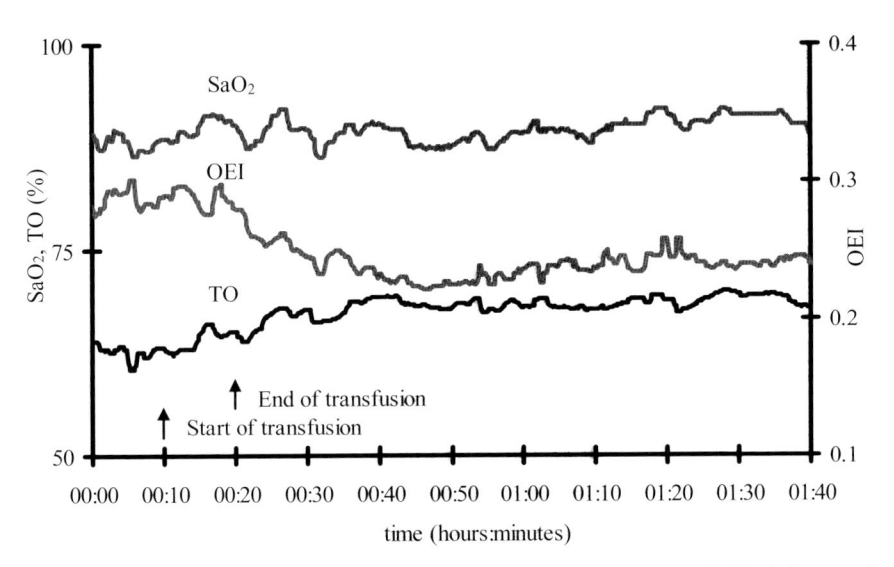

Figure 1. Tissue oxygenation (TO), O_2 extraction index (OEI) and arterial saturation (SaO_2) during transfusion in patient 6.

The 10 preterm neonates had a median gestational age of 27.9 weeks (range 25.1-31.2), birth weight of 840g (690-1310), and age of 29 days (2-45). The transfused amount of packed red cells was 16.0ml (9-25) or 13.3ml/kg body weight (10-18). The mean hematocrit of the transfused packed red cells was 0.64±0.1%. The mean arterial blood pressure was 37mmHg (29-58) prior to transfusion and 39mmHg (31-62, $p<0.05$) after. The heart rate was 169 beats/min (150-182) prior and 168.5 beats/min (142-180, $p>0.05$) after. The hematocrit increased from 28% (25-38) to 38.1% (33-47%, $p<0.005$).

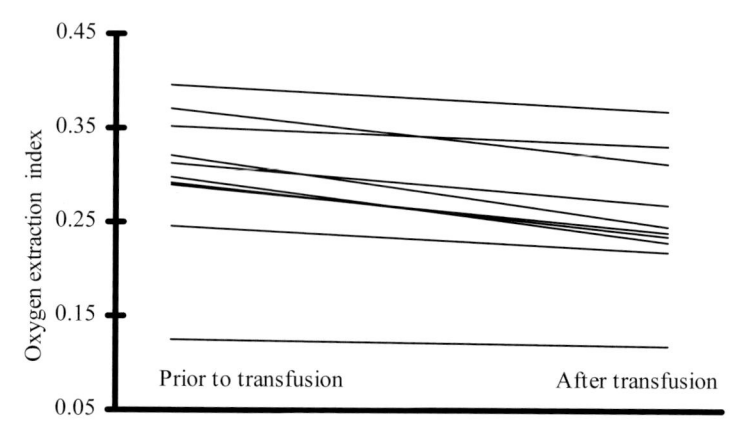

Figure 2. Changes in oxygen extraction index prior to and after packed red cell transfusion.

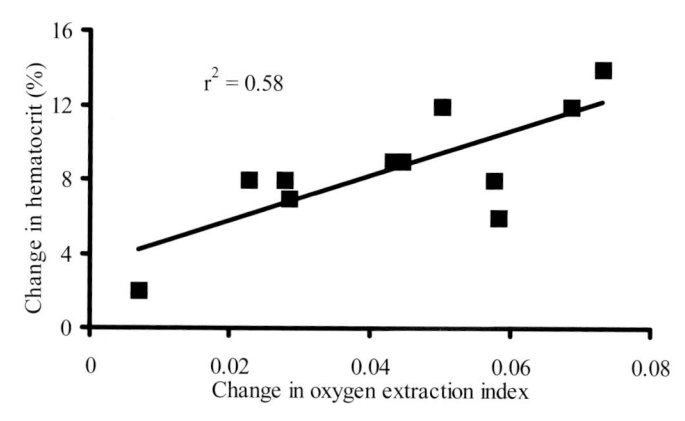

Figure 3. Correlation of change in hematocrit and oxygen extraction index.

Table 1. Tissue oxygenation (TO), arterial saturation (SaO_2), oxygen extraction index (OEI) for the ten infants prior and after transfusion in mean ± standard deviation.

Patient	Prior to transfusion			After transfusion		
	TO %	SaO_2 %	OEI	TO %	SaO2 %	OEI
1	65.3±0.9	94.9±1.0	0.32±0.01	68.0±1.4	93.0±1.2	0.27±0.01
2	71.9±0.7	95.4±0.8	0.25±0.01	72.0±0.6	92.1±1.3	0.22±0.01
3	58.6±0.5	97.0±1.3	0.40±0.01	61.8±1.0	97.8±2.0	0.37±0.00
4	66.7±0.9	94.9±4.7	0.30±0.03	73.9±0.5	95.8±1.2	0.23±0.01
5	62.8±0.9	99.7±1.2	0.37±0.01	64.6±0.6	94.0±2.0	0.31±0.01
6	62.7±1.0	88.3±1.3	0.29±0.01	69.5±0.6	91.3±0.6	0.24±0.01
7	61.9±2.0	91.0±2.6	0.32±0.01	66.4±1.1	88.1±1.5	0.25±0.01
8	60.7±1.3	93.8±1.6	0.35±0.01	63.5±2.4	94.8±3.1	0.33±0.01
9	64.8±1.4	91.6±1.7	0.29±0.01	68.3±1.0	89.2±2.1	0.23±0.01
10	80.2±0.2	91.8±0.4	0.13±0.00	80.9±0.2	91.8±0.5	0.12±0.01

A typical example of the measurements of TO, OEI and SaO_2 for infant 6 is shown in Fig. 1. The median TO increased from 63.8% (58.6-80.2) to 68.1% (61.8-80.9, p<0.005), the SaO_2 did not significantly change from 94.3% (88.3-99.6) to 92.5% (88.1-97.8, p>0.05) and OEI (Fig. 2) decreased from 0.31 (0.13-0.40) to 0.24 (range 0.12-0.37, p<0.005). The decrease in OEI correlated positively with the weight matched amount of packed red cell transfusion (r^2=0.40, p<0.05) and with the increase in hematocrit (r^2=0.58, p<0.005) (Fig. 3). The values for the individual patients are given in table 1.

The mean TO for healthy adult subjects was 61.7±3.7% with a TRV of 2.2±0.9%. The median difference in TO between the first and the second measurement in the infants was 4.3% and its standard deviation was 0.97% and 0.93%, respectively.

4. DISCUSSION

Anemia is the inadequacy of hemoglobin-determined O_2 availability to meet tissue requirements, and the peripheral fractional O_2 extraction correlates with the degree of anemia[7]. Currently hematocrit or total hemoglobin are used to evaluate if an infant needs to be transfused, but other markers that better reflect the adequacy of the oxygen delivery to the tissues are needed[7,8]. Other conditions such as a low cardiac output also result in an inadequate oxygen delivery to the tissue and are difficult to quantify with noninvasive methods. The knowledge of the tissue oxygenation in different organs would help to assess the patient's clinical state, and his need for medical interventions such as volume replacement, inotropic support, or red blood cell transfusions. To evaluate the ability of the TO to detect changes in tissue oxygenation we studied anemic preterm infants requiring red blood cell transfusion for clinical reasons.

Before applying the method to the preterm infant the TRV of the method was tested and was considerably lower than the measured change in TO in infants. Another important factor, the system noise in steady state of the clinical measurement reflected by the standard deviation of TO averaged over 10 minutes for every individual patient was considerably smaller than the expected changes in TO.

The red blood cell transfusion led to an increase in TO and a decrease in OEI detected by the NIRS, which represents an increase in tissue oxygenation due to increased organ oxygen delivery. Despite the large interindividual range of TO and OEI, the increase of TO and the decrease in OEI was found in all our patients and positively correlated to the amount of transfused red cells and to the increase in hematocrit. The decrease in OEI either due to an increase of oxygen carrying capacity or an increase in cardiac output secondary to the volume load of the blood transfusion[9,10], clearly indicates the desired effect, i.e. a better tissue oxygenation was achieved by the transfusion.

It is important to realize that the algorithm calculates a TO that is not identical to the venous saturation[6], but represents a mixture of arterial and venous oxygen saturation according to the proportion of their compartments. Therefore changes in TO are more important than its actual value. Provided O_2 consumption remains constant, O_2 extraction will decrease when O_2 delivery is increased. The OEI is only valid under the assumption that TO represents only hemoglobin within the vascular compartment, and that the influence of myoglobin on the estimation of TO can be neglected. The contribution of myoglobin to the NIRS data although controversial is probably in the range of 10%[13]. However the correlation of TO and OEI with the hematocrit is not affected by myoglobin and confirms the results obtained by the venous occlusion method[7]. A very important factor affecting the fractional O_2 extraction[7] was the concentration of fetal hemoglobin (HbF) due to its higher O_2 affinity. A higher concentration of HbF results in a higher fractional O_2 extraction. Since we did not measure HbF concentration, its influence on our results is unknown. By transfusing predominantly adult hemoglobin the relative concentration of HbF may have dropped and the OEI may also be reduced after transfusion.

Movement artifacts are an important problem, particularly in muscular tissue. In sedated or intubated patients these artifacts pose no problems. However in conscious children with muscle contractions the results are difficult to interpret. By comparing the pulse rate measured by the pulse oximeter at the same limb to the heart rate from the electrocardiogram we were able to identify and exclude movement artifacts

5. CONCLUSION

Although our results are still preliminary we conclude: The OEI obtained by NIRS may allow to detect changes in tissue oxygenation in preterm infants.

6. REFERENCES

1. R. C. Bone, C. L. Sprung, and W. J. Sibbald, Definitions for sepsis and organ failure, *Crit Care Med* **20**, 724-726 (1992).
2. R. D. Bland, W. C. Shoemaker, E. Abraham, and J. C. Cobo, Hemodynamic and oxygen transport patterns in surviving and nonsurviving postoperative patients, *Crit Care Med* **13**, 85-90 (1985).
3. M. E. Astiz, and E. C. Rackow, Assessing perfusion failure during circulatory shock, *Crit Care Clin* **9**, 299-312 (1993).
4. M. Wolf, N. Brun, G. Greisen, M. Keel, K. von Siebenthal, and H. Bucher, Optimising the methodology of calculating the cerebral blood flow of newborn infants from near infrared spectrophotometry data, *Med Biol Eng Comput* **34**, 221-226 (1996).
5. M. Wolf, H. U. Bucher, V. Dietz, M. Keel, K. von Siebenthal, and G. Duc, How to evaluate slow oxygenation changes to estimate absolute cerebral haemoglobin concentration by near infrared spectrophotometry, *Adv Exp Med Biol* **411**, 495-501 (1997).
6. M. Wolf, P. Evans, H. U. Bucher, V. Dietz, M. Keel, R. Strebel, and K. von Siebenthal, The measurement of absolute cerebral haemoglobin concentration in adults and neonates, *Adv Exp Med Biol* **428**, 219-227 (1998).
7. S. P. Wardle, C. W. Yoxall, E. Crawley, and A. M. Weindling, Peripheral oxygenation and anemia in preterm babies, *Pediatr Res* **44**, 125-31 (1998).
8. D. C: Alverson, V. H. Isken, and R. S. Cohen, Effect of booster blood transfusions on oxygen utilization in infants with bronchopulmonary dysplasia, *J Pediatr* **113**, 722-6 (1988).
9. C. A. Finch, and C. Lenfant, Oxygen transport in man, *N Engl J Med* **186**, 407-415 (1972).
10. L. G. R. Delima, and J. E. Wynards, Oxygen transport, *Can J Anaesth* **40**, R81-86 (1993).
11. M. Ferrari, T. Binzoni, and V. Quaresima, Oxidative metabolism in muscle, *Philos Trans R Soc Lond B Biol Sci* **352**, 677-83 (1997).

NON-INVASIVE MEASUREMENT OF THE SUPERFICIAL CORTICAL OXYGEN PARTIAL PRESSURE

Chris Woertgen, Jan Warnat, Alexander Brawanski, and Georg Liebsch[1]

Abstract: We present a non invasive fluorescein based method to measure and visualise the partial oxygen pressure of the rat cortex in a 2D picture. We studied 10 Wistar rats. A trepanation was done over the hemisphere and the dura was opened. A PMMA cylinder with a calibrated optical membrane was fixed over the surface of the brain. The CCD camera with the light source is placed over the cylinder. This allows the generation of two-dimensional maps of the pO_2 pressure. Using the white light picture we defined regions of interest (ROI) in an artery, vein, parenchyma and an overall ROI. For every ROI a mean emission value was calculated. We increased, stepwise, the FiO_2 from 30% up to 100%. Thereafter we established ventilation with an FiO_2 of 30% and induced a stepwise hypo- and hyperventilation. The ROI`s showed significantly different pO_2 values. The apO_2 showed a good correlation to the pO_2 in the ROIs. This new set up seems to give reliable absolute pO_2 values of the brain surface. This method seems to be able for the first time to give a non invasive pO_2 map of the brain surface reflecting oxygenation and ventilation effects.

1. INTRODUCTION

Oxygen delivery to the brain is a key point in the treatment of patients with intracranial pathologies. In the clinical setting, direct measurement of oxygen partial pressure of the brain is only possible by insertion of electrodes into the brain, for example by a Clarke type electrode (Licox, IntegraNeuroScience, San Diego, CA). Besides the brain penetration, the disadvantages of this method are the position of the electrode in the white matter and the possibility of false high measurements if the electrode is in the neighbourhood of a vessel. Other indirect methods of oxygen delivery measurements, like

[1] Chris Woertgen, Jan Warnat, Alexander Brawanski, Dept. of Neurosurgery, University of Regensburg, 93042 Regensburg, Germany; Gregor Liebsch, Biocam GmbH, 93053 Regensburg, Friedensstr. 30, Germany

P. Liss et al. (eds.), *Oxygen Transport to Tissue XXX*, DOI 10.1007/978-0-387-85998-9_26, **167**
© Springer Science+Business Media, LLC 2009

NIRS, jugular bulb spectroscopy, xenon CCT, PET and MRI measurements, are not accepted widely or are very costly and give no online data to monitor patients on the intensive care unit.

In the following we present a technique for the direct measurement of the partial oxygen pressure of the cortical surface. The measurement is based on the detection of oxygen-dependent quenching of the sensor luminescence [1,2] and has been evaluated for the measurement of the oxygenation of melanoma on skin surfaces. [3] In this study, the device is placed directly on the cortex. It consists of a planar opto-chemical sensor foil, a light conductor, a pulsed LED-based light source and a CCD camera. This setup allows time-resolved two-dimensional luminescence lifetime imaging of the sensor and the calculation of two dimensional maps of partial oxygen pressure distribution over the cortex under the sensor.[1,2,4] Normal camera images of the measured area can be obtained simultaneously.

We studied the feasibility of the method and tried to determine major topics for further technical development towards a laboratory and clinical tool.

2. MATERIAL AND METHDS

We studied 10 Wistar rats (Charles River) with a mean body weight of 376 ±33g, which were sedated, endotracheally intubated and ventilated with a gas mixture of isoflurane, oxygen and nitrous oxide. The isoflurane was set to 1.2 – 1.5%. The oxygen fraction was varied between 5% and 98.5% throughout the experiment. For the preparation and initial phase of the experiment gas fractions were 1.2 – 1.5% isoflurane, ~30% O_2 and ~70% NO. The left femoral artery was catheterized for arterial blood pressure measurement and blood gas analysis. After cannulation of the artery, the head was fixed in a stereotactic frame. Arterial blood pressure and temperature were monitored during the experiment. The blood gases were checked after every experimental step looking for arterial pO_2, pCO_2, Hb and haematocrit. Then a right-sided trepanation was done over the total hemisphere. The dura was opened without injuring the brain surface. A PMMA cylinder with the optical membrane (diameter 12 mm) was placed and fixed over the surface of the brain, having contact to the tissue with minimal pressure. The CCD camera (SensiMOD, PCO, Kehlheim, Germany) with the light source was placed over the PMMA cylinder and focused on the tissue. Then a white light colour picture was projected onto a TV screen. The gated CCD camera and a pulsed LED excitation light (405 nm) source were used for time-resolved monitoring of the emission intensity (645 nm). Either the fluorescence picture or the colour picture could be used to identify anatomical structures such as veins. The principle of this method is oxygen-dependent quenching of phosphorescence. The sensor, a platinum(II)–octaethyl–porphyrin, is immobilized in a transparent membrane. This allows the non-invasive generation of two-dimensional maps of surface pO_2 following excitation by an LED array (see also Fig. 1). The sensor membrane was calibrated in vitro to calculate the oxygen pressure values. The measurement itself takes 200ms and after every measurement the fluorescence picture is stored on a personal computer using special software for data evaluation. In this setting data evaluation was carried out later after finishing the animal experiments.

For every pO_2 measurement we defined four regions of interest (ROI). Using the colour picture we defined a ROI in an artery, vein, parenchyma and an overall ROI. The ROIs were chosen out of the same region. The ROIs of the artery, vein and parenchyma consisted of 100 pixels, the overall ROI of 2500 pixels. For every ROI a mean emission value was calculated and due to the calibration of the optical membrane a partial oxygen pressure for the ROI was calculated. The formula is : pO_2 (mmHg) = ((0.95/ ((1.225/R) − 0.05)) − 1) / 0.0615. R is the measured light intensity value.

After surgery and installation of the measurement setup, a steady state with respect to blood pressure and temperature as well as FiO_2 at 30% was established. After this, we checked the blood gases and made a first pO_2 measurement. We then increased the FiO_2 stepwise by 5-10% up to 100%. After every increase of the FiO_2 a steady state was established (5 minutes) and then we looked at the blood gases and a brain tissue pO_2 measurement was done. After this measurement we established ventilation with an FiO_2 of 30% and induced a stepwise hypo- and hyperventilation to manipulate the pCO_2 and during hyperventilation increased the FiO_2 to 100%. Finally, in stable animals we decreased the FiO_2 stepwise to 10%. After every change of ventilation, blood gases and brain surface pO_2 measurements were done. After finishing the measurements the animals were sacrificed.

After calculation of the correlation between the values, we formed groups with distinct arterial oxygen (apO_2) and carbon dioxide partial pressure ($apCO_2$). For the apO_2 we defined groups with apO_2 below 80mmHg, between 80 and 120 mmHg and above 120 mmHg. For the $apCO_2$ we defined a hyperventilation group with $apCO_2$ below 30 mmHg, a normo-ventilation group with $apCO_2$ between 30 and 40 mmHg and a hypoventilation group with $apCO_2$ values above 40 mmHg.

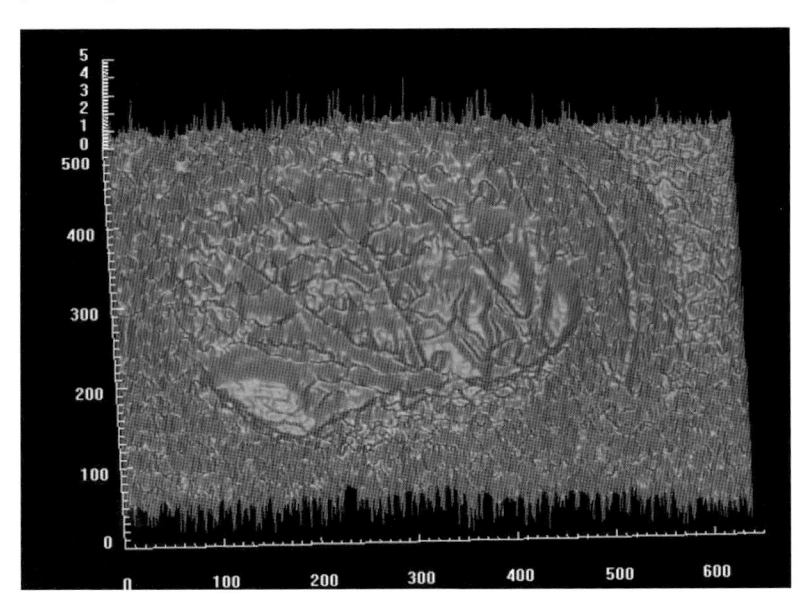

Figure 1: Three-dimensional presentation of the oxygen map (apO_2 of 320 mmHg).

For statistical analysis we used Sigma Stat version 2.0 (SPSS Inc.). For the difference between groups we used Kruskal-Wallis One Way Analysis of Variance on Ranks, for the correlation between numerical values we used the Spearman Rank Order Correlation. For the calculation of differences between groups we also used the Chi square test. To isolate a group or groups that differ from the others we used a pairwise multiple comparison procedure and the Tukey Test. The significance level was defined at $p < 0.05$.

All experiments were approved by the regional authorities according to the German animal care regulations which are in accordance with the international guidelines for animal care and use in scientific experiments (AZ 54-2531.1-30/05).

3. RESULTS

The ROIs showed significantly different pO_2 values. The pO_2 in the artery was 20.3, in the vein 17.1, in the parenchyma 9.1 and in the overall ROI 14 (mmHg, mean, $p<0.001$, n=203-213). Each group differed significantly from each other ($p < 0.05$). See Table 1.

Table 1: Partial oxygen pressure at different regions of interest and different ventilation conditions.

	pO_2 ROI artery	pO_2 ROI vein	pO_2 ROI parenchyma	pO_2 ROI overall
Count	212	203	213	212
pO_2 mmHg (mean)	20.3	17.1	9.1	14
SD	10.7	8.2	7.5	8.1
min, max	-0.7 / 68.5	-0.8 / 40.2	-0.8 / 38.5	-0.9 / 51.5
pO_2 ROI artery		$p < 0.05*$	$p< 0.05*$	$p < 0.05*$
pO_2 ROI vein			$p < 0.05*$	$p < 0.05*$
pO_2 ROI parenchyma				$p < 0.05*$
FiO$_2$ cc, p-value	0.65;p<0.0001	0.53;p<0.0001	0.4;p<0.0001	0.52;p<0.0001
apO$_2$ cc, p-value	0.72;p<0.0001	0.57;p<0.0001	0.45;p<0.0001	0.58;p<0.0001
apCO$_2$ cc, p-value	0.09;p=0.18	0.15;p<0.05	0.08;p=0.22	0.14;p<0.05
apO$_2$ < 80	8.3 (1.3)**	8.1 (1)**	3.7 (1.1)****	5.6 (1) **
apO$_2$ 80 – 120	18.7 (1.5)	15.7 (1.2)	8.4 (1.2)	12.5 (1.1)
apO$_2$ > 120	25.6 (1)	20.3 (0.7)	11.5 (0.8)	17.4 (0.7)
apCO$_2$ < 30	16.3 (1.7) n.s.	12.9 (1.3)***	6.9 (1.3) n.s.	10.7 (1.3) n.s.
apCO$_2$ 30 – 40	16.6 (1.1)	14.3 (0.8)	7.7 (0.9)	10.9 (0.9)
apCO$_2$ > 40	19.8 (0.9)	17 (0.7)	8.9 (0.7)	13.9 (0.7)

apO$_2$: arterial oxygen partial pressure; apCO$_2$: arterial CO$_2$ partial pressure;
pO$_2$: partial oxygen pressure; ROI: region of interest, pO$_2$ values given in mmHg (mean);
min: minimum, max: maximum; cc: correlation coefficient;
n.s.: not significant between groups; SEM standard error of mean is given in parenthesis
* $p < 0.05$: significant difference between groups; ** $p < 0.003$ between apO$_2$ group each;
*** $p < 0.03$ between apCO$_2$ < 30 versus (vs.) > 40
**** $p < 0.001$ between apO$_2$ < 80 vs. > 120 and p = 0.008 between < 80 vs. 80-120

The FiO$_2$ was significantly correlated to the pO$_2$ in the ROIs. The best correlation was found in the artery ($r = 0.65$). Nearly similar r–values showed the correlation of the FiO$_2$ to the ROI vein and overall ($r = 0.53$ and $r = 0.52$), the lowest r-value was calculated looking for the ROI parenchyma (r = 0.4). The arterial pO_2 (apO_2) showed a good correlation to the pO_2 in the ROIs of the cortex. The best correlation factor showed the apO_2 to the artery r = 0.72, followed by the overall and vein ROI (r = 0.58 and r = 0.57, respectively). The lowest correlation between apO_2 and the cortex showed the ROI parenchyma with r = 0.45. The correlation between the $apCO_2$ and the pO_2 of the cortex was rather low ranging from r = 0.08 to r = 0.15. See also Table 1.

Taking all measurements into account we compared groups with different apO_2 according to our definitions. Here we saw significantly different pO_2 in the ROI`s of the cortex. There was a significant difference in the $pO_2$ of the ROI`s between the groups with apO_2 < 80 mmHg, apO_2 between 80 – 120 and above 120 mmHg (8.3, 18.7 and 25.6 mmHg, mean, p < 0.003). Also, the pO_2 in the vein and overall ROI differed significantly between the apO_2 groups (p < 0.003), see Table 1. In the ROI parenchyma the mean pO_2 of the apO_2 group 80 – 120 did not differ significantly from the apO_2 group > 120, the remaining mean pO_2 values differed significantly, see Table 1.

With regard to the $apCO_2$ groups, we only found a significant difference between pO_2 values in the $apCO_2$ group < 30 and > 40 mmHg (ROI vein 12.9 and 17 mmHg, mean, p < 0.03). The remaining groups did not differ significantly, see Table 1.

4. DISCUSSION

Up to now, partial oxygen pressure measurement in human brain has only been possible by insertion of probes into the tissue.[5-9] The influence of this tissue injury on the quality of data remains unclear. A further point of uncertainty is the influence of the sensor position in relation to vessels, e.g. arteries.[5,7] The technique introduced here has the advantage of the non-invasiveness of the measurement and, in addition, the region of interest can be chosen by anatomical structures. Another different point is the location of the measurement.[7] The established electrodes like Licox, Neurotrend and Raumedic are positioned into the white matter at a depth of 2-3cm below the dura.[7] The presented technique registers the oxygen partial pressure on the surface of the brain, especially the grey matter. The two locations have different anatomical structures and represent different tasks.[7,9] The best location for oxygen measurement is still under discussion [7], but we are convinced that measurement on the surface has some advantages, for example, to conduct measurements on different sites of the cortical surface (vein, artery, parenchyma) and at various distances from the microvessels.

To test the general feasibility of our technique we manipulated the FiO_2 and therefore the apO_2. Using the Spearman rank test we saw a good connection of the FiO_2 and the apO_2 with the pO_2 values in the different ROIs. The best correlation was seen between the apO_2 and the pO_2 ROI artery (r: 0.72). The correlation between the pO_2 of the parenchyma was in both the lowest (r: 0.4 and 0.45, Tab.1). The apO_2 values correlated better than the FiO_2 to the pO_2 in the ROIs. Menzel et al. using a Licox $ptiO_2$ probe in pigs, also found a correlation of r = 0.67 between the apO_2 and the brain tissue pO_2 after manipulation of the FiO_2.[10] In concordance with other authors we saw that this new technique had a good connection between the pO_2 in the different ROI and the apO_2. Comparing the different oxygenation groups we found significant different pO_2 values in all ROI`s except for the ROI parenchyma which was not significantly different for the groups of apO_2 > 120 and apO_2 80 – 120 (Table.1).

Vovenko for the first time conducted measurements in the brain surface in rats combining the techniques of intravital microscopy and pO_2 measurements by needle oxygen microelectrodes (diameter 1μm).[8] He saw a decrease along the arterial microvasculature in order of branching from 81.2 mmHg SD 6.2 to 61.5 mmHg SD 12. The mean pO_2 in the capillaries was 57.9 mmHg SD 10.6, in the venules 38.2 mmHg SD 12.3. The arterial pO_2 in this setting was stable between 90 and 120 mmHg, the pCO_2 was also stable at 40 mmHg. Due to our experimental setup we had a large range of arterial pO_2 and pCO_2 and saw distinct lower pO_2 values on the cortex. In the ROI artery we saw a pO_2 of 20.3 mmHg SD 10.7, in the vein a mean value of 17.1 mmHg SD 8.2. A reason for the absolute difference of the pO_2 values could be that Vovenko and colleagues measured directly on the vessels and our sensor foil measures at a distinct distance from the structure, at least separated by the pia mater and the subarachnoid space.[8] The pO_2 values in the paper of Vovenko at a distance of 20 μm from the vessel are in the same range as our measurements. [8,11] Also Duling et al. saw a transmural PO_2 gradient across the wall of large arterioles, in their study the averaged gradient was 0.9 mmHg / μm.[12]

The $apCO_2$ showed only a weak correlation to the pO_2 in the ROI`s. A change in the $apCO_2$ had only an influence on the ROI vein regarding the values below 30mmHg and above 40mmHg CO_2. There are no comparable data in the literature.

As arterial pO_2 rises, the cortical measures rise, too. The correlation coefficients and p-values show clearly that these are closely coupled parameters. In other words, changes in paO_2 result in highly significant changes in cortex measures indicating the good reactivity of the measurement system. Changes in arterial pCO_2 had no effect on cortical pO_2 regardless of likely changes of CBF. Interestingly, changes in vascular diameter were noticeable (data not shown). It must be said that our findings merely reflect normal physiological findings due to the invasive approach in this feasibility study. However, exactly these questions such as CO_2 reactivity, effects of CBF and CPP, influence of anaesthetics, cortex oxygenation under pathophysiological conditions, the possibility of investigating arteriovenous gradients and so forth may be addressed in a unique and very detailed fashion using this measurement principle.

5. CONCLUSION

To our knowledge, this is the first report of nearly real-time time acquisition of two-dimensional cortical pO_2 maps. The spatial and temporal resolution of the technique is already good, and can be improved by further technical development. Single measurements are very well reproducible. Changes in cortical oxygen partial pressure can be detected immediately and appear to be reasonable which is shown by the strong correlation with the systemic arterial pO_2. Both the cortical oxygenation and the corresponding normal view of the measured area can be displayed and analyzed. Consequently specific measurements over visual detectable structures like arterial and venous vessels are available. The system is semi-invasive, because an access to the cortex surface is necessary, but lesions of the brain by probe insertion are avoided. This opens a wide field for applications as a clinical monitoring tool for patients with TBI or SAH and for laboratory investigations.

Our data are clearly preliminary. Further development of the setup and evaluation studies in the lab and in the clinical setting have to be done.

6. REFERENCES

1. G. Liebsch, I. Klimant, C. Krause, OS. Wolfbeis, Fluorescent imaging of pH with optical sensors using timedomain dual lifetime referencing, Anal.Chem **73**, 354-4363 (2001).
2. G. Liebsch, I. Klimant, B. Frank, G. Holst, OS. Wolfbeis, Luminescence Lifetime Imaging of Oxygen, pH, and Carbon Dioxide Distribution Using Optical Sensors, Applied Spectroscopy **54**, 548-59 (2000).
3. P. Babilas, G. Liebsch, V. Schacht, I. Klimant, OS. Wolfbeis RM. Szeimies, C. Abels, In vivo phosphorescence imaging of pO_2 using planar oxygen sensors. Microcirculation **12**, 477-87 (2005).
4. G. Holst, O. Kohls, I. Klimant, B. König, M. Kühl, T. Richter, A modular luminescence lifetime imagingsystem for mapping oxygen distribution in biological samples. Sensors and Actuators **51**,163-70 (1998).
5. A.J. Johnston, L.A. Steiner, A.K. Gupta, D.K. Menon, Cerebral oxygen vasoreactivity and cerebral oxygenreactivity, British Journal of Anaesthesia **90**, 774-786, (2003).
6. D.W. Lübbers, H. Baumgärtl, Heterogeneities and profiles of oxygen pressure in brain and kidney as examples of the pO_2 distribution in the living tissue, Kidney International, **51**, 372-380, (1997).
7. J.C. Rose, T.A. Neill, J.C. Hemphil III, Continuous monitoring of the microcirculation in neurocritical care: an update on brain tissue oxygenation, Current Opinion in Critical Care **12**, 97-102, (2006).
8. E. Vovenko, Distribution of oxygen tension on the surface of arterioles, capillaries and venules of brain cortex and in tissue in normoxia: an experimental study on rats, Pflügers Arch-Eur J Physiol **437**, 617-623, (1999).
9. M.J. Purves, The Physiology of the Cerebral Circulation, (Cambridge University Press, 1972).
10. M. Menzel, E.M.R. Doppenberg, A. Zauner, J. Soukoup, D. Henze, T. Clausen, A. Rieger, R. Bullock, J. Radke, Oxygen reactivity in patients after severe head injury – a prognostic tool ? Zentralblatt für Neurochirurgie **61**, 181 -187, (2000).
11. K.P. Ivanov, I.B. Sokolova, E.P. Vovenko, Oxygen transport in the rat brain cortex at normobaric hyperoxia. Eur J Appl Physiol **80**, 582-587, (1999).
12. B.R. Duling, W. Kushinsky, M. Wahl, Measurements of the perivascular PO_2 in the vicinity of the pial vessels of the cat. Pflügers Arch **383**, 29-34, (1979).

ONE SENSOR FITS ALL – A NEW APPROACH IN MONITORING BRAIN PHYSIOLOGY

H. Doll , N. Davies, S.L.E Douglas, F. Kipfmueller, M. Maegele, R. Pauly, G. Woebker, A.N. Obeid, and H. Truebel[*].

Abstract: Oxygen plays a pivotal role as a nutrient to the brain. Monitoring partial pressure of oxygen (ptO_2) has been shown to correlate with outcome after brain injury if certain tissue-ptO_2-goals can be achieved. Oxford Optronix has recently developed a new fiber-optic based sensor (MPBS) with a large tissue sampling volume and long-term stability up to 10 days. Direct comparison of the MPBS sensor with the Licox™ system was performed using an in-vitro and in-vivo model. No statistically significant differences between the MPBS and the Licox™ sensor in different settings were found. The response times to a sudden drop in ptO_2 was faster for the MPBS than for the Licox™ probes (time of 80% signal change; 65±11 vs 110±14s; p<0.05).

1. INTRODUCTION

Traumatic brain injury (TBI) is the leading cause of death and disability in young adults[1] in industrialized countries. The annual incidence rate reported in Germany is 7.3 per 100,000 with an overall mortality rate of 45.8%[2]. TBI undoubtedly represents a highly relevant medical and socioeconomic burden for modern society[3,4]. In a multi-injury pattern, TBI is crucial in respect to survival and functional outcome[5]. If the sequelae (e.g. arterial hypotension, brain swelling) are neither prevented nor treated then brain damage will be exacerbated with a concomitant increase in morbidity and mortality[6]. Major improvements in patient outcome can be achieved by rapid resuscitation and closely monitoring changes in brain physiology in a specialized trauma facility[7,8]. Since the primary insult has already occurred, the secondary damage needs to be urgently addressed[9,10]. Current treatment guidelines for patients with TBI are focused on the normalization of intracranial pressure (ICP) and cerebral perfusion pressure (CPP). The control of ICP and thereby CPP is aimed to provide nutrients to the diseased brain

[*] H. Doll, F. Kipfmueller, G. Woebker, H. Trubel (corresponding author), Department of Pediatrics, HELIOS Klinikum Wuppertal, Witten/Herdecke University, Wuppertal, Germany, 42283. M. Maegele, Department of Traumatology and Orthopedic Surgery, Witten/Herdecke University, Collogne, Germany, 51058. N. Davies, S.L.E. Douglas, R. Pauly, A. Obeid, Oxford Optronics, Oxford, UK, OX14 4SA

P. Liss et al. (eds.), *Oxygen Transport to Tissue XXX*, DOI 10.1007/978-0-387-85998-9_27,
© Springer Science+Business Media, LLC 2009

and optimize outcome of the penumbra[11]. Amongst the nutrients in the brain, oxygen plays a pivotal role in providing oxidative metabolism[12]. Since the CPP is a global hemodynamic measurement and local blood flow is under the influence of mechanisms different from global hemodynamics, CPP does not provide enough information of the local situation[13]. Measuring local blood flow (=doppler blood flow unit (DBF)) therefore appears to be a desirable parameter for the intensivist. Currently, monitoring devices either measure pressure (e.g. Codman Microsensor, (Codman, Johnson & Johnson Company, Raynham, MA, USA)) or brain tissue partial pressure of oxygen (ptO_2) and temperature (e.g. Licox™, Integra Lifesciences, Plainsboro, NJ, USA). Current guidelines[11] focus on the optimization of ICP and CPP but studies have shown that management of tissue partial pressure of oxygen (ptO_2)[14-16], which might be low (=inadequat supply) or normal/high (=severe depression with reduced O_2-consumption), and temperature (T)[17,18] in patients with TBI can improve outcome. A monitoring device that records all four variables (ptO_2, T, ICP, DBF) with one sensor therefore seems highly desirable.

Oxford Optronix Ltd (Oxford, UK) manufactures a device that can measure ptO_2 (based on luminescense quenching), T, and laser Doppler based cerebral blood flow for experimental use in animals. Millar Inc. (Houston, USA) manufactures a device (VPN10) for measuring ICP based on micro-electronic mechanical technology. By combining these different technologies, a prototype, multi-parametric brain sensor (MPBS) has been produced that provides continuous measurements of ptO_2, temperature, and ICP (which allows the calculation of CPP). Aspects of this new technology have been evaluated *in vitro* (part 1) and *in vivo* (part 2).

2. MATERIALS AND METHODS

2.1. *In Vitro* Test

The in vitro test-rig comprised a standard filter housing with a membrane supporting glass frit and rubber bungs fitted with copper gas tubing. The housing was filled with (viscosified) water polyvinyl pyrrolidone (PVP K-90 10% wt/wt in water; ISP Europe) in order to damp convection currents in the liquid layer. To evaluate the effect of spatial averaging over an artificially created oxygen gradient, two types of ptO_2 sensors were compared; a new fibre-optic-based ptO_2 sensor with an extended oxygen measuring area (MPBS, Oxford Optronix Ltd, Oxford U.K.)) and the Licox™ catheter ptO_2 sensor (Licox™, Integra Lifesciences, Plainsboro, NJ, USA). In both cases, the stated sampling area of each sensor was $13mm^2$. The temperature was controlled by placing the setup in a water bath. Output data from the various monitors were recorded simultaneously and in real-time using a multi-channel data acquisition system (PowerLab 8SP, ADInstruments, Australia) running under Chart™ for Windows™ (Ver.5.02, ADInstruments, Australia). Readings were taken at five 1mm stepped positions in the induced oxygen gradient. Simultaneous measurements from the two types of sensors (with attached reference probes) were obtained in the oxygen gradient column. The lateral separation was 2.4mm. Each data point representing at least two minutes of averaged readings taken at a sampling rate of 2Hz. Additional measurements consisted of a reduction of the ptO_2 in the test system by replacing oxygen with nitrogen.

2.2. In-Vivo Test

2.2.1. Animals

Two adult Large-White pigs of approx. 40 kg bodyweight were kept in standard housing with free access to food and drink ad libitum. The study was conducted in accordance with German national guidelines for the committee of animal care.

2.2.2. Surgical Preparation

After initial sedation, intravenous anaesthesia was maintained using Ketamine and Midazolam. Muscle relaxation was achieved with i.v. Pancuronim as required. Central venous and arterial lines were positioned for the continuous infusion of saline fluids (5ml/kg/h) and the measurement of arterial blood pressure. Oxygen saturation (using pulse oximetry), ECG and cardiac output were also monitored continuously.

Following instrumentation, two burr holes (1cm lateral and 1cm posterior to the Bregma) were drilled in the skull. The aperture in the dura was minimized in order to reduce cerebrospinal fluid leakage and environmental air-inflow. The MPBS was inserted 2.5cm sub-durally on the left side of the animal. In addition to the MPBS, a laser Doppler flowmeter probe (MSF100XP, Oxford Optronix Ltd, Oxford, UK) was inserted on the ipsilateral side to the same depth (2.5cm). On the contra-lateral side, a reference ICP sensor (Codman Microsensor, Raynham, MA, USA) was located together with a reference ptO_2 oxygen electrode (Licox™, Integra Neurosciences, Plainsboro, NJ, USA) at the same depth (2.5cm) and through a single burr hole. All sensors were positioned securely at a fixed depth in the parenchyma so as to avoid traumatic brain injury or hemorrhage through uncontrolled movements. Continuous data for temperature (MPBS), DBF, ptO2 (Licox™ and MPBS) and ICP (MPBS only) were recorded with Chart v4.2.4 software (PowerLab 8SP, ADInstruments, Australia) and exported for further analysis into MS Excel. It was not possible to continuously record the output from the reference ICP Microsensor (Codman) as there was no analogue voltage output available from the monitor. Instead, reference ICP values were noted from the built-in display at regular intervals. Similarly, readouts from a cardiac output monitor (PiCCO, Pulsion Medical Systems, Germany) were also noted in the Chart v4.2.4 software.

After a period of stabilization, the respiratory rate for each pig was slowly decreased from 26 BPM to 10 BPM. The trend of the $paCO_2$ was followed with regular measurements of blood gases. Oxygen saturation was not affected by this type of hypoventilatory CO_2 challenge since the FiO_2 was maintained at 60%. Hypoventilatory CO_2 challenge was halted when the arterial blood gas pH decreased below 7.1.

3. RESULTS

In part 1, an *in vitro* comparison between the MPBS (ptO_2-) sensor and a conventional (Clark electrode based) Licox™ sensor was carried out in a controlled model of an oxygen gradient. In a direct comparison of the two different ptO_2 sensors during a stepwise up/down movement in the ptO_2 gradient model no significant difference was observed. In respect to the response time, both sensor systems display an increase in ptO_2 when the sensors were simultaneously lowered towards the oxygen inlet. Conversely, a drop in the ptO_2 was observed when the sensors are withdrawn.

A significant difference in the response times of the two sensors was seen when the sensors were exposed to a sudden change in partial pressure of O_2. The Licox™ system gave a response time drop of approximately 110s reduction in ptO_2 between 90% and 10% of the starting level whereas the response time of the MPBS was around 80s.

Sensor comparison was carried out in an animal model with additional recording of DBF and ICP as well as temperature. During the CO_2 challenge (ventilation rate was dropped to 10 BPM), the doppler blood flow rises (see Fig. 1) which is followed by an increase in ICP (not shown). From the same initial starting level, both ptO_2 sensors show an increase during the CO_2 challenge. Interestingly, the ptO_2 from the MPBS levels out at a higher ptO_2 value than that obtained from the Licox™ (see also Fig. 2).

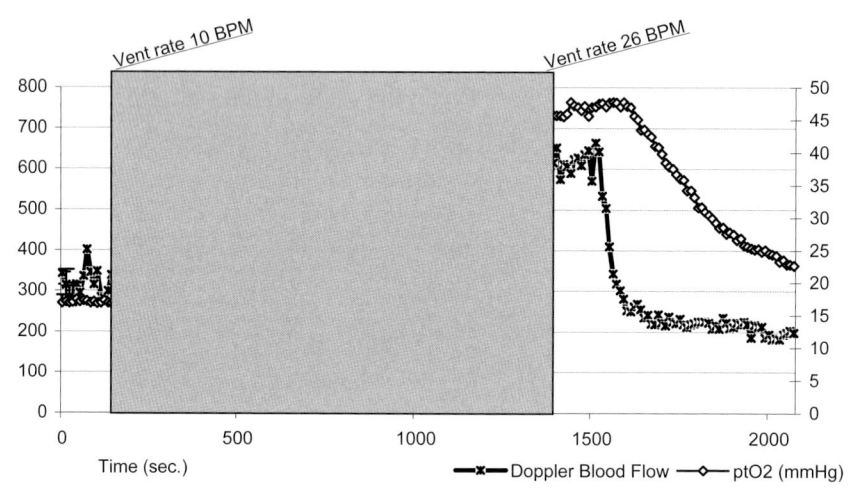

Figure 1. Original sensor registration of animal #2 during hypoventilatory CO_2 challenge (gray shaded area). Doppler Blood Flow (Oxylite), and ptO_2 (MPBS - ptO_2). Vent rate 10 BPM: Ventalatory rate 10 breath/min.

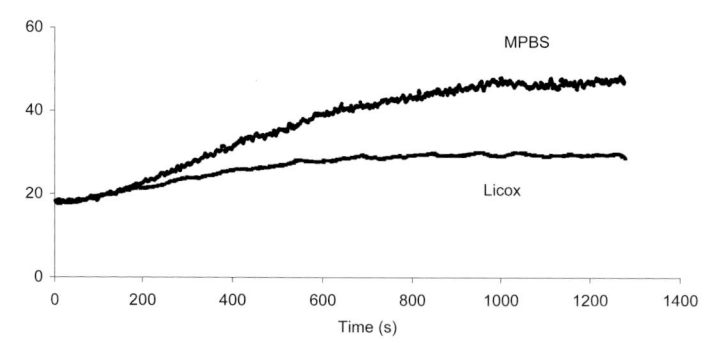

Figure 2. During the CO_2 challenge the MPBS measures higher values than the Licox™ system (data from animal #2 which corresponds with the registration shown in Figure 1).

Figure 2 shows data from the CO_2 challenge in which the values were imported into Excel and displayed versus time. It can be seen that, under hyperoxic conditions (in this case a supernormal arterial oxygen tension at an inspired oxygen fraction of $FiO_2=0.6$; arterial $ptO_2 > 300mmHg$) during a period of increased blood flow to the brain, the ptO_2 measurement from the MPBS, again stabilises at a higher, maximal, ptO_2 at the end of the CO_2 challenge than that obtained from the Licox™.

4. DISCUSSION

These experiments present in vitro and in vivo data for a new, multi parameter brain sensor (MPBS) that is able to continuously measure ptO_2, ICP and T (with DBF added) from a single, implantable device. The *in vitro* data demonstrate that the large area sensing capability of the MPBS is equivalent to the 'gold standard' Licox™ sensor. Large area sensing is considered important because such devices are less spatially sensitive to localized fluctuations in ptO_2. In the *in vivo* pilot study, the prototype MPBS was found to be easy-to-handle, introduce and fix in the brain via a standard burrhole.

The results from this study demonstrate the ability of the MPBS to measure ptO_2 and ICP in a manner similar to the reference (Licox™) electrode and reference (Codman) ICP sensor. Of interest, is the fact that the in-vivo response time for a change in ptO_2 is slightly faster for the MPBS. In a clinical setting, this may be of importance e.g. in providing an opportunity for early intervention of brain-saving treatment modalities.

In vivo data were predominately obtained during artificially induced (increased) pCO_2 challenges. Since cerebral vascular tone is dependent on levels of pCO_2, cerebral blood flow (CBF=DBF) and hence ptO_2 also increases when pCO_2 increases. These increases in ptO_2 were detected with both the MPBS as well as the reference Licox™. However, the Licox™ showed significantly lower values when compared with the MPBS at the end of several maneuvers. The fact that during the hypoventilatory CO_2 challenge the ptO_2 values from the MPBS (Fig. 2) are higher than those obtained from the Licox™ reference can be explained by several factors: For example, the MPBS sensor may have been closer to a large blood vessel (particularly as they were inserted on different sides and the position was not anatomically confirmed). Alternatively, the well-known 'stirring effect' of the polarographic-based Licox™ electrode may also contribute to 'false negatives' by its consumption of oxygen in the area of interest.

Of particular note, however, is the fact that both systems showed almost identical starting (baseline) values. Future experiments should be directed at confirming this finding and comparing the two systems in the hypoxic range (e.g. ischemic stroke model). According to the literature, measurements (and treatment) of low levels of ptO_2 may help to guide therapy for better patient outcome.

For DBF measurements, no reference system was available, but the DBF changed as physiologically expected during the CO_2 challenge leading to concomitant increases in ptO_2 and ICP as reported above. This result is also expected due to the fact that the intact brain autoregulates blood flow in the vasculature. Since this physiological feedback mechanism is often depressed or lost in the injured brain, direct measurements of CBF in the diseased area are of vital importance.

In summary, these first in-vivo tests of the MPBS demonstrate that the MPBS records similar baseline in-vivo measurements of ptO_2. In the hyperoxic range, the MPBS readings for ptO_2 were consistently higher than those measured with the Licox™ system. Temperature and DBF effects were also recorded and were entirely consistent with those expected physiologically during CO_2 challenge.

5. FUTURE PLANS

Future animal studies should be directed at comparing the performance of the MPBS versus reference systems for measuring ICP, ptO$_2$ and T (as well as blood flow) in a broader range of clinical scenarios. These should include models of induced TBI, ischemic stroke and venous stasis/thrombosis. MPBS performance during the application of 'conventional' as well as perhaps 'new' treatment strategies might also be tested during these scenarios.

6. REFERENCES

1. Langlois JA, Rutland-Brown W, Wald MM. The epidemiology and impact of traumatic brain injury: a brief overview. J Head Trauma Rehabil. 2006;21(5):375-8.
2. Maegele M, Engel D, Bouillon B, et al. Incidence and Outcome of Traumatic Brain Injury in an Urban Area in Western Europe over 10 Years. Eur Surg Res. 2007;39(6):372-9.
3. Ghajar J. Traumatic brain injury. Lancet. 2000;356(9233):923-9.
4. Murray CJ, Lopez AD. Global mortality, disability, and the contribution of risk factors: Global Burden of Disease Study. Lancet. 1997;349(9063):1436-42.
5. Sarrafzadeh AS, Peltonen EE, Kaisers U, Kuchler I, Lanksch WR, Unterberg AW. Secondary insults in severe head injury--do multiply injured patients do worse? Crit Care Med. 2001;29(6):1116-23.
6. Manley G, Knudson MM, Morabito D, Damron S, Erickson V, Pitts L. Hypotension, hypoxia, and head injury: frequency, duration, and consequences. Arch Surg. 2001;136(10):1118-23.
7. The Brain Trauma Foundation. The American Association of Neurological Surgeons. The Joint Section on Neurotrauma and Critical Care. Initial management. J Neurotrauma. 2000;17(6-7):463-9.
8. Fakhry SM, Trask AL, Waller MA, Watts DD. Management of brain-injured patients by an evidence-based medicine protocol improves outcomes and decreases hospital charges. J Trauma. 2004;56(3):492-9; discussion 9-500.
9. Werner C, Engelhard K. Pathophysiology of traumatic brain injury. Brit J Anaesth. 2007;99(1):4-9.
10. Jeremitsky E, Omert L, Dunham CM, Protetch J, Rodriguez A. Harbingers of poor outcome the day after severe brain injury: hypothermia, hypoxia, and hypoperfusion. J Trauma. 2003;54(2):312-9.
11. Guidelines for the management of severe traumatic brain injury. J Neurotrauma. 2007;24 Suppl 1:S1-106.
12. van den Brink WA, van Santbrink H, Steyerberg EW, et al. Brain oxygen tension in severe head injury. Neurosurgery. 2000;46(4):868-76; discussion 76-8.
13. DeBacker D, in: Monitoring microcirculation: The next frontier?/ 25 Years of Progress and Innovation in Intensive Care Medicine, edited by Kuhlen R, Moreno R, Ranieri M, Rhodes A (Medizinisch Wissenschaftliche Verlagsgesellschaft, Berlin, 2007), pp. 301-305.
14. Stiefel MF, Spiotta A, Gracias VH, et al. Reduced mortality rate in patients with severe traumatic brain injury treated with brain tissue oxygen monitoring. J Neurosurg. 2005;103(5):805-11.
15. Cruz J. The first decade of continuous monitoring of jugular bulb oxyhemoglobinsaturation: management strategies and clinical outcome. Crit Care Med. 1998;26(2):344-51.
16. Nortje J, Gupta AK. The role of tissue oxygen monitoring in patients with acute brain injury. Brit J Anaesth. 2006;97(1):95-106.
17. McIlvoy LH. The effect of hypothermia and hyperthermia on acute brain injury. AACN clinical issues. 2005;16(4):488-500.
18. Alderson P, Gadkary C, Signorini DF. Therapeutic hypothermia for head injury. Cochrane Database Syst Rev. 2004(4):CD001048.

EVALUATION OF NIRS DATA BASED ON THEORETICAL ANALYSIS OF OXYGEN TRANSPORT TO CEREBRAL TISSUE

Kazunori Oyama, Toshihiro Kondo, Hidefumi Komatsu, and Toshihiko Sugiura[*]

Abstract: NIRS has been widely utilized for monitoring oxygen concentration of cerebral blood flow (CBF). However, meanings of signals measured by NIRS still have many unclear points. One of the factors is that the physiological mechanism of coupling between neuronal activity, metabolism and CBF is not clarified enough. In this study, we evaluate NIRS data based upon numerical simulation of oxygen transport to cerebral tissue. With a 2-dimensional mathematical model of oxygen transport from an arteriole to its surrounding tissue, we simulate the activity-dependent oxygenation changes. On the basis of calculated oxygen tension distribution, we derive quantities of two kinds of hemoglobin in the arteriole by using the oxygen dissociation curve, and theoretically decompose each hemoglobin change into its factors. This decomposition has revealed that NIRS data can reflect two types of physiological phenomena: a qualitative change caused by oxygen dissociation and a quantitative change caused by an increase of CBF. These results indicate that cellular oxygen consumption can be reflected more in the time courses of deoxygenated hemoglobin than those of oxygenated hemoglobin. It will be desirable to focus not only on oxygenated hemoglobin but also on deoxygenated hemoglobin when conducting evaluation of a brain function.

1. INTRODUCTION

In recent years, NIRS (near-infrared spectroscopy) has been widely utilized for monitoring oxygen concentration of cerebral blood flow. Compared to BOLD-fMRI, NIRS can measure narrower vessels, such as arteriole and capillary, which reflect neural activity more directly.[1] Though NIRS can measure two kinds of hemoglobin (oxygenated hemoglobin and deoxygenated hemoglobin), only oxygenated hemoglobin has received

[*] Department of Mechanical Engineering, Keio University, 3-14-1, Hiyoshi, Kouhoku-ku, Yokohama, Japan. E-mail:sugiura@mech.keio.ac.jp

P. Liss et al. (eds.), *Oxygen Transport to Tissue XXX*, DOI 10.1007/978-0-387-85998-9_28,

attention as a criterion for judgment of a brain function. However, importance of early increase of deoxygenated hemoglobin before the rise of CBF has been shown by experiments using animals[2] and humans.[3] In numerical simulation correlation between O_2 consumption and CBF in steady state was examined with Krogh's cylindrical model,[4,5] and with a model which took into account regional structure of microvascular networks.[6] However, changes in O_2 consumption are usually accompanied by transient change in CBF due to its adjusting function, which was simulated by Zheng et al.,[7] though their model does not represent spatial distribution of O_2 partial pressure (pO_2) in tissue.

In this study, we evaluate signals measured by NIRS based on numerical simulation of O_2 transport to cerebral tissue.[8] Transient response is simulated with our model including adjusting function of CBF. Consideration of spatial distribution of pO_2 in tissue is one of features of our study, though we use a 2-dimensional model for simplicity as a first step. On the basis of calculated pO_2 distribution, we derive quantities of two kinds of hemoglobin in the arteriole by using the O_2 dissociation curve, and decompose the hemoglobin quantity changes into two different factors. Finally, we discuss the meanings of the obtained results.

2. MODELING AND FOMULATION

The authors simulate O_2 transport from the arteriole to its surrounding tissue with a 2-dimensional mathematical model.[8] Figure 1 shows an analytical model and the coordinate system. The length in the axial direction is L, the arteriole radius is r, and the distance between the center of the arteriole and the far end of the tissue in the radial direction is R_0.

CBF increases to compensate for energy demand caused by regional neural activity, which is called the adjusting function of CBF. In brain microcirculation, increase of CBF is caused by a vasodilating action of a vascular smooth muscle which covers the arteriole. In this study, we assume that pO_2 is a factor of vasodilating action and decrease in pO_2 at the vessel wall is proportional to increase in the vessel radius r, as shown in Eq. (1).

$$r(p_w) = \begin{cases} r_{max} & (p_w \le p_l) \\ \dfrac{p_s - p}{p_s - p_l}(r_{max} - r_c) + r_c & (p_l < p_w \le p_s) \\ r_c & (p_s < p_w) \end{cases} \tag{1}$$

where p_w denotes the minimum value of pO_2 at the vessel wall. p_s and p_l are the threshold and the critical value of pO_2, respectively. r_c is the vessel radius at rest, and r_{max} is the maximum of the vessel radius.

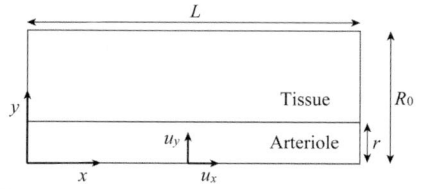

Figure 1. 2-D analytical model for O_2 transport.

O$_2$ in the blood vessel is transported by advective flow and diffusion. On the other hand, O$_2$ in the tissue is transported only by diffusion and O$_2$ consumption occurs only in the tissue. The nondimensional governing equation in the vessel is shown in Eq. (2) and that in the tissue is shown in Eq. (3). Nondimensional parameters are shown in Eq. (4). Nondimensional variables are as follows. $x^* = x/r_c$, $y^* = y/r_c$, $r^* = r/r_c$, $t^* = (k_1/r_c^2)t$, $p^* = (\alpha_1/c_0)p$.

$$\frac{\partial p^*}{\partial t^*} = \left(\frac{\partial^2 p^*}{\partial x^{*2}} + \frac{\partial^2 p^*}{\partial y^{*2}} \right) - Pe \left(1 - y^{*2} \right) r^{*2} \frac{\partial p^*}{\partial x^*} \tag{2}$$

$$\frac{\partial p^*}{\partial t^*} = k \left(\frac{\partial^2 p^*}{\partial x^{*2}} + \frac{\partial^2 p^*}{\partial y^{*2}} \right) - \alpha f^* \tag{3}$$

$$Pe = \frac{\Delta P r_c^3}{4\eta L k_1}, \quad k = \frac{k_2}{k_1}, \quad \alpha = \frac{\alpha_1}{\alpha_2}, \quad f^* = \frac{r_c^2}{k_1 c_0} f \tag{4}$$

3. EVALUATION OF HEMOGLOBIN QUANTITIES

To compare with NIRS data, we have to derive hemoglobin quantities from calculated pO_2. The ratio of oxygenated hemoglobin in a certain blood volume can be calculated by degree of O$_2$ saturation S which is expressed by mathematical expression (Eq. 5) of O$_2$ dissociation curve.[8] O$_2$ dissociation of hemoglobin depends on pO_2.

$$S(x, y, t) = \frac{(pO_2(x, y, t)/p_{50})^n}{1 + (pO_2(x, y, t)/p_{50})^n} \tag{5}$$

Based on the calculated pO_2 distribution, regional hemoglobin quantities in a blood vessel are obtained as shown in Eqs. (6)-(7).

$$\delta q_{oxyHb}(x, y, t) = c_{totalHb} S(x, y, t) \delta V \tag{6}$$

$$\delta q_{deoxyHb}(x, y, t) = c_{totalHb}(1 - S(x, y, t)) \delta V \tag{7}$$

where $\delta q_{oxyHb}(x,y,t)$ and $\delta q_{deoxyHb}(x,y,t)$ denote regional quantities of oxygenated and deoxygenated hemoglobin, respectively. In order to calculate volume of the vessel from 2-D pO_2 distribution, a cubic shaped vessel is considered. We assume that the cross section of the vessel is to be πr^2 ($=\pi r^* r$), which is equivalent to the cross-sectional area of the cylinder with radius r. By integrating regional hemoglobin quantities, overall hemoglobin quantities in a blood vessel $q_{oxyHb}(t)$ and $q_{deoxyHb}(t)$ are obtained as presented in Eqs. (8)- (9).

$$q_{oxyHb}(t) = \int \delta q_{oxyHb} \approx c_{totalHb} \int S \delta V \tag{8}$$

$$q_{deoxyHb}(t) = \int \delta q_{deoxyHb} \approx c_{totalHb} \int (1-S)\delta V \tag{9}$$

It is assumed that there are two factors in changes of two kinds of hemoglobin. One factor is an increase of hemoglobin quantities in a blood vessel with a rise of CBF and another factor is qualitative change of hemoglobin caused by O_2 dissociation reaction. Incoming hemoglobin quantity changes are expressed in Eqs. (10)-(11).

$$\Delta q_{oxyHb}(t)_{in} = c_{totalHb} S_0 \Delta V \tag{10}$$

$$\Delta q_{deoxyHb}(t)_{in} = c_{totalHb}(1-S_0)\Delta V \tag{11}$$

where $\Delta q_{oxyHb}(t)_{in}$ and $\Delta q_{deoxyHb}(t)_{in}$ denote incoming hemoglobin quantities of oxygenated and deoxygenated hemoglobin, respectively. S_0 is degree of O_2 saturation of incoming blood and ΔV is the volume of incoming blood.

The goal of this subsection is to obtain reacted hemoglobin quantities which can be derived by subtracting incoming hemoglobin quantities from whole hemoglobin quantities. Equations (12)-(13) show the reacted hemoglobin quantities.

$$\Delta q_{oxyHb}(t)_{re} = \Delta q_{oxyHb}(t) - \Delta q_{oxyHb}(t)_{in} \tag{12}$$

$$\Delta q_{deoxyHb}(t)_{re} = \Delta q_{deoxyHb}(t) - \Delta q_{deoxyHb}(t)_{in} \tag{13}$$

where $\Delta q_{oxyHb}(t)_{re}$ and $\Delta q_{deoxyHb}(t)_{re}$ denote reacted hemoglobin quantities of oxygenated and deoxygenated hemoglobin, respectively.

The authors have simulated a case using the above equations. Parameters employed in this calculation are shown in Table 1.

Table 1. Parameters employed in this calculation

Parameters	Values
r_c (resting arteriole radius)	0.02 mm
r_{max} (maximum of arteriole radius)	0.022 mm
L (arteriole length)	1.5 mm
k_1 (diffusivity of O_2 in blood)	1.5×10^{-3} mm^2/s
k_2 (diffusivity of O_2 in tissue)	2.4×10^{-3} mm^2/s
α_1 (solubility coefficient for O_2 in blood)	2.8×10^{-3} mM/mmHg
α_2 (solubility coefficient for O_2 in tissue)	3.9×10^{-3} mM/mmHg
f_c (O_2 consumption rate at rest)	3.72×10^{-2} mM/s
p_0 (pO_2 in inflow blood)	90 mmHg
p_s (threshold pO_2 in vessel reaction)	50 mmHg
p_l (critical pO_2 in vessel reaction)	30 mmHg
p_{50} (O_2 tension at half heme saturation)	26.4 mmHg
$c_{totalHb}$ (total Hb concentration in blood)	2.24 mM
n (Hill component)	2.65
η (blood viscosity of arteriole)	0.04 g/mm/s
Δp (difference of the pressure between arteriole and venous ends of it)	90 mmHg

4. RESULTS AND DISCUSSIONS

Figure 2 shows time courses of changes in overall hemoglobin quantities in a blood vessel. To examine the transient response in this model, the authors have simulated a case in which neuronal activity lasts for one second in an active area so that O_2 consumption rate to be 10 times as high as that at rest. To see the mechanism of early increase of deoxygenated hemoglobin, rather high O_2 consumption ratio was given. This result indicates that deoxygenated hemoglobin increases during a task and gradually decreases. On the other hand, the response of oxygenated hemoglobin shows an initial decrease followed by a positive peak. We consider that the biphasic change of oxygenated hemoglobin arises from a time delay in CBF response to changes in $pO2$.

To interpret the significance of these signals, each hemoglobin change has been theoretically decomposed into its factors. Figure 3(a) indicates the quantitative changes of deoxygenated and oxygenated hemoglobin. The levels of both kinds of hemoglobin increase with a rise of CBF. On the other hand, Figure 3(b) illustrates qualitative changes caused by O_2 dissociation. Due to the decrease of pO_2 after task onset, the blood vessel falls into hypoxic condition, which causes increase of deoxygenated hemoglobin and decrease of oxygenated hemoglobin. Even after task offset, increase of deoxygenated hemoglobin lasts and does not return to its baseline immediately. Therefore, it is believed that the hypoxic condition shifts to the downstream of the blood vessel by advective flow and it takes time to return to the steady state.

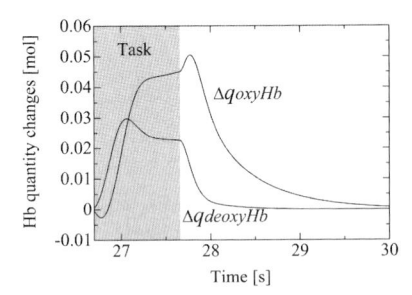

Figure 2. Response of hemoglobin dynamics to 1 second stimulation throughout arteriole.

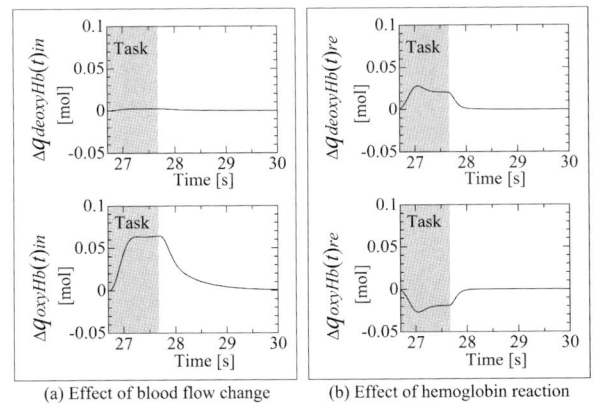

(a) Effect of blood flow change   (b) Effect of hemoglobin reaction

Figure 3. Decomposition of hemoglobin changes into their factors.

Contrary to our results, most of NIRS experiments have shown a fall of deoxygenated hemoglobin. Possible factors for the fall can be strong advection and weak diffusion due to high blood flow velocity in large veins which are usually included in areas covered by NIRS. The former causes a large fall of deoxygenated hemoglobin and the latter causes a small increase of deoxygenated hemoglobin. As a whole, the fall of deoxygenated hemoglobin can easily appear in NIRS signals. Whereas, an early deoxygenation have been reported by experiments using animals[2] and humans.[3] It can be caused by weak advection and strong diffusion due to low blood flow velocity in small vessels which transport O_2 to activated tissue. This early deoxygenation is thus more close to neuronal activities and NIRS can detect it, if it is measured carefully.[9] To simulate this response in a small vessel, we adopted our model with rather high O_2 consumption rate and with low blood flow velocity.

5. CONCLUSIONS

To understand meanings of physiological signals measured by NIRS, two kinds of hemoglobin changes have been simulated using a 2-dimensional mathematical model.

One of the important findings obtained through this study is that two kinds of hemoglobin changes, measured by NIRS, have different physiological meanings. Change in deoxygenated hemoglobin shows oxygen dissociation closely related to neuronal activity, while change in oxygenated hemoglobin shows CBF in general.

These results indicate that hemoglobin dynamics are highly affected by an increase of oxygenated hemoglobin. Tissue oxygen consumption, namely neuronal activity, can be reflected more in the time courses of deoxygenated hemoglobin than those of oxygenated hemoglobin. It will be desirable to focus not only on oxygenated hemoglobin but also on deoxygenated hemoglobin when conducting evaluation of a brain function.

6. REFERENCES

1. T. Kato, Principle and technique of NIRS-Imaging for human brain FORCE: fast-oxygen response in capillary event, International Congress Series, 1270, 85-90 (2004).
2. Offenhauser N., Thomsen K., Caesar K. and Lauritzen M., Activity-induced tissue oxygenation changes in rat cerebellar cortex: interplay of postsynaptic activation and blood flow, J Physiol. 565(pt1), 279-94 (2005).
3. T. Akiyama, T. Ohira, T. Kawase and T. Kato, TMS orientation for NIRS functional motor mapping, Brain Topogr., 19(1-2), 1-9 (2006).
4. Hudetz A. G., Mathematical model of oxygen transport in the cerebral cortex, Brain Res. 817 (1-2), 75-83 (1999).
5. Wang C. H. and Popel A. S., Effect of red blood cell shape on oxygen transport in capillaries, Math Biosci. 116 (1), 89-110 (1993).
6. Tanishita K., Masamoto K., Negishi T., Takizawa N. and Kobayashi H., in Organ Microcirculation edited by Ishii H et al., (Springer, Tokyo, 2004), pp. 13-20.
7. Y. Zheng, J. Martindale, D. Johnston, M. Jones, J. Berwick and J. Mayhew, A model of the hemodynamic response and oxygen delivery to brain, NeuroImage 16, 617-637 (2002).
8. T. Kondo, K. Oyama, H. Komatsu, and T. Sugiura, Numerical simulation of oxygen transport in cerebral tissue, Proceedings of ISOTT 2007 (to be submitted).
9. T. Yamamoto, and T. Kato, Paradoxical correlation between signal in functional magnetic resonance imaging and deoxygenated hemoglobin content in capillaries: a new theoretical explanation, Phys Med Biol. 47, 1121-1141 (2002).

MYOCARDIAL CAPILLARY NET AND BLOOD CONSTITUENTS IN STREPTOZOTOCIN (STZ)-INDUCED DIABETIC RATS

Tomiyasu Koyama[*] and Akira Taka[**]

Abstract: Type 1 diabetes was induced in Wistar rats by injection of streptozotocin (STZ). Changes in the myocardial capillary network were examined using the double-staining enzymatic method for alkaline phosphatase (AP) and dipeptidylpeptidase IV (DPPIV) This method allows the identification of the arteriolar (AP-containing) and the venular (DPPIV-containing) portions of the capillary network. In addition, blood plasma was analysed. The AP- and AP/DPPIV-containing capillary portions increased significantly, accompanied by a decrease in the DPPIV-containing portions in 60 days. A significant increase in AP was observed in the plasma. The capillary domain areas of each capillary portion were larger in the STZ-injected group than in the controls. It appears that oxygen transport to the subendocardial myocardial tissues may be decreased in the STZ group. In rats fed with Saji-supplemented chow there was a decrease in plasma AP, with increases in hemoglobin, hematocrit and vitamin C, suggesting a partial improvement of metabolic function and oxygen supply in these diabetic Wistar rats.

1. INTRODUCTION

Cardiac insufficiency is one of the main complications in diabetes. Several causes for the development of insufficiency have been postulated in relation to microvessels. These include excessive deposition of collagen surrounding capillaries, increased activity of the plasminogen activator inhibitor, PAI-1 (reported in the OLETF rat model of non-insulin dependent diabetes melitus[1] and a decrease in angiogenic factors as found in diabetic patients[2]. It also seems probable that hyperglycemia may affect the expression of several enzymes.

A significant increase in alkaline phosphatase (AP)-containing capillaries has been observed in the study on OLETF rats[1]. However, the converse result, that is a decrease in AP despite an overall increase in capillaries, has been reported in the hearts of STZ-diabetic rats[3]. Different types of diabetes in different strains of rat may cause different

[*] T.Koyama, Hokkaido Universityu, Sapporo Japan, tomkoyamajp@yahoo.co.jp,   [**]A. Taka, Sapporo-Aoba School of Holistic Medicine

P. Liss et al. (eds.), *Oxygen Transport to Tissue XXX*, DOI 10.1007/978-0-387-85998-9_29,   **187**
© Springer Science+Business Media, LLC 2009

modifications to the capillary networks and enzyme expression. In the present experiments we have studied changes in myocardial capillaries in the STZ-induced diabetic Wistar rats.

It has been established by a number of different techniques that the capillaries in the subendocardium that are stained blue, red, and violet by the double-staining method represent, respectively, alkaline phosphatase-containing arteriolar portions, venular portions containing dipeptidylpeptidase IV (DPPIV) and intermediate portions containing both enzymes[4-6]. In the present studies, using this method, particular attention has been paid to the capillaries expressing AP. It is known that expression of proliferating cell nuclear antibody is low in AP-sensitive capillary portions[7], suggesting AP acts to suppress endothelial cell proliferation, resulting in the arterialization and calcification of blood vessels.

In addition AP in the plasma was studied since it is multi-functional and widely distributed in the whole body. Blood plasma was also analysed for some other factors. Further, in an attempt to find a herbal medicine that might be active against diabetes, the effects of supplementing rat chow with Saji (*Hippophae rhamnoides*) were studied. Preliminary results will be presented.

2. METHODS

The present study conformed with the guidelines of Hokkaido University for the care and use of laboratory animals.

STZ was dissolved in a 0.1M citrate buffer solution (pH 4.3) (100mg/ml). Six 7-week-old male Wistar rats were injected with STZ (50mg/kg) in the tail vein; five controls received 0.5 ml citrate buffer. Sixty days after the injections, rats were anesthetized with i.p. Nembutal (50mg/kg) and blood samples from the abdominal vein collected in heparinized cylinders for analysis (see below).

Hearts were removed, divided horizontally, placed in O.C.T. compound® and quickly frozen in liquid nitrogen. Frozen cross-sections of the left ventricle were sectioned at 16 μm with a cryotome and stained for AP and DPPIV using the double-staining method[5] with a slight change in timing. In STZ-diabetic tissues the normal 25 min reaction time produced intense blue coloration throughout the whole section; the staining time was therefore reduced to 5 minutes, after which sections were rinsed several times in water for visualization and counting of AP- or DPPIV-containing capillaries. The number of each type of capillary portion was counted in four visual fields in each of four sections of the subendocardium from each heart. The polygonal capillary domain area (CDA) which represents the tissue area perfused by each capillary was calculated by the method proposed by Hoofd et al.[4]. About 100 to 150 capillaries were present in a visual field viewed at x400, hence the four visual fields counted in each of the four sections from each of the five rats in a group yielded 3000 to 4000 values for statistical analysis of CDAs. The proportion of the three of capillary types, and the CDAs for each, were statistically analysed by the Mann-Whitney test (means ±S.D. p<0.05)

The blood plasma was analysed by routine methods: AP by the JSCC standard method, glucose, total cholesterol (TC) and triglyceride (TG) by enzymatic methods; vitamin C (VC) by HPLC, and hemoglobin (Hb) and hematocrit (HCT) by an automated hemoanalyzer. The values for the control and treated groups were compared by Student's unpaired t-test (means ±S.D. p<0.05).

In a preliminary study on the effects of the traditional herbal medicine, powdered Saji fruit was obtained from North Eastern China. It was added to dry rat chow at a ratio of 3.5g/kg, mixed to a dough with water and dried in solid blocks of about 5 g. STZ- and buffer-injected rats were fed ad libitum on supplemented or unsupplemented blocks for 49 days. Blood samples were then obtained from the abdominal vein under anesthesia and analysed as described above.

3. RESULTS

Table 1 shows basic parameters for STZ-treated and control rats at the end of the 60-day experiment. Body weight (BW), heart weight (HW) and left ventricular weight (LVW) were significantly lower in the STZ group than in the controls. The ratio of LVW to BW (LVW/BW) as well as blood glucose levels were significantly higher in the STZ group.

Table 1. Physical paramaters of control and diabetic rats

Group	BW (g)	HW (mg)	LVW(mg)	LVW/BW (mg/g)	Glucose (mg/dl)
Control	526 ± 19	1250 ± 20	930 ± 30	1.78 ± 0.10	148 ± 14
STZ	$330 \pm 50^*$	$890 \pm 120^*$	$670 \pm 80^*$	$2.05 \pm 0.11^*$	$401 \pm 65^*$

* $P<0.01$

Despite the reduction in the incubation period for the double-staining method, the whole visual field, showing capillaries and cross-sections of cardiomyocytes, was still faintly blue in sections from the STZ group. The differently coloured capillaries could, however, be distinguished in these sections.

The numbers of the different capillary portions in the ventricular subendocardium from control and STZ groups are given in Table 2. This shows that STZ treatment increased the number of blue-stained and violet-stained capillary portions. In both the control and STZ groups the proportions of the different capillary portions were in the order blue<violet<red but there was a marked difference in the numbers in the two groups. In the controls the proportion of red-stained capillaries was about 2.5 times the proportion of those stained violet. In the STZ group the proportion of red-stained capillary portions was only 1.1 times that of the violet portions as a result of the significant increase in the latter.

The CDAs were in the order: blue > violet > red for the capillary portions in both the control and STZ groups. This is consistent with the finding[4,5] that the arteriolar capillary portions (blue) have a higher oxygen tension. In the STZ group CDAs for each capillary portion were significantly larger than those in the control group.

Table 3 gives the results of the hematological analyses in another rat group. STZ injection caused large changes in the hematological parameters. Compared with controls, the level of glucose was significantly higher, but VC was significantly lower. The plasma concentration of AP was particularly high in the STZ group, being nearly 5 times that in controls.

Table 2. Capillary parameters in ventricular subendocardium from diabetic rats

Group (number of rats)	Capillary color (expressed enzymes)	Capillary proportion (/mm^2)	Capillary domain area (μm^2)
Control (n=5)	Blue (AP)	274 ± 72[†]	617 ± 175[†]
	Violet (AP+DPPIV)	464 ± 104[†]	569 ± 261[†]
	Red (DPPIV)	1173 ± 241	519 ± 143
	Total	1911 ± 206	
STZ (n=6)	Blue (AP)	382 ± 178[†]	642 ±174[†]
	Violet(AP+DPPIV)	686 ± 223*	584 ±155*[†]
	Red (DPPIV)	763 ± 193**	542 ± 133*
	Total	1831 ± 168	

*P<0.01, **P<0.001 against control rats
[†]P<0.01 against red capillary within each group.

Table 3. Effects of STZ and Saji administration on blood plasma constituents

STZ: mg/kg	Saji	TC: mg/dl	AP IU/l	Glucose mg/dl	Hb g/dl	Hct %	VC mic/ml
0	0 (n=3)	72.7±9.3	790±58	221±43	16.0±1.2	47.8±3.7	4.7±0.34
0	3.5 (n=3)	71.3±3.2	541±25	174±17	16.3±0.85	49.6±0.6	3.8±0.76
	p	ns	0.002	ns	ns	Ns	ns
50	0 (n=5)	74.0±6.0	3722±1387	521±49	16.9±0.20	51.8±0.41	0.98±0.35
50	3.5 (n=5)	63.8±3.2	3281±740	527±49	18.0±0.35	54.1±0.91	3.08±0.59
	p	0.02	ns	ns	0.032	0.05	0.03
	p*	ns	0.001	0.001	ns	Ns	0.001

Saji stands for Saji supplementation to rat chow in g/kg. (n=) means the number of studied rats.
p: p-value for the difference between Saji-supplementation 0 and 3.5 g/kg.
p*: p-value for the difference between STZ 0 and 50 without Saji-supplmentation (0-Saji supplementation).

Supplementation of the chow with Saji produced changes in both the control and STZ groups. While Saji supplementation produced a significant decrease in plasma AP in controls, the trend towards a decrease seen in STZ-treated rats did not reach significance. As Table 3 shows, Saji supplementation did, however, significantly increase VC, Hb and Hct and a significant decrease in TC in the STZ group.

4. DISCUSSION

Expression of the angiogenic factors for endothelial cells, HIF-1a and VEGF, is intensified during the first few weeks after STZ administration, and then decreases[8]. Accordingly, capillary density may first increase, then decrease. It would be informative, if, after STZ injection, changes in total capillary density could be studied by time-lapse photography. The present measurements made 60 days after STZ injection, showed a trend towards a decrease in the total capillary density, with marked changes in the proportion of the different capillary portions and in their CDAs. Increases in the blue- and violet-stained capillary portions suggested an increase in AP expression in capillaries, consistent with the significant increase in plasma AP. It is unclear why this increase in AP-containing capillaries was observed in the present experiments while a decrease was reported by Okruhlicova et al.[3].

Some comment is needed on the finding in the STZ group that, unless the incubation period was reduced, there was a very strong reaction for AP throughout the tissue sections. This probably reflects the high level of AP found in blood plasma of the STZ group. This in turn may result from the hyperglycemia in these rats. It has been shown that AP is expressed in vascular smooth muscle cells during hyperglycemia as a result of changes in phenotypic expression[9]. A similar phenotypic change may occur in endothelial cells in the STZ-diabetic rats.

AP is multi-functional but its biological role in blood vessels is unknown. Suggestions in the literature include a role in the maintenance of the blood-brain barrier[10] and suppression of the proliferation of endothelial cells[11]. Since AP is closely related to proliferation and maturation of vascular smooth muscle[12] its expression may limit increases in capillarity. It is possible that the increase in AP expression may be among the factors that cause cardiac complications in diabetes.In the STZ group there was an increase in the CDAs of the violet and red-stainable capillary portions despite the decrease in LVW. This change in the myocardial capillary network will reduce oxygen transport to the cardiac tissues, although there may be increased arterialization of capillaries.

5. REFERENCES

1. T. Sugawara, S. Fujii S. et al., Coronary capillary remodelling in noninsulin- dependent diabetic rats:amelioration by inhibition of angiotensin converting enzyme and its potential clinical implications, Hypertens. Res. **24**:75-81 (2001).
2. H. Marfella, K. Esposito et al. Myocardial infarction in diabetic rats: role of hyperglycaemia on infarct size and early expression of hypoxia-inducible factor 1, Diabetes **53**:2383-91 (2004).
3. L. Okruhlicova, N. Tribulova et al., Ultrastructure and histochemistry of rat myocardial capillary endothelial cells in response to diabeties and hypertension, Cell Res. **15**, 532-8 (2005).
4. L.Hoofd, Z. Turek, K. Kubat et al. Variability of intercapillary distance estimated on histological sections of rat heart. In: Oxygen Transport to Tissue VII, edited by F. Kreuzer, S.M. Cain, Z. Turek and T.K. Goldstick, Plenum Press, New York, (1985), pp239-247.
5. S. Batra, and K. Rakusan, Geometry of capillary networks in hypertrophied rat heart. Microvasc Res **41**, 29-40 (1991).
6   T. Koyama, Z Xie, M. Gao, J. Suzuki and S. Batra. Adaptive changes in the capillary network in the left ventricle of rat heart. Jpn J Physiol, **48**, 229-241 (1998)..
7. T. Koyama, Z.L. Xie, and J. Suzuki, Coronary ischemia / reperfusion increases proliferating cell nuclear antigen in vascular endothelial cells in rat hearts. Angiogenesis, **5**, 119-124 (2002).
8. J. C. Chavez, K. Almhannna, L. N. Berti-Mattera, Transient expression of hypoxia-inducible factor-1 alpha and target genes in peripheral nerves from diabetic rats, Neuroscience Lett. **347**, 179-82 (2005).

9. N. X. Chen and S. M. Moe, Arterial calcification in diabetes, Curr. Diab. Rep. **3**(1), 28-32, (2003).

10. W. Risau, R. Hallmann and U. Albrecht, Differentiation-dependent expression of proteins in brain endothelium during development of the blood-brain barrier. Dev. Biol. **117**, 537-545 (1986)

11. Lindhorn, Schulz-Hector, Loss of myocardial capillary endothelial-cell alkaline phosphatase (ALP) activity in primary endothelial cell culture, Cell Tissue Res. **291,** 497-505 (1998).

12. R. A. Rowan, D. S. Maxwell, An ultrastructural study of vascular proliferation and vascular alkaline phosphatase activity in the developing cerebral cortex of the rat, Am. J. Anat. **160**, 257-65 (1981).

EFFECTS OF PROINSULIN C-PEPTIDE ON OXYGEN TRANSPORT, UPTAKE AND UTILIZATION IN INSULINOPENIC DIABETIC SUBJECTS – A REVIEW

Lina Nordquist[*] and Sara Stridh[*]

Abstract: Exogenous C-peptide administration has beneficial effects in many of the tissues commonly affected by diabetic complications. Diabetes-induced circulatory impairments such as decreased blood flow are prevented by C-peptide, possibly via Ca^{2+}-mediated effects on nitric oxide release. C-peptide also improves diabetes-induced erythrocyte deformability, which likely improves oxygen availability and uptake in affected tissues. Furthermore, C-peptide prevents diabetic neuropathy via improvements of endoneural blood flow and by preventing axonal swelling. In the kidney, C-peptide normalizes the diabetes-induced increase in oxygen consumption via inhibition of the Na^+/K^+-ATPase. Surprisingly, C-peptide has also been shown to prevent complications in patients with type II diabetes. Taken together, these results may indicate that C-peptide treatment has the potential to reduce the prevalence of diabetic complications. In this paper, the current knowledge regarding these beneficial effects of C-peptide administered to diabetic subjects will be reviewed briefly.

1. INTRODUCTION

Intensive treatment of hyperglycemia effectively prevents diabetic complications[1]. However, diabetic patients with close to normal blood glucose levels still have increased risk to develop retinopathy, neuropathy, and nephropathy. The proinsulin connecting peptide, C-peptide, is produced by the pancreatic beta cells during normal insulin synthesis and released to the circulation in equimolar amounts to insulin. Thus, an impaired insulin synthesis results in equal impairment of C-peptide production. However, C-peptide is not included in the common anti-diabetic treatment for type I diabetic

[*] Department of Medical Cell Biology, Uppsala University, BMC, PO 571, 751 23 Uppsala, Sweden

P. Liss et al. (eds.), *Oxygen Transport to Tissue XXX*, DOI 10.1007/978-0-387-85998-9_30,   **193**
© Springer Science+Business Media, LLC 2009

patients even though C-peptide has been shown to improve function in several of the tissues commonly affected by diabetes. This paper will shortly present the current knowledge regarding the beneficial effects of C-peptide on oxygen transport, uptake, and utilization when administered to patients and animal models of insulinopenic diabetes mellitus.

2. CIRCULATORY BENEFITS OF C-PEPTIDE

More than thirty years ago, it was first reported that diabetes can impair the microcirculation and cause relative tissue hypoxia[2]. Type I diabetic patients are known to display reduced maximal oxygen uptake[3]. Shortly after induction of diabetes in lambs, coronary vascular resistance is elevated[4], and it has been shown that diabetic animals have increased total peripheral resistance and are more prone to develop hypertension[5]. In type I diabetes, C-peptide restores diabetes-impaired skeletal muscle perfusion[6, 7], skin capillary red blood cell velocity[8], and myocardial blood perfusion[9]. These effects may not be restricted to type I diabetes since correlations have been found between C-peptide levels and the presence of coronary artery disease, peripheral vascular disease, and autonomic neuropathy in type II diabetic patients[10].

2.1. NO-mediated vasodilation

Nitric oxide (NO) is produced by the three different NO synthase (NOS) isoforms and is a potent vasodilator. Diabetes is associated with decreased NO bioavailabilty and/or production[11, 12]. In type I diabetic patients, the vasodilatory effect of C-peptide is mediated via induction of NO release from endothelial (e)-NOS[7]. The cellular mechanism of C-peptide-induced NO release includes increased intracellular calcium levels, which stimulates both eNOS and inducible NOS[13, 14]. Surprisingly, in myocardial ischemia-reperfusion, C-peptide has been shown to exert cardio protective effects through NO release[15], indicating that the NOS stimulating effect of C-peptide is not restricted to the diabetic state.

2.2. Effects on erythrocyte deformability

Erythrocyte deformability is a component in the regulation of vasodilation via the release of adenosine triphosphate (ATP)[16]. ATP released from the erythrocyte stimulates eNOS and results in NO release from vascular endothelial cells[17]. The reduced vascular tone results in increased blood perfusion and increased tissue oxygenation. In diabetes, erythrocyte deformability and aggregation are increased[18], with decreased capillary blood flow and oxygen availability in the tissue. C-peptide levels correlate to erythrocyte deformability[19], and exogenous C-peptide administration ameliorates this impairment in diabetic patients[20]. In addition, C-peptide increases oxygen uptake in exercising forearm muscles of diabetic patients, possibly by increasing capillary recruitment in the activated muscle[6].

Na^+/K^+-ATPase is a ubiquitous ion exchanging enzyme situated in most cell membranes. Decreased Na^+/K^+-ATPase activity in erythrocytes is associated with diabetes complications[19], which has been suggested to mediate some of the commonly occurring diabetic complications[21]. There are reports showing that C-peptide levels correlate with erythrocyte Na^+/K^+-ATPase activity in type I as well as type II diabetic subjects[19, 22]. Thus, it is possible that the beneficial effects of C-peptide on erythrocyte deformability and tissue oxygenation are mediated via the increased Na^+/K^+-ATPase activity[20].

3. C-PEPTIDE-INDUCED NEUROPROTECTION

C-peptide prevents and even improves autonomic and somatic nerve function in patients with type I diabetic neuropathy[23]. These improvements, manifested as circulatory improvements, are also mediated via induction of NO release and increased Na^+/K^+-ATPase. Patients with diabetic neuropathy display impaired neural NO activity and endoneural blood flow, and have a reduced maximal oxygen uptake[24], which results in tissue hypoxia. Impaired neural Na^+/K^+-ATPase activity leads to increased intra-axonal Na^+, which causes axonal swelling[25]. As axonal degeneration develops, nerve conduction velocity decreases and becomes progressively less reversible[23]. Exogenous C-peptide administration preserves neural Na^+/K^+-ATPase activity in diabetic rats[26] and prevents neuronal apoptosis in type I diabetic patients[27]. Stimulation of eNOS-mediated NO production appears to contribute to the C-peptide-induced effects on diabetic neuropathy, causing vasodilation and subsequently increases the endoneural blood flow[23].

4. RENOPROTECTIVE EFFECTS OF C-PEPTIDE

NO is reduced in the kidney cortex of diabetic rats[28]. In renal mesangial cells, C-peptide increases the expression of the vasopressin-activated calcium-mobilizing receptor-1 (VACM-1), thereby causing Ca^{2+} influx and increased eNOS activity[29]. C-peptide also increases intracellular Ca^{2+} in proximal tubular cells from opossum[30] and in human tubular cells[29]. Several studies have reported renoprotective effects of C-peptide, both in diabetic subjects as well as in animal models of insulinopenic diabetes mellitus. C-peptide decreases glomerular hyperfiltration in diabetic patients[31], and long-term C-peptide treatment improves renal function[32].

Renal hypoxia is a known pathway leading to end-stage renal failure[33]. In the kidney, 80% of the oxygen is consumed by the Na^+/K^+-ATPase[34]. Na^+/K^+-ATPase activity is increased in the diabetic kidney, with subsequent elevated oxygen consumption[35]. It is possible that the increased Na^+/K^+-ATPase activity can limit oxygen availability in the diabetic kidney. It has recently been shown that C-peptide normalizes oxygen consumption in proximal tubular cells isolated from diabetic rats, via Na^+/K^+-ATPase inhibition[36].

5. C-PEPTIDE AND RETINOPATHY

Diabetic retinal microangiopathy is characterized by increased extracellular matrix protein deposition and thickening of the capillary basement membrane[37, 38]. The current knowledge regarding the effects of C-peptide on diabetes-induced retinopathy is scarce, but a recent study suggests that C-peptide has effects on angiogenesis[39]. There are indications that diabetes induces up-regulation of total fibronectin, as well as oncofetal fibronectin in the retina[40], which is believed to be involved in angiogenesis and is normally not found in mature tissue[41, 42]. Retinas of diabetic rats have increased expression of oncofetal fibronectin compared to control rats, which is completely prevented by C-peptide treatment[39]. Chakrabarti and co-workers concluded that normalization of the diabetes-induced up-regulation of oncofetal fibronectin in diabetic retinas may suggest an important role of C-peptide for the development of retinopathy and microangiopathy.

6. CONCLUDING REMARKS

C-peptide affects oxygen homeostasis via several mechanisms. In diabetes mellitus, C-peptide administration induces NO-mediated vasodilation and improves erythrocyte deformability. In addition, C-peptide improves endoneural blood perfusion and normalizes oxygen consumption by proximal tubular cells in the diabetic subjects. Recent studies even suggest effects on angiogenesis. In summary, the C-peptide-mediated effects on oxygen transport, uptake, and utilization suggest it is likely that C-peptide-substitution to insulinopenic diabetics reduces several of the most commonly occurring diabetic complications including nephropathy, neuropathy and retinopathy. Thus, it is likely that exogenous C-peptide administration of these patients would reduce the prevalence of these devastating effects of diabetes.

7. REFERENCES

1. The effect of intensive treatment of diabetes on the development and progression of long-term complications in insulin-dependent diabetes mellitus. The Diabetes Control and Complications Trial Research Group, *N Engl J Med* **329**(14), 977-86 (1993).
2. J. Ditzel, and E. Standl, The problem of tissue oxygenation in diabetes mellitus, *Acta Med Scand Suppl* **578**(59-68 (1975).
3. T. Jensen, E. A. Richter, B. Feldt-Rasmussen, H. Kelbaek, and T. Deckert, Impaired aerobic work capacity in insulin dependent diabetics with increased urinary albumin excretion, *Br Med J (Clin Res Ed)* **296**(6633), 1352-4 (1988).
4. J. C. Lee, and S. E. Downing, Coronary dynamics and myocardial metabolism in the diabetic newborn lamb, *Am J Physiol* **237**(2), H118-24 (1979).
5. T. D. Bell, G. F. DiBona, Y. Wang, and M. W. Brands, Mechanisms for renal blood flow control early in diabetes as revealed by chronic flow measurement and transfer function analysis, *J Am Soc Nephrol* **17**(8), 2184-92 (2006).
6. B. L. Johansson, B. Linde, and J. Wahren, Effects of C-peptide on blood flow, capillary diffusion capacity and glucose utilization in the exercising forearm of type 1 (insulin-dependent) diabetic patients, *Diabetologia* **35**(12), 1151-8 (1992).
7. B. L. Johansson, J. Wahren, and J. Pernow, C-peptide increases forearm blood flow in patients with type 1 diabetes via a nitric oxide-dependent mechanism, *Am J Physiol Endocrinol Metab* **285**(4), E864-70 (2003).
8. T. Forst, T. Kunt, T. Pohlmann, K. Goitom, M. Engelbach, J. Beyer, and A. Pfutzner, Biological activity of C-peptide on the skin microcirculation in patients with insulin-dependent diabetes mellitus, *J Clin Invest* **101**(10), 2036-41 (1998).
9. A. Hansen, B. L. Johansson, J. Wahren, and H. von Bibra, C-peptide exerts beneficial effects on myocardial blood flow and function in patients with type 1 diabetes, *Diabetes* **51**(10), 3077-82 (2002).
10. R. Sari, and M. K. Balci, Relationship between C peptide and chronic complications in type-2 diabetes mellitus, *J Natl Med Assoc* **97**(8), 1113-8 (2005).
11. T. Traupe, P. C. Nett, B. Frank, L. Tornillo, R. Hofmann-Lehmann, L. M. Terracciano, and M. Barton, Impaired vascular function in normoglycemic mice prone to autoimmune diabetes: role of nitric oxide, *Eur J Pharmacol* **557**(2-3), 161-7 (2007).

12. T. Thum, D. Fraccarollo, M. Schultheiss, S. Froese, P. Galuppo, J. D. Widder, D. Tsikas, G. Ertl, and J. Bauersachs, Endothelial nitric oxide synthase uncoupling impairs endothelial progenitor cell mobilization and function in diabetes, *Diabetes* **56**(3), 666-74 (2007).
13. H. Li, L. Xu, J. C. Dunbar, C. B. Dhabuwala, and A. A. Sima, Effects of C-peptide on expression of eNOS and iNOS in human cavernosal smooth muscle cells, *Urology* **64**(3), 622-7 (2004).
14. T. Wallerath, T. Kunt, T. Forst, E. I. Closs, R. Lehmann, T. Flohr, M. Gabriel, D. Schafer, A. Gopfert, A. Pfutzner, J. Beyer, and U. Forstermann, Stimulation of endothelial nitric oxide synthase by proinsulin C-peptide, *Nitric Oxide* **9**(2), 95-102 (2003).
15. L. H. Young, Y. Ikeda, R. Scalia, and A. M. Lefer, C-peptide exerts cardioprotective effects in myocardial ischemia-reperfusion, *Am J Physiol Heart Circ Physiol* **279**(4), H1453-9 (2000).
16. R. S. Sprague, M. L. Ellsworth, A. H. Stephenson, and A. J. Lonigro, ATP: the red blood cell link to NO and local control of the pulmonary circulation, *Am J Physiol* **271**(6 Pt 2), H2717-22 (1996).
17. R. S. Sprague, J. J. Olearczyk, D. M. Spence, A. H. Stephenson, R. W. Sprung, and A. J. Lonigro, Extracellular ATP signaling in the rabbit lung: erythrocytes as determinants of vascular resistance, *Am J Physiol Heart Circ Physiol* **285**(2), H693-700 (2003).
18. D. E. McMillan, N. G. Utterback, and J. La Puma, Reduced erythrocyte deformability in diabetes, *Diabetes* **27**(9), 895-901 (1978).
19. D. D. De La Tour, D. Raccah, M. F. Jannot, T. Coste, C. Rougerie, and P. Vague, Erythrocyte Na/K ATPase activity and diabetes: relationship with C-peptide level, *Diabetologia* **41**(9), 1080-4 (1998).
20. T. Kunt, S. Schneider, A. Pfutzner, K. Goitum, M. Engelbach, B. Schauf, J. Beyer, and T. Forst, The effect of human proinsulin C-peptide on erythrocyte deformability in patients with Type I diabetes mellitus, *Diabetologia* **42**(4), 465-71 (1999).
21. D. Raccah, C. Fabreguetts, J. P. Azulay, and P. Vague, Erythrocyte Na(+)-K(+)-ATPase activity, metabolic control, and neuropathy in IDDM patients, *Diabetes Care* **19**(6), 564-8 (1996).
22. T. Forst, D. D. De La Tour, T. Kunt, A. Pfutzner, K. Goitom, T. Pohlmann, S. Schneider, B. L. Johansson, J. Wahren, M. Lobig, M. Engelbach, J. Beyer, and P. Vague, Effects of proinsulin C-peptide on nitric oxide, microvascular blood flow and erythrocyte Na+,K+-ATPase activity in diabetes mellitus type I, *Clin Sci (Lond)* **98**(3), 283-90 (2000).
23. A. A. Sima, W. Zhang, and G. Grunberger, Type 1 diabetic neuropathy and C-peptide, *Exp Diabesity Res* **5**(1), 65-77 (2004).
24. C. B. Kremser, N. S. Levitt, K. M. Borow, J. B. Jaspan, C. Lindbloom, K. S. Polonsky, and A. R. Leff, Oxygen uptake kinetics during exercise in diabetic neuropathy, *J Appl Physiol* **65**(6), 2665-71 (1988).
25. D. A. Greene, and S. A. Lattimer, Impaired energy utilization and Na-K-ATPase in diabetic peripheral nerve, *Am J Physiol* **246**(4 Pt 1), E311-8 (1984).
26. W. Zhang, M. Yorek, C. R. Pierson, Y. Murakawa, A. Breidenbach, and A. A. Sima, Human C-peptide dose dependently prevents early neuropathy in the BB/Wor-rat, *Int J Exp Diabetes Res* **2**(3), 187-93 (2001).
27. Z. G. Li, and A. A. Sima, C-peptide and central nervous system complications in diabetes, *Exp Diabesity Res* **5**(1), 79-90 (2004).
28. F. Palm, D. G. Buerk, P. O. Carlsson, P. Hansell, and P. Liss, Reduced nitric oxide concentration in the renal cortex of streptozotocin-induced diabetic rats: effects on renal oxygenation and microcirculation, *Diabetes* **54**(11), 3282-7 (2005).
29. J. Shafqat, L. Juntti-Berggren, Z. Zhong, K. Ekberg, M. Kohler, P. O. Berggren, J. Johansson, J. Wahren, and H. Jornvall, Proinsulin C-peptide and its analogues induce intracellular Ca2+ increases in human renal tubular cells, *Cell Mol Life Sci* **59**(7), 1185-9 (2002).
30. N. M. Al-Rasheed, F. Meakin, E. L. Royal, A. J. Lewington, J. Brown, G. B. Willars, and N. J. Brunskill, Potent activation of multiple signalling pathways by C-peptide in opossum kidney proximal tubular cells, *Diabetologia* **47**(6), 987-97 (2004).
31. B. L. Johansson, K. Borg, E. Fernqvist-Forbes, A. Kernell, T. Odergren, and J. Wahren, Beneficial effects of C-peptide on incipient nephropathy and neuropathy in patients with Type 1 diabetes mellitus, *Diabet Med* **17**(3), 181-9 (2000).
32. T. Forst, T. Kunt, A. Pfutzner, J. Beyer, and J. Wahren, New aspects on biological activity of C-peptide in IDDM patients, *Exp Clin Endocrinol Diabetes* **106**(4), 270-6 (1998).
33. M. Nangaku, Chronic hypoxia and tubulointerstitial injury: a final common pathway to end-stage renal failure, *J Am Soc Nephrol* **17**(1), 17-25 (2006).
34. N. A. Lassen, O. Munck, and J. H. Thaysen, Oxygen consumption and sodium reabsorption in the kidney, *Acta Physiol Scand* **51**(371-84 (1961).
35. A. Korner, A. C. Eklof, G. Celsi, and A. Aperia, Increased renal metabolism in diabetes. Mechanism and functional implications, *Diabetes* **43**(5), 629-33 (1994).

36. L. Nordquist, Fasching, A., Palm, F C-peptide-Induced Reduction in Oxygen Consumption in Isolated Proximal Tubular Cells from Diabetic Rats is Inhibited by Pretreatment with Ouabain, but not L-NAME. In: International Society on Oxygen Transport to Tissue; 2007; Uppsala, Sweden; 2007.

37. E. C. Carlson, and N. J. Bjork, SEM and TEM analyses of isolated human retinal microvessel basement membranes in diabetic retinopathy, *Anat Rec* **226**(3), 295-306 (1990).

38. E. Beltramo, S. Buttiglieri, F. Pomero, A. Allione, F. D'Alu, E. Ponte, and M. Porta, A study of capillary pericyte viability on extracellular matrix produced by endothelial cells in high glucose, *Diabetologia* **46**(3), 409-15 (2003).

39. S. Chakrabarti, Z. A. Khan, M. Cukiernik, W. Zhang, and A. A. Sima, C-peptide and retinal microangiopathy in diabetes, *Exp Diabesity Res* **5**(1), 91-6 (2004).

40. Z. A. Khan, M. Cukiernik, J. R. Gonder, and S. Chakrabarti, Oncofetal fibronectin in diabetic retinopathy, *Invest Ophthalmol Vis Sci* **45**(1), 287-95 (2004).

41. P. Castellani, G. Viale, A. Dorcaratto, G. Nicolo, J. Kaczmarek, G. Querze, and L. Zardi, The fibronectin isoform containing the ED-B oncofetal domain: a marker of angiogenesis, *Int J Cancer* **59**(5), 612-8 (1994).

42. T. V. Karelina, and A. Z. Eisen, Interstitial collagenase and the ED-B oncofetal domain of fibronectin are markers of angiogenesis in human skin tumors, *Cancer Detect Prev* **22**(5), 438-44 (1998).

REDUCED OXYGENATION IN DIABETIC RAT KIDNEYS MEASURED BY T2* WEIGHTED MAGNETIC RESONANCE MICRO-IMAGING

Jenny Edlund[*], Peter Hansell[*], Angelica Fasching[*], Per Liss[†], Jan Weis[†], Jerry D. Glickson[‡] and Fredrik Palm[*§]

Abstract: By applying invasive techniques for direct measurements of oxygen tension, we have reported decreased kidney oxygenation in experimental diabetes in rats. However, the non-invasive MRI technique utilizing the BOLD effect provides several advantages with the possibility to perform repetitive measurements in the same animals and in human subjects. In this study, we applied a modified single gradient echo micro-imaging sequence to detect the BOLD effect in kidneys of diabetic rats and compared the results to normoglycemic controls.

All measurements were performed on inactin-anaesthetized adult male Wistar Furth rats. Diabetes was induced by streptozotocin (45 mg/kg) 14 days prior to MRI-analysis. Sixteen T2*-weighted image records (B_0=1.5 T) were performed using radiofrequency spoiled gradient echo sequence with 2.6 ms step increments of TE (TE_1=12 ms), while TR (75 ms) and bandwidth per pixel (71.4 Hz) were kept constant. T2* maps were computed by mono-exponential fitting of the pixel intensities.

Relaxation rates R2* (1/T2*) in cortex and outer stripe of the outer medulla were similar in both groups (cortex for controls 22.3±0.4 vs. diabetics 23.1±1.8 Hz and outer stripe of outer medulla for controls 24.9±0.4 vs. diabetics 26.4±1.8 Hz; n=4 in both groups), whereas R2* was increased in the inner stripe of the outer medulla in diabetic rats (diabetics 26.1±2.4 vs. controls 18.8±1.4 Hz; n=4, P<0.05).

This study demonstrates that experimental diabetes in rats induces decreased oxygenation of the renal outer medulla. Furthermore, the proposed T2*-weighted MR micro-imaging technique is suitable for detection of regional changes in kidney oxygenation in experimental animal models.

[*] Department of Medical Cell Biology, BMC, PO Box 571, Uppsala University, 751 23 Uppsala, Sweden
[†] Department of Oncology, Radiology and Clinical Immunology, University Hospital, 751 23 Uppsala, Sweden
[‡] Department of Radiology, University of Pennsylvania, Philadelphia, PA., USA
[§] Department of Medicine, Georgetown University, Washington, DC., USA

P. Liss et al. (eds.), *Oxygen Transport to Tissue XXX*, DOI 10.1007/978-0-387-85998-9_31,
© Springer Science+Business Media, LLC 2009

1. INTRODUCTION

The early phase after the onset of experimental diabetes in rats is closely associated with altered oxygen metabolism, mainly as a result of the profound hyperglycemia[1]. We have consistently reported that streptozotocin (STZ)-induced diabetic rats have reduced oxygen availability in the kidney secondary to increased oxygen consumption[2]. It has been hypothesized that the reduced oxygen availability in the diabetic kidney is a key mediator for the development of diabetic nephropathy and subsequent progressive renal dysfunction.

In our previous studies we have used Clark-type microelectrodes to monitor oxygen tension in different parts of the kidney during different experimental set-ups[3]. This invasive technique is reliable and reproducible but limits studies of a repetitive nature in the same animal. The non-invasive nature of the magnetic resonance imaging (MRI) makes it suitable for the clinical settings, but needs to be established and validated in animal models. Blood oxygenation level dependent (BOLD) MRI is sensitive to intravascular oxygenation and therefore has the potential to measure physiological relevant changes in oxygenation of biological tissues[4].

In this study, we used the BOLD effect sensitive spectroscopic micro-imaging technique to detect diabetes-induced alterations in the intrarenal oxygenation in rats.

2. MATERIALS AND METHODS

Age-matched male Wistar Furth rats weighing 230–250 g were purchased from M&B (Ry, Denmark). Animals had free access to tap water and standard rat chow (R3, Ewos, Södertälje, Sweden) throughout the study. All experiments were performed in accordance with the NIH guidelines for use and care of laboratory animals and approved by the Animal Care and Use Committee for Uppsala University.

2.1. Diabetes induction and surgical procedures

Diabetes mellitus was induced by an injection of STZ (45 mg/kg BW dissolved in saline; Sigma-Aldrich, St. Louis, MO) in the tail vein of four rats. Animals were considered diabetic if blood glucose concentrations increased to ≥ 15 mmol/l within 24 hours after STZ-injection and remained elevated. Blood glucose concentrations were determined with test reagent strips (MediSense, Bedford, MA) from blood samples obtained from the cut tip of the tail.

Fourteen days after allocation to the study, all rats were anaesthetized by an intraperitoneal injection of thiobutabarbital sodium (Inactin, 120 mg/kg BW for non-diabetic rats and 80 mg/kg BW for diabetic rats; Sigma-Aldrich). Tracheotomy was performed to ensure free airways.

2.2. Magnetic Resonance Imaging

The measurements were performed with a circular receiver coil with a diameter of 53 mm on a Gyroscan NT 1.5 T MR imaging system (Philips, Amsterdam, The Netherlands). BOLD effect sensitive T2* measurements of the rat kidneys were performed using magnetic resonance spectroscopic micro-imaging sequence as previously described[5]. The technique consisted of a standard 2D, radiofrequency spoiled fast gradient echo sequence in a partial echo version (partial factor=0.625). Sixteen T2*-weighted image records were performed with 2.6 ms step increments of time echo (TE_1=12 ms), while time repetition (75 ms) and bandwidth per pixel (71.4 Hz) were kept constant. Resolution in plane was 0.2x0.2 mm, slice thickness 1 mm. Measurement time for the whole spectroscopic acquisition was 15 minutes and 22 seconds. T2* distribution was computed by mono-exponential fitting of pixel intensities. Note that spectroscopic imaging property of our sequence enabled calculation of pure water, fat and chemical shift artifact-free micro-images[5].

2.3. Statistical analysis

All values are expressed as means±SEM. Student's t-test was used to compare means of two groups. Statistical analysis was performed using Graph Pad Prism software (Graph Pad Inc., San Diego, CA, USA), and P<0.05 was considered statistically significant.

3. RESULTS

STZ-injection resulted in a sustained hyperglycemia in all injected rats (20.9±1.0 vs. 5.8±0.7 mmol/l for control animals; n=4 each). The diabetic animals gained less weight compared to the normoglycemic controls (251±8 vs. 290±9 g; n=4 each).

Diabetic rats had increased R2* relaxation rates in the inner stripe of the outer medulla compared to controls (Fig. 1 and 2).

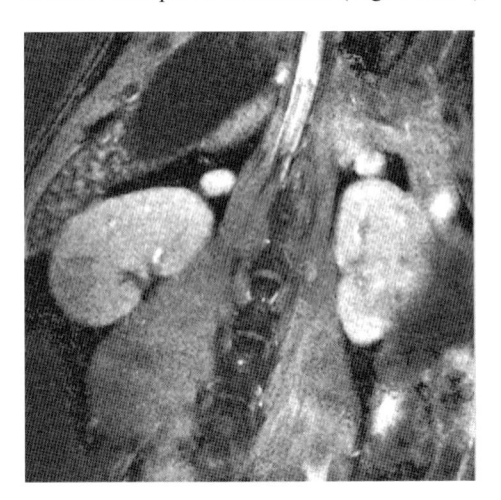 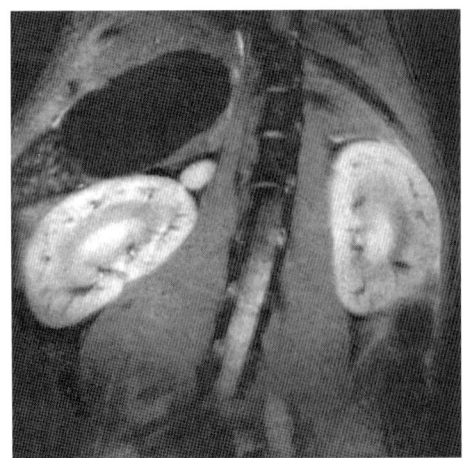

Figure 1. T2*-weighted image of normoglycemic control (left) and streptozotocin-diabetic (right) Wistar Furth rats. The inner (darker) stripe of the outer medulla of the diabetic rats displayed increased R2* relaxation rates.

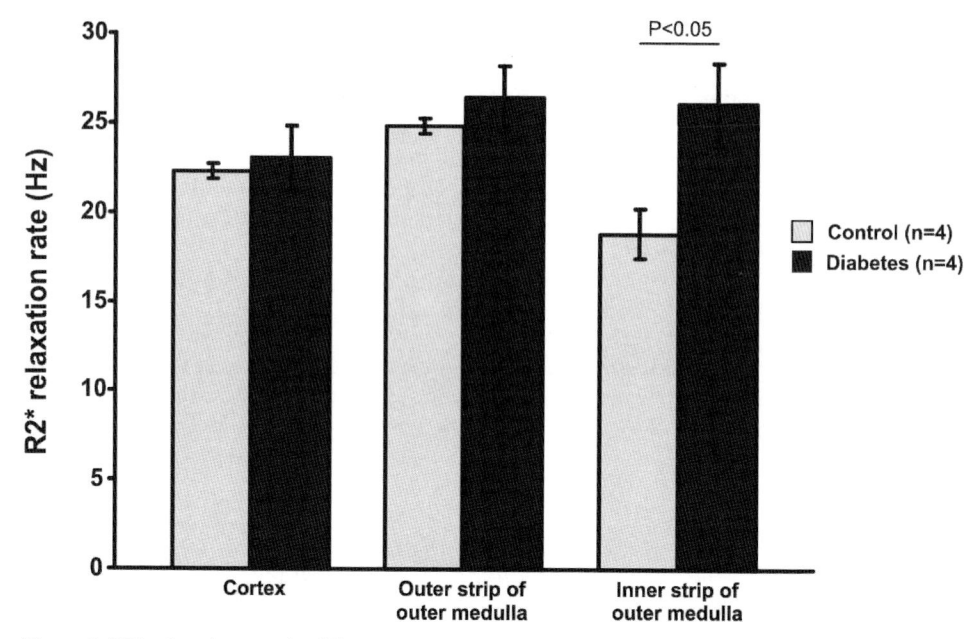

Figure 2. R2* relaxation rates in different parts of the kidneys of normoglycemic control and streptozotocin-diabetic Wistar Furth rats.

4. DISCUSSION

The main finding of the present study is that the previously reported reduced oxygen availability in the diabetic rat kidney[6-8] can also be detected by a non-invasive magnetic resonance micro-imaging method. BOLD effect sensitive T2*-weighted measurements of the rat kidneys has previously been performed using multiple-gradient echo sequence[9]. In this study, we used spectroscopic micro-imaging technique[10]. By increasing the echo time it is possible to acquire both T2* (BOLD) and spectroscopic information. The subsequent data processing allows reconstruction of the proton (^{1}H) spectra in each voxel of the measured slice. The spectrum consists of water and/or fat line in our case. In addition to R2* maps, it is possible to compute water, fat and chemical-shift artifact free micro-images. A significant feature of the described method is an improvement of the signal-to-noise ratio by narrowing the receiver bandwidth without the undesired consequence of increased chemical shift artifacts.

Utilizing the spectroscopic imaging sequence has similar drawbacks as standard BOLD techniques, predominantly severe influences by blood flow alterations and factors altering the hemoglobin dissociation curve[3]. The effect of alterations in the blood flow and/or vascular pattern have to be taken into consideration when utilizing the BOLD technique for oxygenation measurements, especially in tumors with a chaotic and constantly changing blood vessel pattern. However, the kidney has an extremely well organized vasculature and highly regulated internal blood flow, which potentially increases the usefulness of BOLD measurements. Furthermore, we have previously shown that STZ-diabetic rats have decreased intrarenal oxygen availability, despite similar intrarenal blood flow. Therefore, the detected difference in BOLD signal between control and diabetic rats might reflect a true difference in intravascular oxygen availability in the inner stripe of the outer medulla. However, our previous reports using the invasive microsensor technique have also detected reduced tissue oxygen availability in the more superficial kidney cortex and the outer stripe of the outer medulla, which could not be detected using the BOLD measurement. The most likely explanations for these discrepancies are the higher blood perfusion rate in combination with the relatively high baseline oxygen tension in these regions of the kidney. One may speculate that the increase in oxygen extraction from the blood in the diabetic kidney is below the detection level of the BOLD measurement, and therefore undetectable by this technique. Also previous studies using various BOLD effect sensitive MRI sequences have reported a lack of increase in BOLD signal from different pathological states known to reduce the oxygen tension in the kidney cortex, including diabetes and spontaneously hypertensive rats.

5. SUMMARY

Diabetes-induced decrease in intravascular oxygenation of the inner stripe of the outer medulla can be detected using T2*-weighted MRI. Our findings also show that this method is unable to detect differences in oxygenation in the kidney cortex and the outer stripe of the inner medulla under the present conditions of diabetes. However, the non-invasive nature of this technique makes it suitable for investigating if similar oxygenation changes in the inner parts of the kidney occur in diabetic patients.

6. ACKNOWLEDGEMENTS

This work was funded by The Swedish Research Council, The Marcus and Amalia Wallenberg Foundation, The Linné Foundation for Medical Research, The Swedish Diabetes Association, The Swedish Society for Medical Research, and NIH K-99 grant (DK-077858) to FP.

7. REFERENCES

1. F. Palm, Intrarenal oxygen in diabetes and a possible link to diabetic nephropathy, *Clin Exp Pharmacol Physiol* **33**(997-1001 (2006).
2. P. V. Prasad, Functional MRI of the kidney: tools for translational studies of pathophysiology of renal disease, *Am J Physiol Renal Physiol* **290**(5), F958-74 (2006).
3. L. Li, P. Storey, D. Kim, W. Li, and P. Prasad, Kidneys in hypertensive rats show reduced response to nitric oxide synthase inhibition as evaluated by BOLD MRI, *J Magn Reson Imaging* **17**(6), 671-5 (2003).
4. J. Weis, I. Frollo, and L. Budinsky, Magnetic field distribution measurement by the modified FLASH method, *Zeitschr Naturforsch* **44a**(1151-1154 (1989).
5. J. Weis, A. Ericsson, and A. Hemmingsson, Chemical shift artifact-free microscopy: spectroscopic microimaging of the human skin, *Magn Reson Med* **41**(5), 904-8 (1999).
6. F. Palm, D. G. Buerk, P. O. Carlsson, P. Hansell, and P. Liss, Reduced nitric oxide concentration in the renal cortex of streptozotocin-induced diabetic rats: effects on renal oxygenation and microcirculation, *Diabetes* **54**(11), 3282-3287 (2005).
7. F. Palm, J. Cederberg, P. Hansell, P. Liss, and P. O. Carlsson, Reactive oxygen species cause diabetes-induced decrease in renal oxygen tension, *Diabetologia* **46**(8), 1153-1160 (2003).

8. F. Palm, P. Hansell, G. Ronquist, A. Waldenstrom, P. Liss, and P. O. Carlsson, Polyol-pathway-dependent disturbances in renal medullary metabolism in experimental insulin-deficient diabetes mellitus in rats, *Diabetologia* **47**(7), 1223-1231 (2004).
9. F. A. Howe, S. P. Robinson, D. J. McIntyre, M. Stubbs, and J. R. Griffiths, Issues in flow and oxygenation dependent contrast (FLOOD) imaging of tumours, *NMR Biomed* **14**(7-8), 497-506 (2001).
10. E. A. dos Santos, L. P. Li, L. Ji, and P. V. Prasad, Early changes with diabetes in renal medullary hemodynamics as evaluated by fiberoptic probes and BOLD magnetic resonance imaging, *Invest Radiol* **42**(3), 157-62 (2007).

IDENTIFICATION AND DISTRIBUTION OF UNCOUPLING PROTEIN ISOFORMS IN THE NORMAL AND DIABETIC RAT KIDNEY

Malou Friederich[*†], Lina Nordquist[*†], Johan Olerud[*], Magnus Johansson[‡], Peter Hansell[*], and Fredrik Palm[*§]

Abstract: Uncoupling protein (UCP)-2 and -3 are ubiquitously expressed throughout the body but there is currently no information regarding the expression and distribution of the different UCP isoforms in the kidney. Due to the known cross-reactivity of the antibodies presently available for detection of UCP-2 and -3 proteins, we measured the mRNA expression of UCP-1, -2 and -3 in the rat kidney in order to detect the kidney-specific UCP isoforms. Thereafter, we determined the intrarenal distribution of the detected UCP isoforms using immunohistochemistry. Thereafter, we compared the protein levels in control and streptozotocin-induced diabetic rats using Western blot. Expressions of the UCP isoforms were also performed in brown adipose tissue and heart as positive controls for UCP-1 and 3, respectively.

UCP-2 mRNA was the only isoform detected in the kidney. UCP-2 protein expression in the kidney cortex was localized to proximal tubular cells, but not glomerulus or distal nephron. In the medulla, UCP-2 was localized to cells of the medullary thick ascending loop of Henle, but not to the vasculature or parts of the nephron located in the inner medulla. Western blot showed that diabetic kidneys have about 2.5-fold higher UCP-2 levels compared to controls.

In conclusion, UCP-2 is the only isoform detectable in the kidney and UCP-2 protein can be detected in proximal tubular cells and cells of the medullary thick ascending loop of Henle. Furthermore, diabetic rats have increased UCP-2 levels compared to controls, but the mechanisms underlying this increase and its consequences warrants further studies.

[*] Department of Medical Cell Biology, BMC, PO 571, Uppsala University, 751 23 Uppsala, Sweden
[†] These authors contributed equally to this study.
[‡] Department of Pathology, University of California, San Francisco, CA., USA
[§] Department of Medicine, Georgetown University, Washington, DC., USA

P. Liss et al. (eds.), *Oxygen Transport to Tissue XXX*, DOI 10.1007/978-0-387-85998-9_32,

1. INTRODUCTION

Normal cellular function is dependent on the availability of adenosine triphosphate (ATP), which is mainly produced by mitochondrial oxidative phosphorylation. Oxidative phosphorylation includes electrons that are donated from NADH and $FADH_2$, both produced in the Krebs cycle, to the first of four complexes in the mitochondrial electron transport chain. Electrons are thereafter transferred through the four complexes to oxygen, being the ultimate electron acceptor. This process is accompanied by protons being pumped across the inner mitochondrial membrane, which gradually creates a proton gradient. ATP-synthase, sometimes referred to as the fifth complex, allows for a controlled transport of protons back across the membrane during the production of ATP from adenosine diphosphate (ADP) and inorganic phosphate.

Sustained hyperglycemia is closely associated with increased production of reactive oxygen species (ROS) and subsequently increased oxidative stress[1-3]. The mitochondrial electron transport chain has been hypothesized to be a major source of superoxide radicals[3]. Reducing the high mitochondrial membrane potential can reduce the formation of superoxide radicals, which can be achieved by uncoupling of the mitochondrial electron transport chain. Mitochondrial uncoupling refers to the phenomenon of maintained mitochondrial respiration without any formation of ATP, which also can be referred to as proton leak. The basal proton leakage across the mitochondrial membrane is about 15% of the total mitochondrial respiration[4]. However, the basal proton leakage should be separated from any induced leakage intended to reduce the formation of ROS. A possible mechanism for increased mitochondrial uncoupling is to increase the mitochondrial content of uncoupling proteins (UCP), which has been shown to decrease the formation of ROS[5-8]. UCP-1 to -5 are members of a large family of mitochondrial anion carriers. The first discovered UCP was UCP-1, which also was called thermogenin. UCP-1 is predominantly found in brown adipose tissue (BAT), where its main purpose is to create heat in order to sustain non-shivering thermogenesis in for instance hibernating animals[9]. Lately, UCP-2 and -3 have gained interest due to their uncoupling ability, which reduces the mitochondrial formation of ROS. UCP-2 and -3 share 59% and 57% sequence similarity with UCP-1, respectively. It is still debated if UCP-2 and -3 really have the capability to uncouple mitochondria in vivo, but several studies have reported the involvement of these proteins for the regulation of mitochondrial ROS formation[10-14].

About 30-40% of all patients with insulinopenic diabetes develop diabetic nephropathy, eventually resulting in end-stage renal disease requiring dialysis or transplantation. Animal studies have shown that oxidative stress is closely associated with diabetic nephropathy[15]. We have previously proposed that UCP can be protective against diabetes-induced ROS production[16]. Indeed, we reported increased UCP-2 protein levels in kidneys from diabetic rats, both the kidney cortex and medulla. These results were obtained by Western blot with antibodies directed against UCP-2. The current literature states that both UCP-2 and -3 are ubiquitously expressed throughout the body and there is currently no information regarding the expression and distribution of the different UCP isoforms in the kidney. Due to the known cross-reactivity of the antibodies presently available for detection of UCP-2 and -3 proteins Western blot is not the preferred method for investigating this issue. We designed a set of experiments to identify the UCP isoforms expressed in control and diabetic rat kidneys. First, we used real-time polymerase chain reaction (RT-PCR) to identify which UCP mRNA is detectable and thereafter utilized immunohistochemistry to localize which cells are expressing the proteins and thereafter measured the protein levels of the detected UCP isoform in control and diabetic kidney tissue.

2. MATERIALS AND METHODS

Age-matched male Wistar Furth rats weighing 250-300g were purchased from B&K Universal (Sollentuna, Sweden). They had free access to tap water and standard rat chow (R3, Ewos, Södertälje, Sweden) throughout the study. All experiments were performed in accordance with the NIH guidelines for use and care of laboratory animals and approved by the Animal Care and Use Committee for Uppsala University.

2.1. Diabetes induction and surgical procedures

Diabetes mellitus was induced by an injection of streptozotocin (STZ; 45 mg/kg BW dissolved in saline; Sigma-Aldrich, St. Louis, MO) in the tail vein. Animals were considered diabetic if blood glucose concentrations increased to $\geq$ 15 mmol/l within 24 hours after STZ-injection and remained elevated. Blood glucose concentrations were determined with test reagent strips (Medisense, Bedford, MA) from blood samples obtained from the cut tip of the tail. All rats were anaesthetized with an intraperitoneal injection of thiobutabarbital sodium, (Inactin, 120 mg/kg BW for non-diabetic rats and 80 mg/kg BW for diabetic rats; Sigma-Aldrich) and placed on a servo-regulated heating pad. Tracheotomy was performed to ensure sufficient breathing. A polyethylene catheter was placed in a carotid artery and perfused with 20 ml of ice-cold phosphate buffer saline (Medicago AB, Uppsala, Sweden), and the right renal vein was cut open in order to facilitate complete perfusion of the kidneys.

2.2. Semi quantitative Real time PCR

Total RNA was isolated with the guanidinium-based lysis buffer method with RNAquous-4 PCR Kit (Ambion, Austin, TX) and treated with DNaseI. RT reactions were performed using Superscript III first strand cDNA synthesis (Invitrogen, Carlsbad, CA). Amplification was obtained with a Lightcycler system (Roche-Diagnostic, Lewers, UK) using DyNAmo™ Capillary SYBR® Green qPCR Kit (Finnzymes, Espoo, Finland). Beta-actin was used as house keeping gene. For primer sequences used, see Table 1. PCR products were run on 1.8% agarose gel for size identification.

Table 1. Primer sequences used in the present study[a].

Gene	Sense primer (5′-3′)	Anti-sense primer (5′-3′)	Product size (bp)
UCP-1	GTGAAGGTCAGAATGCAAGC	AGGGCCCCCTTCATGAGGTC	199
UCP-2	GCATTGGCCTCTACGACTCT	CTGGAAGCGGACCTTTACC	151
UCP-3	TGCAGCCTGTTTTGCTGATCT	GGGTTCTCCCCTTGGATCTG	80
B-actin	CCACCGATCCACACAGACTACTTG	GCTCTGCGTCCTAGCACC	76

[a]Primers obtained from MWG Biotech, Ebersberg, Germany.

2.3. Immunohistochemistry

Kidneys were fixed with 4% formaldehyde, dehydrated and embedded in paraffin. 5μm sections were stained as previously described[17]. Sections (5 μm) were deparaffinized and immersed in ethanol with concentration gradients and were then heated in citrate solution (0.01 M, pH 6.0) for antigen retrieval. Endogenous peroxidase activity was blocked using 3% H_2O_2 and non-specific binding was prevented by blocking with normal goat serum (Santa Cruz Biotechnology, Heidelberg, Germany). Thereafter, the sections were incubated with an antibody against UCP-2 in a 1:50 dilution overnight at 4°C. The sections were rinsed with TBST and incubated with a biotinylated secondary antibody against goat IgG (Santa Cruz Biotechnology; 1:500). After rinsing with TBST, the sections were incubated with an avidin biotin enzyme reagent (ABC Elite Kit; Vector laboratories, Burlingame, CA, USA). Labeling was visualized using a peroxide substrate solution with 0.8 mM DAB and 0.01% H_2O_2. The sections were subsequently counterstained with hematoxylin and mounted.

2.4. Western blot

Samples were homogenized in 700 μl buffer (1.0% NP40, 0.5% sodium deoxycholate, 0.1% SDS, 10 mM NaF, 80mM Tris, pH 7.5) containing enzyme inhibitors (Phosphatase inhibitor cocktail-2; 10 μl/ml; Sigma-Aldrich, and Complete Mini; 1 tablet/1.5 ml; Roche Diagnostics, Mannheim, Germany). Samples were run on 12.5% Tris-HCl gels with Tris/glycine/SDS buffer and the proteins detected, after transfer to nitrocellulose membranes, using goat anti-rat UCP-2 antibody (1:1000; Santa Cruz Biotechnology, Santa Cruz, CA) and HRP-conjugated secondary antibody (rabbit anti-goat, 1:10,000; Kirkegaard and Perry Laboratories, Gaithersburg, MD) by an ECL-camera (Kodak image station 2000; New Haven, CT). Beta-actin was detected using mouse anti-rat beta-actin antibody (1:10,000; Sigma Aldrich) and secondary HRP-conjugated goat-anti mouse antibody (1:60,000; Kirkegaard and Perry Laboratories).

2.5. Statistical analysis

All values are expressed as means±SEM. Student's t-test was used to compare means of two groups. Statistical analysis was performed using Graph Pad Prism software (Graph Pad Inc., San Diego, CA, USA), and $P<0.05$ was considered statistically significant.

3. RESULTS

The STZ-injected rats developed a sustained hyperglycemia compared to controls (23.5±1.0 vs. 6.1±0.8 mmol/l; n=5/group).

mRNA expression of UCP- 1 was exclusively found in BAT (Fig. 1; left panel), UCP-2 mRNA was found in BAT, heart and kidneys (Fig 1; middle panel), and UCP-3 mRNA was found in both BAT and heart (Fig. 1; right panel). Highest expression of UCP-2 mRNA within the kidney was found in the medulla (Fig. 1 middle panel). UCP-2 was increased in diabetic kidneys. The absence of UCP-1 and -3 in the kidney was confirmed by visualizing the PCR products on an agarose gel (Fig. 2).

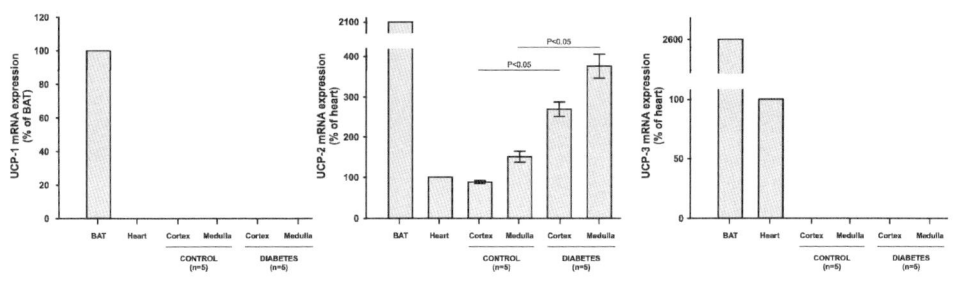

Figure 1. UCP-1 (left panel), UCP-2 (middle panel) and UCP-3 (right panel) mRNA expressions in brown adipose tissue (BAT) and heart in normoglycemic Wistar Furth rats and kidney cortex and medulla of normoglycemic control and diabetic Wistar Furth rats.

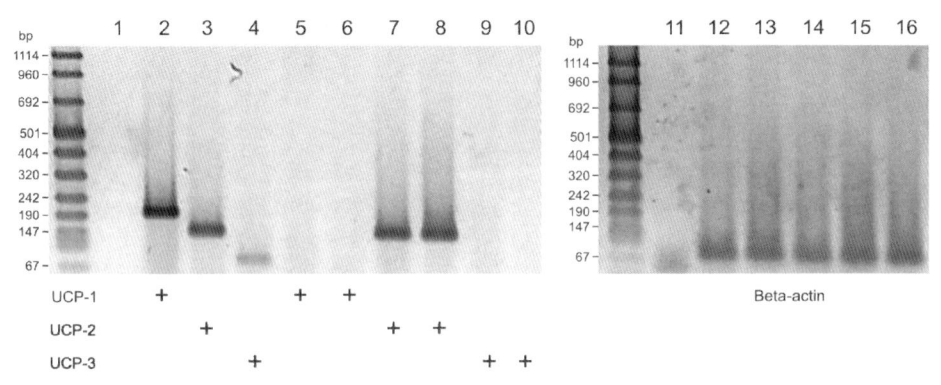

Figure 2. Left panel: PCR products from negative control (lane 1) positive controls (2:BAT, 3:kidney, 4: heart) and kidney tissue (5, 7, 9: normoglycemic controls and 6, 8, 10: diabetics). Right panel: Beta-actin content of negative control (11), BAT (12), heart (13) and kidney samples (14-16).

Immunohistochemistry revealed UCP-2 protein expression in proximal tubular cells in the kidney cortex (Fig. 3A) and cells of medullary thick ascending loop of Henle (Fig. 3B-D). No staining was detected in glomerulus, distal nephron, vascular bundles of the outer medulla or cells of the nephron located in the inner medulla (Fig. 3A-D).

4. DISCUSSION

The main finding of the present study is that UCP-2 is the only isoform detectable in the kidney, whereas neither UCP-1 nor UCP-3 mRNA could be detected using RT-PCR. Furthermore, proximal tubular cells and cells of the medullary thick ascending loop of Henle stained positive for UCP-2 protein within the kidney. Also, increased UCP-2 mRNA expression in kidneys of diabetic rats correlated with increased UCP-2 protein expression.

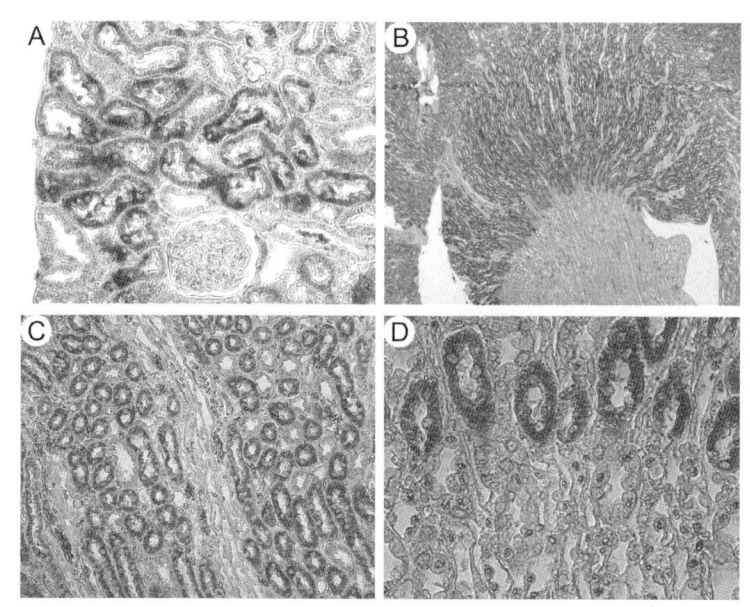

Figure 3. Immunohistochemistry directed against UCP-2 in kidneys of Wistar Furth rats. *A*. Staining of proximal tubular cells, but not glomerulus or cells in the distal nephron. *B*. Staining of outer medullary region. *C*. Staining of cells of the medullary thick ascending loop of Henle, but not vascular bundles. *D*. High magnification of mTAL cells positive for UCP-2, whereas no staining occurs in other cells of the loop of Henle.

Diabetic rats had 2.5-fold higher UCP-2 protein levels in the kidney compared to normoglycemic controls (Fig. 4).

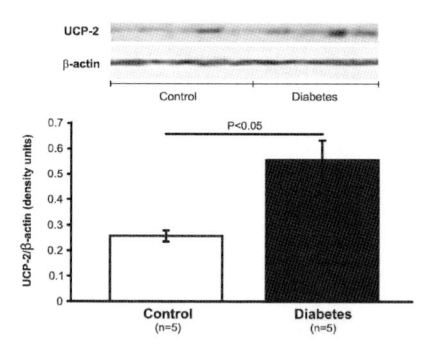

Figure 4. UCP-2 protein expression in kidneys from normoglycemic controls and diabetic Wister Furth rats.

As expected, UCP-1 mRNA was highly expressed in BAT, whereas UCP-3 mRNA could be detected in both BAT and heart tissue. We have previously shown that UCP-2 protein expression is increased in both kidney cortex and medulla of diabetic rats[16]. We now confirmed that the increased protein expression correlates to increased mRNA levels, suggesting increased protein synthesis as a main mechanism for the observed increase.

Immunohistochemistry revealed that UCP-2 is localized to proximal tubular cells, but not glomerulus or distal tubulus in the kidney cortex. In the medulla, UCP-2 is localized to medullary thick ascending loop of Henle cells, but not to the vasculature or tubular cells located in the inner medulla. This is very interesting because both proximal tubular cells and medullary thick ascending loop of Henle cells have a remarkable high amount of mitochondrial content[18]. Also, these two cell types are responsible for most of the intrarenal oxygen consumption related to tubular transport of electrolytes. We have previously reported that oxygen consumption by both these cell types is increased in diabetic rats compared to corresponding normoglycemic controls[15, 19, 20]. This is expected from the present and previous results since more oxygen will be consumed in order to sustain ATP production when mitochondrial uncoupling is induced.

Several possible mechanisms have been proposed, and shown, to be involved in the diabetes-induced increased oxygen consumption, including increased oxidative stress[15], increased polyol pathway activity[19], and decreased renomedullary nitric oxide levels [21]. Since UCPs are known to be activated by superoxide and fatty acids[22, 23], it can be hypothesized that hyperglycemia induces oxidative stress, activating uncoupling proteins, thereby inducing increased oxygen consumption that ultimately results in decreased oxygen availability in the diabetic kidney. Thus, these events may contribute to the development of diabetic nephropathy. However, this hypothesis does not exclude the involvement of other mechanisms. Other ROS sources than the mitochondria, e.g. the membrane bound NADPH oxidase and xanthine oxidase, might very well be important or even more important contributors to the increased oxidative stress commonly observed during hyperglycemia. Reduced nitric oxide bioavailability has also been shown to occur in the diabetic kidney cortex[24].

5. SUMMARY

UCP-2 is the only isoform present in the kidney, whereas neither UCP-1 nor UCP-3 mRNA could be detected anywhere in the kidney. Within the kidney, UCP-2 protein is only localized to proximal tubular cells and medullary thick ascending loop of Henle cells. The present study confirmed previous reports showing increased UCP-2 protein expression in the diabetic kidney, which also correlates to increased mRNA levels.

6. ACKNOWLEDGEMENTS

This work was funded by The Swedish Science Council Medicine (9940, 10840), The Marcus and Amalia Wallenberg Foundation, The Linné Foundation for Medical Research, The Swedish Diabetes Association, and The Swedish Society for Medical Research and NIH K-99 grant (DK-077858) to FP.

7. REFERENCES

1. T. Nishikawa, D. Edelstein, and M. Brownlee, The missing link: a single unifying mechanism for diabetic complications, *Kidney Int Suppl* **77**(S26-30 (2000).
2. R. P. Robertson, Chronic oxidative stress as a central mechanism for glucose toxicity in pancreatic islet beta cells in diabetes, *J Biol Chem* **279**(41), 42351-4 (2004).
3. M. Brownlee, Biochemistry and molecular cell biology of diabetic complications, *Nature* **414**(6865), 813-20 (2001).

4. B. B. Lowell, and G. I. Shulman, Mitochondrial dysfunction and type 2 diabetes, *Science* **307**(5708), 384-7 (2005).
5. S. S. Korshunov, V. P. Skulachev, and A. A. Starkov, High protonic potential actuates a mechanism of production of reactive oxygen species in mitochondria, *FEBS Lett* **416**(1), 15-8 (1997).
6. S. S. Liu, Generating, partitioning, targeting and functioning of superoxide in mitochondria, *Biosci Rep* **17**(3), 259-72 (1997).
7. M. D. Brand, Uncoupling to survive? The role of mitochondrial inefficiency in ageing, *Exp Gerontol* **35**(6-7), 811-20 (2000).
8. T. Nishikawa, D. Edelstein, X. L. Du, S. Yamagishi, T. Matsumura, Y. Kaneda, M. A. Yorek, D. Beebe, P. J. Oates, H. P. Hammes, I. Giardino, and M. Brownlee, Normalizing mitochondrial superoxide production blocks three pathways of hyperglycaemic damage, *Nature* **404**(6779), 787-90 (2000).
9. P. Jezek, Possible physiological roles of mitochondrial uncoupling proteins--UCPn, *Int J Biochem Cell Biol* **34**(10), 1190-206 (2002).
10. A. G. Dulloo, and S. Samec, Uncoupling proteins: their roles in adaptive thermogenesis and substrate metabolism reconsidered, *Br J Nutr* **86**(2), 123-39 (2001).
11. A. Negre-Salvayre, C. Hirtz, G. Carrera, R. Cazenave, M. Troly, R. Salvayre, L. Penicaud, and L. Casteilla, A role for uncoupling protein-2 as a regulator of mitochondrial hydrogen peroxide generation, *Faseb J* **11**(10), 809-15 (1997).
12. C. Duval, A. Negre-Salvayre, A. Dogilo, R. Salvayre, L. Penicaud, and L. Casteilla, Increased reactive oxygen species production with antisense oligonucleotides directed against uncoupling protein 2 in murine endothelial cells, *Biochem Cell Biol* **80**(6), 757-64 (2002).
13. K. S. Echtay, E. Winkler, K. Frischmuth, and M. Klingenberg, Uncoupling proteins 2 and 3 are highly active H(+) transporters and highly nucleotide sensitive when activated by coenzyme Q (ubiquinone), *Proc Natl Acad Sci U S A* **98**(4), 1416-21 (2001).
14. A. M. Vincent, J. A. Olzmann, M. Brownlee, W. I. Sivitz, and J. W. Russell, Uncoupling proteins prevent glucose-induced neuronal oxidative stress and programmed cell death, *Diabetes* **53**(3), 726-34 (2004).
15. F. Palm, J. Cederberg, P. Hansell, P. Liss, and P. O. Carlsson, Reactive oxygen species cause diabetes-induced decrease in renal oxygen tension, *Diabetologia* **46**(8), 1153-60 (2003).
16. M. Friederich, J. Olerud, A. Fasching, P. Liss, P. Hansell, and F. Palm, Uncoupling protein 2 in diabetic kidneys: Increased protein expression correlates to increased non-transport related oxygen consumption, *Adv Exp Med Biol* **In press**((2007).
17. A. Tojo, M. Kimoto, and C. S. Wilcox, Renal expression of constitutive NOS and DDAH: separate effects of salt intake and angiotensin, *Kidney Int* **58**(5), 2075-83 (2000).
18. F. Palm, and P. O. Carlsson, Thick ascending tubular cells in the loop of Henle: regulation of electrolyte homeostasis, *Int J Biochem Cell Biol* **37**(8), 1554-9 (2005).
19. F. Palm, P. Hansell, G. Ronquist, A. Waldenstrom, P. Liss, and P. O. Carlsson, Polyol-pathway-dependent disturbances in renal medullary metabolism in experimental insulin-deficient diabetes mellitus in rats, *Diabetologia* **47**(7), 1223-31 (2004).
20. A. Korner, A. C. Eklof, G. Celsi, and A. Aperia, Increased renal metabolism in diabetes. Mechanism and functional implications, *Diabetes* **43**(5), 629-33 (1994).
21. F. Palm, L. Nordquist, and D. G. Buerk, Nitric oxide in the kidney; direct measurements of bioavailable renal nitric oxide, *Adv Exp Med Biol* **599**(117-23 (2007).
22. K. S. Echtay, D. Roussel, J. St-Pierre, M. B. Jekabsons, S. Cadenas, J. A. Stuart, J. A. Harper, S. J. Roebuck, A. Morrison, S. Pickering, J. C. Clapham, and M. D. Brand, Superoxide activates mitochondrial uncoupling proteins, *Nature* **415**(6867), 96-9 (2002).
23. K. S. Echtay, T. C. Esteves, J. L. Pakay, M. B. Jekabsons, A. J. Lambert, M. Portero-Otin, R. Pamplona, A. J. Vidal-Puig, S. Wang, S. J. Roebuck, and M. D. Brand, A signalling role for 4-hydroxy-2-nonenal in regulation of mitochondrial uncoupling, *Embo J* **22**(16), 4103-10 (2003).
24. F. Palm, D. G. Buerk, P. O. Carlsson, P. Hansell, and P. Liss, Reduced nitric oxide concentration in the renal cortex of streptozotocin-induced diabetic rats: effects on renal oxygenation and microcirculation, *Diabetes* **54**(11), 3282-7 (2005).

IODINATED CONTRAST MEDIA DECREASE RENOMEDULLARY BLOOD FLOW

A possible cause of contrast media-induced nephropathy

Per Liss, Peter Hansell, Per-Ola Carlsson, Angelica Fasching, and Fredrik Palm[*]

Abstract: The renal medulla has been implicated as a key target for contrast media-induced nephropathy (CIN). Although the effects of contrast media (CM) on whole kidney blood flow are well characterized, the effect of CM on renal medullary blood flow has been controversial. It has been reported that an extremely high dose of a high osmolar CM (iothalamate; 2900 mg I/kg bw) injected rapidly increased the renal outer medullary blood flow (OMBF). However, more clinical relevant doses consistently result in a sustained decrease in medullary blood flow. Furthermore, simultaneous measurements using both laser-Doppler flowmetry and hydrogen washout yield similar results of a decrease in OMBF after CM administration. CM induced a transient 28% decrease in the laser-Doppler signal from the outer medulla, while the hydrogen washout rate in the same region was reduced by approximately 50%. Furthermore, CM administration consistently results in decreased medullary oxygen tension (P_{O2}). The renal medulla works already during normal physiological conditions at the verge of hypoxia, and the majority of the studies published so far are in agreement with the hypothesis that CIN may have its origin in a further reduction in blood flow and/or oxygen availability of this region of the kidney.

[*] PL, Department of Oncology, Radiology and Clinical Immunology, Uppsala University, University Hospital, SE 751 85 Uppsala, Sweden. PH, POC, AF, FP, Department of Medical Cell Biology, Uppsala University, Biomedical Center, PO Box 571, SE 751 23 Uppsala, Sweden.

P. Liss et al. (eds.), *Oxygen Transport to Tissue XXX*, DOI 10.1007/978-0-387-85998-9_33, **213**

1. INTRODUCTION

Acute renal failure following intravenous administration of contrast media (CM) for radiographic purposes is a feared complication, accounting for 10–15% of hospital-acquired renal failure[1]. Increased frequency of acute renal failure is observed in patients with already impaired renal function[2, 3]. An especially large group of patients with increased risk comprises diabetics with nephropathy[4, 5]. These patients frequently suffer from other diabetic complications, e.g., atherosclerosis with concomitant angina pectoris or claudicatio intermittens, which increase the necessity for angiography using CM. Several mechanisms have been suggested to explain the pathogenesis of CM-induced nephropathy, including blood flow changes, direct tubular toxicity and intratubular obstruction[6, 7]. Hemodynamic changes may be involved, since a vasodilation occurs in most regional vascular beds following CM administration. However, this phase is absent in the renal medulla and instead replaced by a sustained vasoconstriction (Fig. 1)[8, 9]. Unique for the kidney is the highly specialized vascular architecture, with only about 10% of renal blood directed through *vasa recta*, which results in that the medulla already during normal conditions functions at the brink of hypoxia[10]. Further reduced medullary baseline P_{O2} has been reported during experimental diabetes[11, 12].

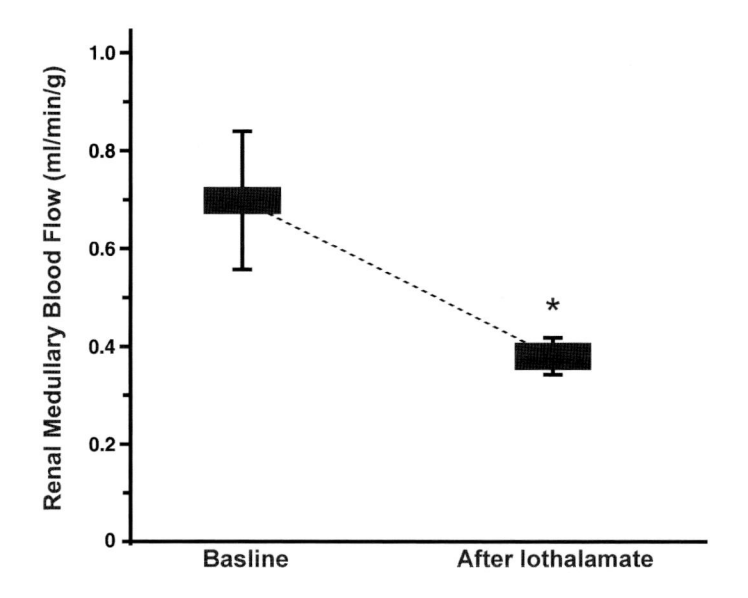

Figure 1. Renomedullary blood flow during baseline and after intravenous injection of iothalamate as estimated from the rate of hydrogen washout. * denotes p<0.05 versus baseline blood flow. Modified from Liss *et al.*[8].

Thus far, the vast majority of data published support the hypothesis that intravenous administration of iodinated CM results in decreased renomedullary blood flow and oxygen availability (Fig. 2)[13]. Several factors have been proposed to explain the changes in renal function occurring after injection of CM. The CM may directly stimulate renal vasoconstrictor receptors, release renal vasoconstrictors, inhibit renal vasodilators, induce formation of oxygen-derived free radicals or change physical factors. Some of the factors potentially involved are the renin-angiotensin system[14], prostaglandins[15], calcium[16], nitric oxide[17], adenosine[18], high osmolarity[19] and the viscosity of the CM[20].

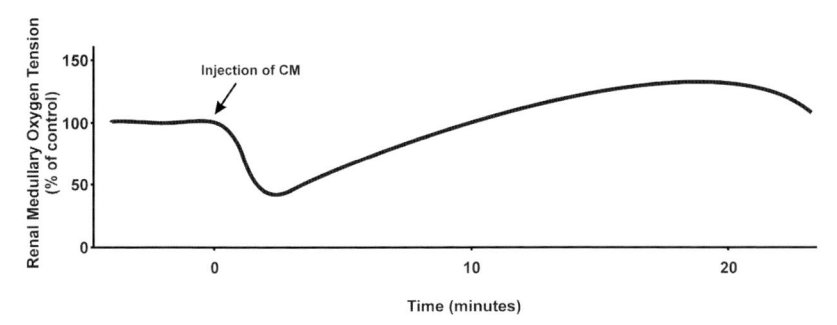

Figure 2. Renomedullary oxygen tension during baseline and after intravenous injection of iodinated CM. Adapted from Liss et al.[13].

2. INFLUENCE OF VASOACTIVE SUBSTANCES ON THE DECREASED MEDULLARY BLOOD FLOW FOLLOWING CM ADMINISTRATION

2.1. Role of Endothelin

Endothelin (ET) is released into the bloodstream following CM injection. Due to its prominent vasoactive properties, the increased risk of acute renal failure after CM injection has been suggested to result from ET-induced renal medullary vasoconstriction[21]. However, several investigators have found no attenuation of CM-induced reduction of whole renal blood flow with use of ET receptor antagonists[22, 23]. The beneficial effect of the ET-A receptor antagonists on CM-induced renal dysfunction does not seem to be primarily mediated on the hemodynamic level but may rather involve tubular transport mechanisms. Results have shown that the depression of outer medullary P_{O2} caused by injection of iopromide is partly mediated over ET-A receptors, while the decrease in outer medullary blood flow is not[24]. ET has previously been shown to directly inhibit Na^+/K^+-ATPase activity in the proximal tubule and inner medullary collecting duct. However, the effect of ET on tubular function is complex and depends on the species examined and dose of ET-1 used[25]. In the dog, low ET-1 doses increase the sodium excretion, while higher doses are markedly antinatriuretic[26]. ET-A receptor antagonists may inhibit the Na^+/K^+-ATPase activity in the thick ascending limb of the loop of Henle and thus prevent the increase in oxygen consumption due to the sodium reabsorption. Furthermore, Rabelink et al.[27] have shown that ET, at a dose that does not alter the blood pressure or renal blood flow, causes significant antidiuresis and antinatriuresis. Interestingly, Khimenko et al.[28] found that ET-1 and the ET-A receptors are involved in polymorphonuclear leukocyte and red blood cell activation in the lung. Heyman et al.[29] have reported trapping of red blood cells in the outer medulla following injection of CM, indirectly indicating a decrease in regional blood flow. We have reported similar findings[30], and it may be speculated that the beneficial effect of ET-A receptor blockade is mediated through a decrease in red blood cell aggregation.

2.2. Role of Adenosine

It has been shown that the depression in outer medullary blood flow and oxygen tension caused by injection of the CM iopromide does not primarily involve the adenosine A1-receptor[31]. Furthermore, these results are in line with those of Oldroyd *et al.*[18] showing that CM-induced depression in renal plasma flow is not attributable to adenosine. However, the adenosine A1-receptor is tonically active during normal physiological conditions and regulate renal hemodynamics, P_{O2} and urine output. Administration of a selective adenosine A1-receptor antagonist resulted in increased cortical blood flow, glomerular filtration rate (GFR) and urinary fluid and electrolyte excretion, whereas outer medullary P_{O2} decreased. This suggests an important tonic activity of adenosine A1-receptors on baseline renal function in normal rats. These actions are in line with studies on the effects of adenosine *per se* with reduced GFR[32], reduced total renal blood flow[33], elevated outer medullary P_{O2}[34] and reduced fluid-electrolyte excretion[35]. The effects on hemodynamics and solute transport fit well with the observations of adenosine receptors on both vascular and tubular structures[36]. The CM-induced reduction in GFR was in that study shown to be mediated by adenosine independent of a change in vascular resistance and possibly secondary to mesangial cell contraction. The mechanism responsible for the reduction in outer medullary P_{O2} after adenosine A1-receptor inhibition is most likely an elevated GFR. This will increase the workload of the tubular system to increase the relative reabsorption of sodium. The metabolic demand of the kidney is in direct linear relation to the reabsorption of sodium; thus, an elevation in the filtered load of sodium will result in an elevation in sodium reabsorption, which increases the oxygen consumption in the cells of the medullary thick ascending limb of the loop of Henle. This will result in a reduction in outer medullary P_{O2}. Although other mechanisms may be involved as well, the given explanation seems most plausible to explain how the adenosine A1-receptor antagonist decreases outer medullary P_{O2} even in the light of an elevated blood perfusion. It is concluded that in the normal rat adenosine A1-receptors are not important mediators in the CM-induced depression of OMBF and outer medullary P_{O2}[31].

3. SUMMARY

Several factors have been proposed to explain the changes in renal function occurring after injection of CM. CM may directly stimulate renal vasoconstrictor receptors, release renal vasoconstrictors, inhibit renal vasodilators, induce formation of oxygen-derived free radicals or change physical factors. These might include the renin-angiotensin system, prostaglandins, catecholamines, nitric oxide, adenosine and endothelin. Furthermore, the high osmolarity and the viscosity of the CM have effects on the renal hemodynamics which may include changes in renal interstitial pressure and red blood cell aggregation. It is likely that many factors are involved in the development of CIN. However, the vast majority of data so far strengthen the hypothesis that CIN may result from reduced blood flow and/or oxygenation of the renal medulla.

4. ACKNOWLEDGEMENTS

Financial support was provided by the Swedish Medical Research Council (projects no 10840 and 15040), the Marcus and Amalia Wallenberg Foundation, the Gunvor and

Josef Anér Foundation, the Knut and Alice Wallenberg Foundation and the Swedish Society for Medical Research.

5. REFERENCES

1. S. H. Hou, D. A. Bushinsky, J. B. Wish, and J. J. Cohen, Hospital-acquired renal insufficiency. A prospective study, *Am J Med* **63**(243-248 (1983).
2. A. S. Berns, Nephrotoxicity of contrast media, *Kidney Int* **36**(730-740 (1989).
3. C. J. Davidson, M. Hlatky, K. G. Morris, K. Pieper, T. N. Skelton, S. J. Schwab, and T. M. Bashore, Cardiovascular and renal toxicity of a nonionic radiographic contrast agent after cardiac catheterization. A prospective trial, *Ann Intern Med* **110**(2), 119-124 (1989).
4. E. M. Lautin, N. J. Freeman, A. H. Schoenfeld, C. W. Bakal, N. Haramiti, A. C. Friedman, J. L. Lautin, S. Braha, S. Sprayregen, and I. Belizon, Radiocontrast-associated renal dysfunction: a comparison of lower-osmolality and conventional high-osmolality contrast media., *Am J Roentgenol* **157**(July), 59-65 (1991).
5. P. S. Parfrey, S. M. Graffiths, B. B.J., M. D. Paul, M. Genge, J. Withers, N. Farid, and P. J. McManamon, Contrast material-induced renal failure in patients with diabetes mellitus, renal insufficiency, or both, *New Engl J Med* **320**(3), 143-149 (1989).
6. C. H. Coggins, and L. S. Fang, Acute renal failure associated with antibiotics, anesthetic agents, and radiographic contrast agents, in: *Acute renal failure,* edited by B. M. Brenner, and J. M. Lazareus (Churchill Livingstone, 1988), pp. 319-352.
7. P. Persson, P. Hansell, and P. Liss, Pathophysiology of contrast medium induced nephropathy., *Kidney Int* **68**(68), 14-22 (2005).
8. P. Liss, K. Aukland, P. O. Carlsson, F. Palm, and P. Hansell, Influence of iothalamate on regional renal hemodynamics and oxygenation in the rat, *Acta Radiol* **46**(8), 823-829 (2005).
9. A. Nygren, H. R. Ulfendahl, P. Hansell, and U. Erikson, Effects of intravenous contrast media on cortical and medullary blood flow in the rat kidney, *Invest Radiol* **23**(753-761 (1988).
10. M. Brezis, and S. Rosen, Hypoxia of the renal medulla - its implication for disease, *N Engl J Med* **332**(10), 647-655 (1995).
11. F. Palm, J. Cederberg, P. Hansell, P. Liss, and P. O. Carlsson, Reactive oxygen species cause diabetes-induced decrease in renal oxygen tension, *Diabetologia* **46**(8), 1153-1160 (2003).
12. F. Palm, P. Hansell, G. Ronquist, A. Waldenstrom, P. Liss, and P. O. Carlsson, Polyol-pathway-dependent disturbances in renal medullary metabolism in experimental insulin-deficient diabetes mellitus in rats, *Diabetologia* **47**(7), 1223-1231 (2004).
13. P. Liss, A. Nygren, N. P. Revsbech, and H. Ulfendahl, Intrarenal oxygen tension measured by a modified Clark electrode at normal and low blood pressure and after injection of x-ray contrast media, *Pflügers Arch* **434**(705-711 (1997).
14. R. W. Katzberg, T. W. Morris, F. A. Burgerner, D. E. Kamm, and H. W. Fischer, Renal renin and hemodynamic responses to selective renal artery catheterization and angiography, *Invest Radiol* **12**(381-388 (1977).
15. G. Lund, S. Einzig, B. A. Rysavy, B. Borgwardt, E. Salomonowitz, A. Cragg, and K. Amplatz, Role of ischemia in contrast-induced renal damage: an experimental study, *Circulation* **69**(4), 783-789 (1984).
16. G. L. Bakris, and C. Burnett, A role for calcium in radiocontrast-induced reductions in renal hemodynamics, *Kidney Int* **27**(465-468 (1985).
17. S. K. Morcos, S. Oldroyd, and J. Haylor, Effect of radiographic contrast media on endothelium derived nitric oxide-dependent renal vasodilatation, *Br J Radiol* **70**(154-159 (1997).
18. S. D. Oldroyd, L. Fang, J. L. Haylor, M. S. Yates, A. M. El Nahas, and S. K. Morcos, Effects of adenosine receptor antagonists on the responses to contrast media in the isolated rat kidney, *Clin Sci (Lond)* **98**(3), 303-11 (2000).
19. R. W. Katzberg, G. Schulman, L. G. Meggs, W. J. H. Caldicott, M. M. Damiano, and N. K. Hollenberg, Mechanism of the renal response to contrast medium in dogs: decrease in renal function due to hypertonicity, *Invest Radiol* **18**(74-80 (1983).
20. J. Ueda, A. Nygren, P. Hansell, and H. R. Ulfendahl, Effect of intravenous contrast media on proximal and distal tubular hydrostatic pressure in the rat kidney, *Acta Radiol* **34**(83-87 (1993).
21. S. N. Heyman, B. A. Clark, N. Kaiser, K. Spokes, S. Rosen, M. Brezis, and F. H. Epstein, Radiocontrast agents induce endothelin release in vivo and in vitro, *J Am Soc Nephrol* **3**(1), 58-65 (1992).
22. J. E. Bird, M. R. Giancarli, J. R. Megill, and S. K. Durham, Effects of endothelin in radiocontrast-induced nephropathy in rats are mediated through endothelin-A receptors, *J Am Soc Nephrol* **7**(1153-1157 (1996).

23. L. Cantley, K. Spokes, B. Clark, E. G. McMahon, J. Carter, and F. H. Epstein, Role of endothelin and prostaglandins in radiocontrast-induced renal artery constriction, *Kidney Int* **44**(1217-1223 (1993).

24. P. Liss, P. O. Carlsson, A. Nygren, F. Palm, and P. Hansell, Et-A Receptor Antagonist BQ123 Prevents Radiocontrast Media-Induced Renal Medullary Hypoxia, *Acta Radiol* **44**(1), 111-7 (2003).

25. D. P. Brooks, and P. Nambi, Blockade of radio-contrast induced nephrotoxicity by the endothelin receptor antagonist, SB 209670, *Nephron* **72**(629-636 (1996).

26. N. C. Sandgaard, and P. Bie, Natriuretic effect of non-pressor doses of endothelin-1 in conscious dogs, *J Physiol* **494**(809-18 (1996).

27. T. J. Rabelink, K. A. Kaasjager, P. Boer, E. G. Stroes, B. Braam, and H. A. Koomans, Effects of endothelin-1 on renal function in humans: implications for physiology and pathophysiology, *Kidney Int* **46**(2), 376-81 (1994).

28. P. L. Khimenko, T. M. Moore, and A. E. Taylor, Blocked ETA receptors prevent ischemia and reperfusion injury in rat lungs, *J Appl Physiol* **80**(1), 203-7 (1996).

29. S. N. Heyman, M. Brezis, F. H. Epstein, K. Spokes, P. Silva, and S. Rosen, Early renal medullary hypoxic injury from radiocontrast and indomethacin, *Kidney Int* **40**(4), 632-642 (1991).

30. P. Liss, A. Nygren, U. Olsson, H. R. Ulfendahl, and U. Erikson, Effects of contrast media and mannitol on renal medullary blood flow and red cell aggregation in the rat kidney, *Kidney Int* **49**(5), 1268-1275 (1996).

31. P. Liss, P. O. Carlsson, F. Palm, and P. Hansell, Adenosine A1 receptors in contrast media-induced renal dysfunction in the normal rat, *Eur Radiol* **14**(7), 1297-302 (2004).

32. F. H. Epstein, M. Brezis, P. Silva, and S. Rosen, Protection against hypoxic injury in renal medulla, *Mol Physiol* **8**(525-534 (1985).

33. W. S. Spielman, S. L. Britton, and M. J. Fiksen-Olsen, Effect of adenosine on the distribution of renal blood flow in dogs, *Circ Res* **46**(3), 449-56 (1980).

34. D. Dinour, and M. Brezis, Effects of adenosine on intrarenal oxygenation, *Am J Physiol* **261**(F787-F791 (1991).

35. P. C. Churchill, and A. Bidani, Renal effects of selective adenosine receptor agonists in anesthetized rats, *Am J Physiol* **252**(2 Pt 2), F299-303 (1987).

36. W. S. Spielman, and L. J. Arend, Adenosin receptors and signaling in the kidney, *Hypertension* **17**(117-130 (1991).

C-PEPTIDE NORMALIZES GLOMERULAR FILTRATION RATE IN HYPERFILTRATING CONSCIOUS DIABETIC RATS

Sara Stridh[*], Johan Sällström[*], Markus Fridén[†], Peter Hansell[*], Lina Nordquist[*] and Fredrik Palm[*†]

Abstract: Tubular electrolyte transport accounts for a major part of the oxygen consumed by the normal kidney. We have previously reported a close association between diabetes and increased oxygen usage, partly due to increased tubular electrolyte transport secondary to glomerular hyperfiltration during the early onset of diabetes. Several studies have shown that acute administration of C-peptide to diabetic rats with glomerular hyperfiltration results in normalized glomerular filtration rate (GFR). In this study, we validated a novel method for precise and repetitive GFR measurements in conscious rats and used C-peptide injection in diabetic rats for evaluation.

First, GFR was determined in normoglycemic control rats before and after C-peptide administration. Thereafter, all rats were made diabetic by an i.v. streptozotocin injection. Fourteen days later, GFR was again determined before and after C-peptide administration. GFR was estimated from plasma clearance curves using a single bolus injection of FITC-inulin, followed by serial blood sampling over 155 min. FITC-inulin clearance was calculated using non-compartmental pharmacokinetic data analysis. Baseline GFR in normoglycemic controls was 2.10±0.18 ml/min, and was unaffected by C-peptide (2.23±0.14 ml/min). Diabetic rats had elevated GFR (3.06±.034 ml/min), which was normalized by C-peptide (2.35±0.30 ml/min).

In conclusion, the used method for estimation of GFR in conscious animals result in values that are in good agreement with those obtained from traditional GFR measurements on anaesthetized rats. However, multiple measurements from the same conscious subject can be obtained using this method. Furthermore, as previously shown

[*] Department of Medical Cell Biology, Uppsala University, Uppsala, Sweden
[†] Division of Pharmacokinetics and Drug Therapy, Department of Pharmaceutical Biosciences, Uppsala University, Sweden
[‡] Department of Medicine, Georgetown University, Washington, DC, USA

P. Liss et al. (eds.), *Oxygen Transport to Tissue XXX*, DOI 10.1007/978-0-387-85998-9_34,
© Springer Science+Business Media, LLC 2009

on anaesthetized rats, C-peptide also normalizes GFR in hyperfiltrating conscious diabetic rats.

1. INTRODUCTION

Tubular transport of electrolytes accounts for approximately 80% of the total oxygen usage in the normal mammalian kidney[1]. Tubular electrolyte transport is predominantly determined by the tubular sodium load, which is derived from the glomerular filtration rate (GFR). Therefore, a close correlation exists between GFR and renal oxygen consumption.

The early onset of diabetes is often associated with glomerular hyperfiltration[2], which has been hypothesized to be involved in the progression of diabetic nephropathy[2, 3]. Although the development of further kidney damage can be slowed down by careful and intense insulin treatment and aggressive anti-hypertensive treatment[4], there is at the present no treatment that can fully reverse a declining kidney function in these patients. There are several hypotheses for the mechanisms leading to the onset and progression of diabetes-induced renal dysfunction. The Diabetes Control and Complication Trial concluded that the degree of hyperglycemia is an important predictor of long-term diabetic renal complications[5]. The early augmented glomerular filtration may play a crucial role in the development of diabetic nephropathy, and inhibition of the diabetes-induced hyperfiltration have beneficial effects on delaying the progression of kidney damage[6].
C-peptide is secreted from the islets of Langerhans together with insulin in equimolar amounts, and has been shown to influence physiological processes[7]. Several studies have also shown renoprotective effects of exogenous C-peptide administration to patients with type-1 diabetes[8, 9], but also to animals with experimental insulinopenic diabetes[9, 10]. Such beneficial effects include decreased glomerular hyperfiltration, albuminuria, renal hypertrophy, and normalized glomerular volume[11]. Therefore, C-peptide has been proposed to be a novel therapeutical treatment with the potential to reduce the development of diabetic nephropathy[12, 13].

GFR measurements are usually done on anaesthetized animals with tedious sampling of both urine and plasma in order to calculate the plasma clearance of a suitable exogenous marker, e.g. inulin or EDTA. However, the anesthesia and subsequent required surgery for placement of infusion and sampling catheters will influence e.g. blood pressure which ultimately will affect the GFR measurement. Furthermore, the invasive nature of this technique normally only allows for GFR measurement at one time point, which results in weak statistics and inclusion of many animals if studying the developmental process over several days or weeks. We therefore, evaluated a novel method for GFR determinations in conscious unrestrained animals and used two different interventions known to influence the GFR, namely induction of diabetes, which causes glomerular hyperfiltration and C-peptide administration to hyperfiltrating diabetic animals, which we previously shown normalizes GFR. Furthermore, instead of using inulin labeled with a radioactive isotope, we used fluorescent fluorescein isothiocyanate (FITC)-inulin since this will result in no radioactive contamination of the cages housing the animals.

2. MATERIALS AND METHODS

2.1. Experimental protocol

Ten adult male Sprague Dawley rats (Møllegaard, Copenhagen, Denmark) had free access to water and standard rat chow (R3, Ewos, Södertälje, Sweden) throughout the study. First, GFR determinations (described in detailed below)[14] were performed on normoglycemic control rats before and after C-peptide administration (0.2 fmol/rat 45 minutes before GFR estimation), in randomized order. Thereafter, diabetes was induced in all rats by a single injection of streptozotocin (50 mg/kg bw; Sigma-Aldrich, St. Louis, MO, USA) in the tail vein. Diabetes (blood glucose above 20 mmol/l) was verified with test reagent strips (MediSense, Bedford, MA, USA) from blood samples obtained from the cut tip of the tail. Two weeks thereafter, GFR was again determined before and after C-peptide, in randomized order. All experiments were performed in accordance with the NIH guidelines for use and care of laboratory animals and approved by the Animal Care and Use Committee for Uppsala University.

2.2. GFR measurements by the single bolus injection method

FITC-inulin (1.5%) was dissolved in saline and filtered through a 0.45 μm syringe filter. In order to remove residual free FITC, the solution was dialyzed in 2000 ml of saline at +4°C overnight using a 1000 Da cut-off dialysis membrane (Spectra/Por® 6 Membrane, Spectrum Laboratories Inc, Rancho Dominguez, CA, USA) and protected from light. The dialyzed inulin was filtered through a 0.22 μm syringe filter before use. Dialyzed FITC-inulin (1 ml) was injected in the tail vein of the conscious rat. The syringe was weighed before and after the injection to calculate the exact dose. Approximately 40 μl of blood was collected from a small cut at the tip of the tail at 1, 3, 7, 10, 15, 35, 55, 75, 95, 125 and 155 minutes after injection of FITC-inulin.

The plasma samples (5 μl) with FITC-inulin were diluted with 70 μl HEPES buffer (pH 7.4), and fluorescence measured after loading the samples onto a black 384-well microplate (Greiner Bio-One GmbH, Kremsmuenster, Austria) using a plate reader (Safire II, Tecan Austria GmbH, Grödig, Austria) with 496 nm excitation and 520 nm emission.

Plasma clearance of FITC-inulin was calculated using non-compartmental pharmacokinetic data analysis as described by Gabrielsson and Weiner[15]. FITC-inulin clearance (CL) is defined as the given i.v. dose divided by the total area under the plasma fluorescence time curve ($AUC_{0-\infty}$):

$$CL = Dose_{iv} / AUC_{0-\infty}$$

The $AUC_{0-\infty}$ was estimated from summation of the trapezoid areas that are formed by connecting the data points, a procedure referred to as the linear trapezoidal rule (Fig. 1). It is essential to calculate the entire area under the curve from time zero to infinity. This necessitates the estimation of the areas before the first measurement and after the last measurement:

$$AUC_{0-\infty} = AUC_{back-extr} + \Sigma trapezoids + AUC_{\lambda}$$

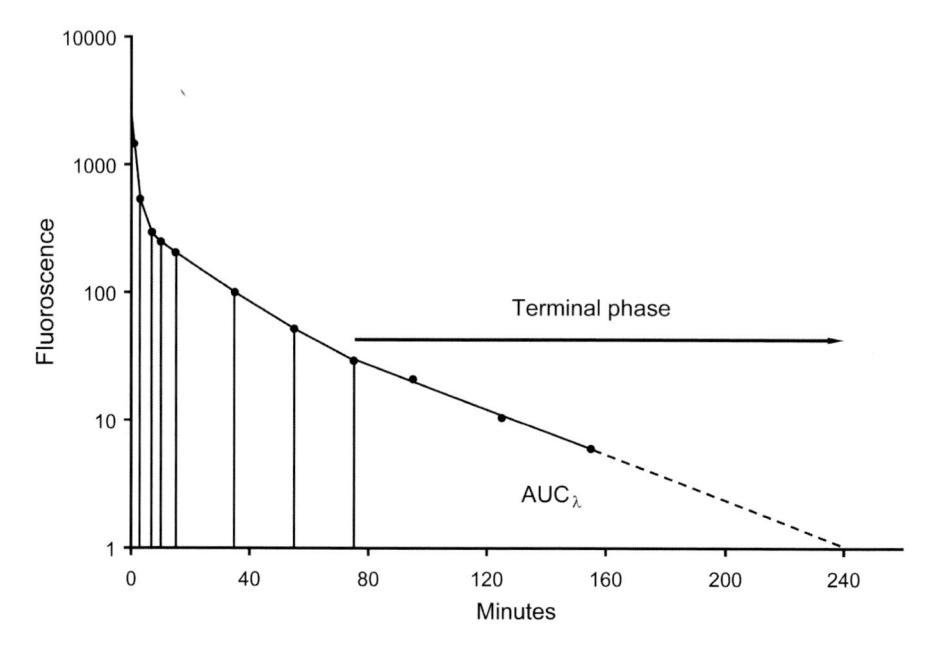

Figure 1. Representative plot of plasma FITC fluorescence versus time. The semi-logarithmic presentation reveals the location of a terminal phase with a constant slope. The slope λ (min^{-1}) was estimated using linear regression analysis of the natural logarithm of fluorescence and time. The prediction line was subsequently used to calculate the area from the first time point of the terminal phase (75 minutes in this figure) to infinity (AUC$_\lambda$), thereby including the residual area after the last time point. AUC$_\lambda$ is obtained by dividing the *predicted* fluorescence at 75 min by the slope. The area from time zero to the first time point (AUC$_{back-extr}$) was calculated by linear back-extrapolation.

2.5. Statistical evaluation

All values are expressed as means±SEM. Multiple comparisons between different groups were performed using repeated one-way analysis of variance (ANOVA) and, when appropriate, followed by Dunnett's Multiple Comparisons Test. For all comparisons, $p<0.05$ was considered statistically significant.

3. RESULTS

The design of the blood sampling schedule was suitable for accurate calculation of GFR as indicated by small residual areas (~3 % of the total area).

The GFR of normoglycemic control rats was unaffected by C-peptide administration (Fig. 2). Diabetic rats had glomerular hyperfiltration, which was normalized by C-peptide administration (Fig. 2).

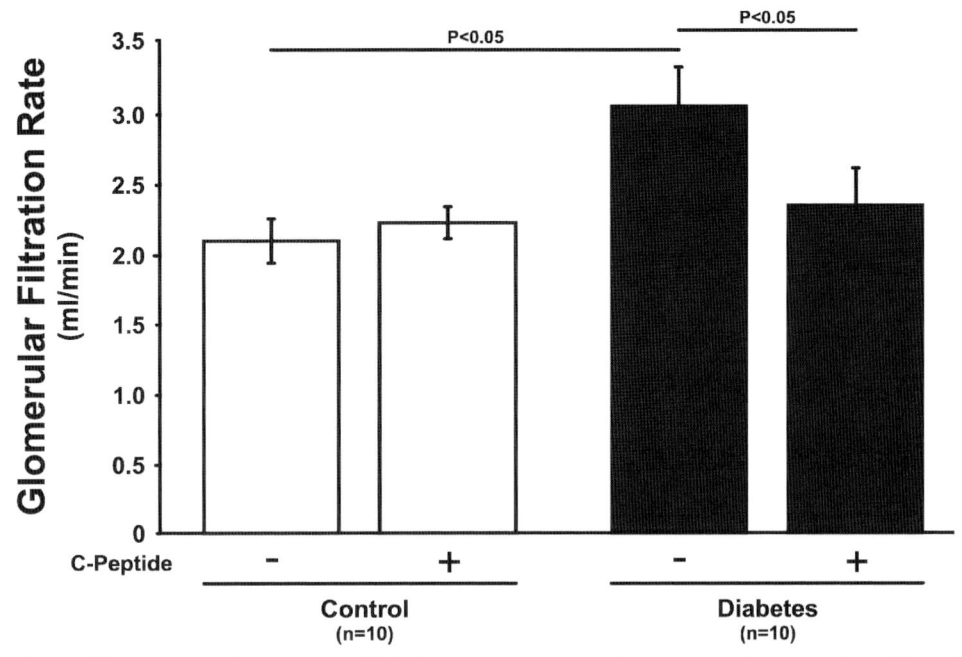

Figure 2 The glomerular filtration rate (GFR) of normoglycemic control rats was not affected by the addition of C-peptide. Diabetic rats displayed elevated GFR, which was normalized by C-peptide administration.

4. DISCUSSION

In this paper, we report that diabetes induces glomerular hyperfiltration, which is normalized by C-peptide administration in conscious unrestrained rats. Our results from conscious rats confirm previous findings that the early onset of STZ-induced diabetes in rats results in glomerular hyperfiltration[16], which is reversed by C-peptide administration[10]. These results clearly show that the applied method is suitable for detecting physiologically relevant changes in GFR in rats.

Plasma clearance of inulin is often used for determinations of GFR because it is freely filtrated in the glomerulus and has no known reabsorption or secretion along the nephron[14]. The classical method for inulin clearance measurements in order to determine GFR involves anaesthetizing the animals and placing infusion catheters in a vein and collection catheters at least in the artery and urinary bladder. These procedures can affect the physiological state of the animals, e.g. blood pressure, which can confound the GFR determination. The present study involved non-compartmental data analysis to estimate GFR based on plasma fluorescence of FITC-inulin following a single i.v. bolus dose and repetitive blood sampling from conscious rats. In contrast to traditional modeling or "curve-fitting", the non-compartmental analysis does not require the assumptions associated with the structure of a model, i.e. a one- or two-compartment model. While there is no consensus on which is the correct model to use[17], a clearance estimation such as GFR does not at all require a model or modeling software as long as a reasonable study design is applied. The benefits from using repetitive GFR determinations in the same animals includes reducing the number of animals needed in order to detect relevant differences, but also minimizing the influence of randomization errors due to individual differences between subjects.

5. CONCLUSION

This method for estimation of GFR in conscious animals results in values that are in good agreement with those obtained from traditional GFR measurements on anaesthetized rats. However, multiple GFR measurements from the same subject can be obtained using this method. As previously shown on anaesthetized rats, diabetes results in glomerular hyperfiltration and C-peptide normalizes GFR in hyperfiltrating conscious diabetic rats.

6. ACKNOWLEDGEMENTS

This work was funded by The Swedish Research Council Medicine (9940, 10840), The Marcus and Amalia Wallenberg Foundation, The Linné Foundation for Medical Research, The Swedish Diabetes Association, and The Swedish Society for Medical Research and NIH K-99 grant (DK-077858) to FP.

7. REFERENCES

1. N. A. Lassen, O. Munck, and J. H. Thaysen, Oxygen consumption and sodium reabsorption in the kidney, *Acta Physiol Scand* **51**(371-84 (1961).
2. M. Marre, M. Hallab, J. Roy, J. J. Lejeune, P. Jallet, and P. Fressinaud, Glomerular hyperfiltration in type I, type II, and secondary diabetes, *J Diabetes Complications* **6**(1), 19-24 (1992).
3. J. W. Yip, S. L. Jones, M. J. Wiseman, C. Hill, and G. Viberti, Glomerular hyperfiltration in the prediction of nephropathy in IDDM: a 10-year follow-up study, *Diabetes* **45**(12), 1729-33 (1996).
4. J. A. Staessen, J. Wang, G. Bianchi, and W. H. Birkenhager, Essential hypertension, *Lancet* **361**(9369), 1629-41 (2003).
5. The effect of intensive treatment of diabetes on the development and progression of long-term complications in insulin-dependent diabetes mellitus. The Diabetes Control and Complications Trial Research Group, *N Engl J Med* **329**(14), 977-86 (1993).
6. K. A. Nath, S. M. Kren, and T. H. Hostetter, Dietary protein restriction in established renal injury in the rat. Selective role of glomerular capillary pressure in progressive glomerular dysfunction, *J Clin Invest* **78**(5), 1199-205 (1986).
7. C. Wojcikowski, V. Maier, K. Dominiak, R. Fussganger, and E. F. Pfeiffer, Effects of synthetic rat C-peptide in normal and diabetic rats, *Diabetologia* **25**(3), 288-90 (1983).
8. J. Wahren, K. Ekberg, B. Samnegard, and B. L. Johansson, C-peptide: a new potential in the treatment of diabetic nephropathy, *Curr Diab Rep* **1**(3), 261-6 (2001).
9. D. Y. Huang, K. Richter, A. Breidenbach, and V. Vallon, Human C-peptide acutely lowers glomerular hyperfiltration and proteinuria in diabetic rats: a dose-response study, *Naunyn Schmiedebergs Arch Pharmacol* **365**(1), 67-73 (2002).
10. L. Nordquist, E. Moe, and M. Sjoquist, The C-peptide fragment EVARQ reduces glomerular hyperfiltration in streptozotocin-induced diabetic rats, *Diabetes Metab Res Rev*, (2006).
11. B. Samnegard, S. H. Jacobson, B. L. Johansson, K. Ekberg, B. Isaksson, J. Wahren, and M. Sjoquist, C-peptide and captopril are equally effective in lowering glomerular hyperfiltration in diabetic rats, *Nephrol Dial Transplant* **19**(6), 1385-91 (2004).
12. J. Wahren, K. Ekberg, and H. Jornvall, C-peptide is a bioactive peptide, *Diabetologia* **50**(3), 503-9 (2007).

13. L. Luzi, G. Zerbini, and A. Caumo, C-peptide: a redundant relative of insulin?, *Diabetologia* **50**(3), 500-2 (2007).

14. Z. Qi, I. Whitt, A. Mehta, J. Jin, M. Zhao, R. C. Harris, A. B. Fogo, and M. D. Breyer, Serial determination of glomerular filtration rate in conscious mice using FITC-inulin clearance, *Am J Physiol Renal Physiol* **286**(3), F590-6 (2004).

15. J. Gabrielsson, and D. Weiner, Non-compartmental Analysis, in: *Pharmacokinetic and Pharmacodynamic Data Analysis: Concepts and Applications,* edited (Swedish Pharmaceutical Press, Stockholm, 2006), pp. 161-180.

16. F. Palm, J. Cederberg, P. Hansell, P. Liss, and P. O. Carlsson, Reactive oxygen species cause diabetes-induced decrease in renal oxygen tension, *Diabetologia* **46**(8), 1153-60 (2003).

17. C. Sturgeon, A. D. Sam, 2nd, and W. R. Law, Rapid determination of glomerular filtration rate by single-bolus inulin: a comparison of estimation analyses, *J Appl Physiol* **84**(6), 2154-62 (1998).

LIPOPROTEIN NANOPLATFORM FOR TARGETED DELIVERY OF DIAGNOSTIC AND THERAPEUTIC AGENTS

Jerry D. Glickson, Sissel Lund-Katz, Rong Zhou, Hoon Choi, I-Wei Chen, Hui Li, Ian Corbin, Anatoliy V. Popov, Weiguo Cao, Liping Song, Chenze Qi, Diane Marotta, David S. Nelson, Juan Chen, Britton Chance and Gang Zheng[*]

Abstract: Low-density lipoprotein (LDL) provides a highly versatile natural nanoplatform for delivery of optical and MRI contrast agents, photodynamic therapy agents and chemotherapeutic agents to normal and neoplastic cells that over express LDL receptors (LDLR). Extension to other lipoproteins ranging in diameter from ~5-10 nm (high density lipoprotein, HDL) to over a micron (chilomicrons) is feasible. Loading of contrast or therapeutic agents has been achieved by covalent attachment to protein side chains, intercalation into the phospholipid monolayer and extraction and reconstitution of the triglyceride/cholesterol ester core. Covalent attachment of folate to the lysine side chain amino groups was used to reroute the LDL from its natural receptor (LDLR) to folate receptors and could be utilized to target other receptors. A semi-synthetic nanoparticle has been constructed by coating magnetite iron oxide nanoparticles (MIONs) with carboxylated cholesterol and overlaying a monolayer of phospholipid to which Apo A1, Apo E or synthetic amphoteric α-helical polypeptides were adsorbed for targeting HDL, LDL or folate receptors, respectively. These particles can be utilized for *in situ* loading of magnetite into cells for MRI monitored cell tracking or gene therapy.

---

[*] Jerry D. Glickson (glickson@mail.med.upenn.edu, ph 215-898-1805, fax 215-573-2113) Rong Zhou, Hoon Choi, Anatoliy Popov, Weiguo Cao, Liping Song, Chenze Qi, Diane Marotta and David S. Nelson, Molecular Imaging Laboratory, Department of Radiology, University of Pennsylvania School of Medicine, Philadelphia, Pennsylvania 19104-6069. Sissel Lund-Katz, Joseph Stokes Jr. Research Institute, The Children's Hospital of Philadelphia, Philadelphia, PA 19104-4318. I-Wei Chen, Department of Materials Science and Engineering, University of Pennsylvania, Philadelphia, PA 19104-6272. Hui Li, Institute for Translational Medicine and Therapeutics, University of Pennsylvania School of Medicine, Philadelphia, PA 19104-6160. Britton Chance, Department of Biochemistry & Biophysics, Johnson Research Foundation, University of Pennsylvania, Philadelphia, Pennsylvania 19104-6089. Ian Corbin, Juan Chen and Gang Zheng, Division of Biophysics and Bioimaging, Ontario Cancer Institute, Department of Medical Biophysics, University of Toronto, Toronto, Canada.

P. Liss et al. (eds.), *Oxygen Transport to Tissue XXX*, DOI 10.1007/978-0-387-85998-9_35, **227**
© Springer Science+Business Media, LLC 2009

1. INTRODUCTION

Lipoproteins are natural nanoparticles spanning a range of diameters from about 10 nm (high density lipoprotein, HDL) to over a micron (chylomicrons).[1] All of them contain a phospholipid monolayer shell and a hydrophobic lipid core consisting of triglycerides and cholesterol esters. Many contain protein components adsorbed onto the phospholipid monolayer. The most commonly studied lipoprotein, low density lipoprotein (LDL), which is ~20 nm in diameter, contains a very large protein component, apo B-100 (550 KD, 4536 amino acid residues) that covers about 40% of its surface area and targets the particle to LDL receptors (LDLRs) in a number of normal tissues such as liver, adrenals and ovaries. LDL and HDL transport cholesterol to and from tissues and play pathological and protective roles, respectively, in cardiovascular disease. Because a number of tumors over express LDLRs, this particle has been utilized as a delivery vehicle for a number of lipophilic drugs.[2] Our laboratories have been developing LDL and more recently other lipoproteins for delivery of optical and NMR probes as well as photodynamic therapy (PDT) agents and cancer drugs to solid tumors. Here we describe some key aspects of this program.

3. METHODS

3.1. Preparation of Lipoproteins.

Methods for preparation of LDL and chemically modified LDL have been previously described.[3] For preparation of the iron oxide encapsulating lipoproteins, iron(III) acetylacetonate was added to phenyl ether with 1,2-hexadecanediol and cholic acid under nitrogen, then heated to 265°C to form cholesterol coated iron oxide particles. The collected oily particles were dispersed in chloroform, to which phosphatidylcholine was added, and the solution was dried. A phosphate buffer solution (PBS) including EDTA and NaN_3 was added to the dried sample and sonicated at 50°C to form oil in water phospholipid micelles containing a shell of phospholipid and a core of cholesterol-coated iron oxide particles. Apolipoprotein A-1 was finally added to the micelle solution under sonication to decorate the phospholipid micelle, forming lipoprotein particles with a total diameter of ~20 nm.

3.2. Synthesis of NIRFs and MRI contrast agents.

Methods have been previously described.[4-6]

3.3. Cell Lines and Animal Models.

Methods have been previously described.[3, 4]

3.4. Confocal Microscopy.

Methods have been previously described.[3-5]

3.5. Redox Scanning.

Methods have been previously described.[4]

3.6. Optical Imaging.

Imaging was performed with a Xenogen imaging system (Palo Alto, CA) under isoflurane anesthesia.

3.7. MRI.

Methods have been previously described.[6]

4. RESULTS AND DISCUSSION

4.1. Lipoproteins: Nature's Nanoparticles

Low density lipoprotein (LDL) is a 22 nm particle (density 1.006-1.063 g/ml) that contains one copy of apo B-100 that is rich in amphipathic α–helices. The hydrophobic faces of these helices are adsorbed onto the outer surface of the phospholipid monolayer, which also contains variable amounts of free cholesterol. The inner surface of the monolayer is adsorbed onto a lipid core consisting of cholesterol esters (~1500 per particle) and triglycerides.

The mechanism of internalization of LDL in cells was delineated through the classic studies of Brown and Goldstein, for which they were awarded the Nobel Prize in Medicine in 1985. Through electrostatic interactions between highly cationic receptor binding sequences on apo B-100 with complementary anionic sequences on the cell surface, the particle binds to LDLRs embedded in clathrin coated pits on the cell surface. The receptor-LDL complex is internalized by endocytosis and digested in lysosomes by acid hydrolysis and enzymatic breakdown into free cholesterol, fatty acids and amino acids; the receptors are cycled back to the cell surface in about 10 minutes. The process can then be repeated with another LDL particle. The turnover time for cell surface LDLRs is about 24 hr. Any probes or drugs incorporated into the LDL particle accumulate within the targeted cell providing a very effective amplification mechanism for drug and imaging probe delivery.

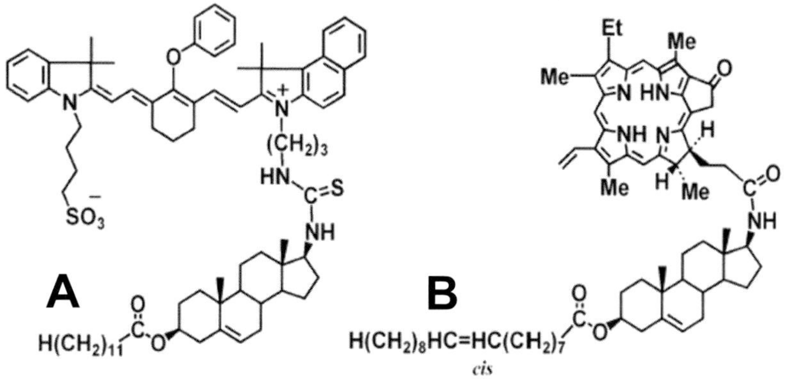

Figure 1. Structure of cholesterol ester conjugates with tricarbocyanine (A) and pyropheophorbide (B).

B16 Melanoma   HepG₂ Tumor

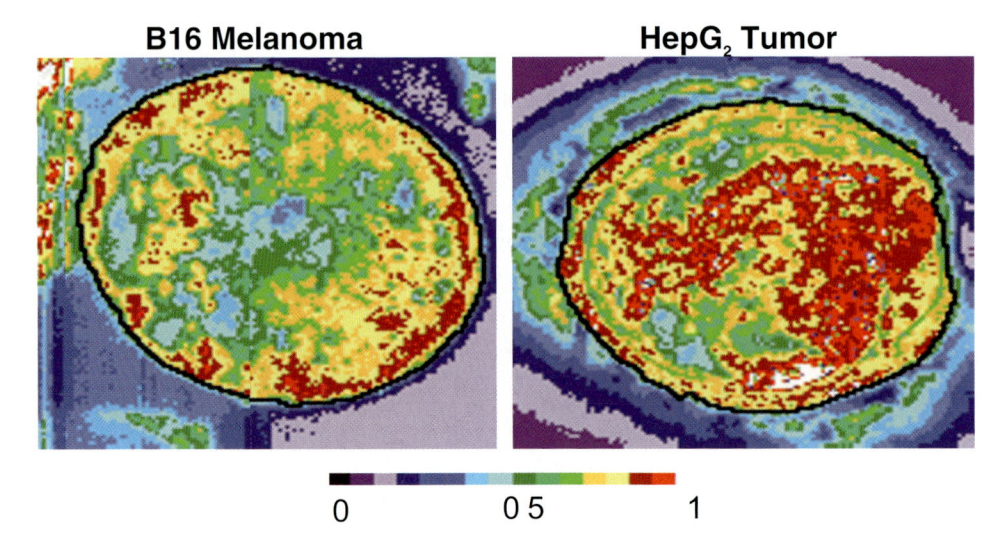

Figure 2. Redox scans of single slices of B16 melanoma and HepG2 hepatoma after i.v. injection of the recombinant LDL-Pyro-CE complexes into mice with subcutaneous tumors.

A number of normal tissues such as liver, adrenals and ovaries contain high levels of LDLRs. Many malignancies also overexpress these receptors, including acute myelogenous leukemia (3-100 fold compared to normal cells), colon cancer (6-fold), adrenal ade-

Surface-loaded DIR-LDL 81 nmol based on DIR (Max bar: 12000, Min bar:2000)

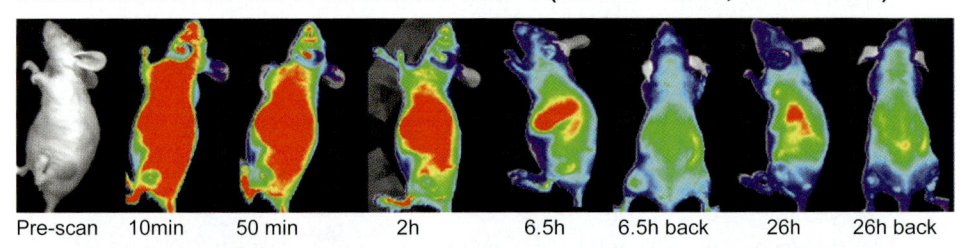

Pre-scan   10min   50 min   2h   6.5h   6.5h back   26h   26h back

Core-loaded DIR-LDL 73 nmol based on DIR (Max bar: 12000, Min bar:2000)

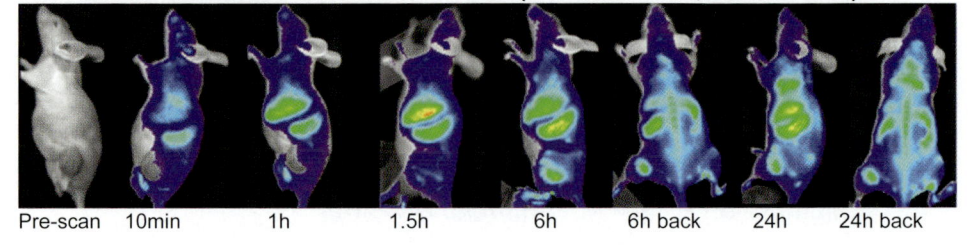

Pre-scan   10min   1h   1.5h   6h   6h back   24h   24h back

Figure 3. Preliminary comparison of effects of surface and core loaded LDL on distribution of the dye delivered by i.v. injection of LDL labeled with DiR.

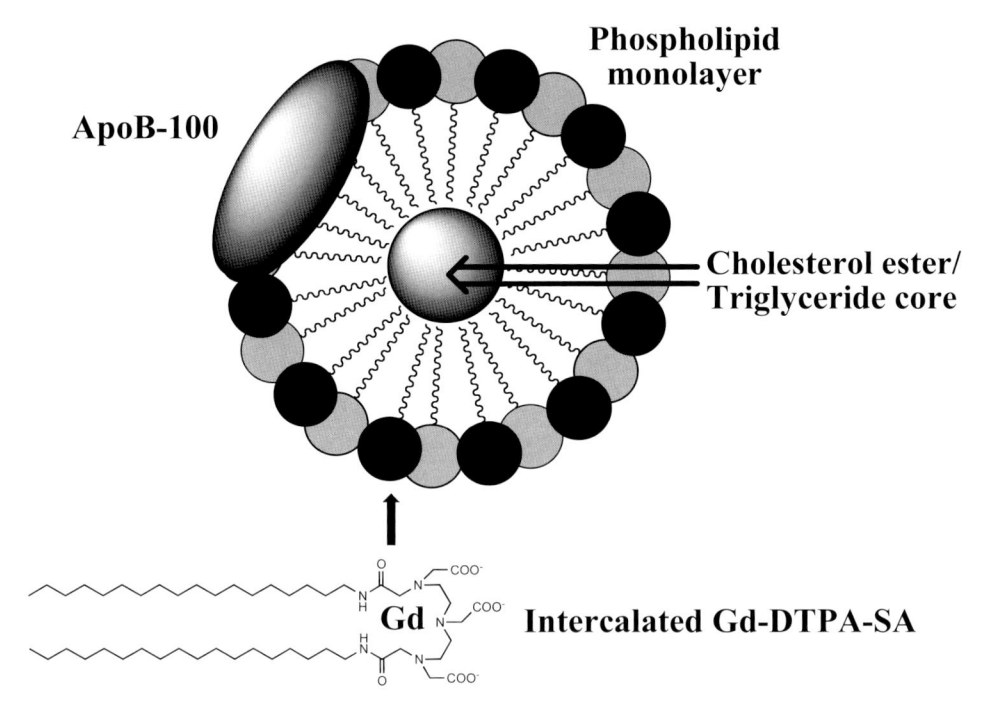

Figure 4. Schematic representation of surface loading of LDL with bis-stearylamide Gd complex for MRI detection.

noma (8-fold) and pancreatic, lung, brain and prostate cancer.[7] Among the laboratory models of human cancer B16 melanoma and HepG2 hepatoma are rich in LDLRs.[8, 9]

Three methods have been used to bind imaging or therapeutic agents to LDL - protein loading, core loading and surface loading. Each has advantages and disadvantages; we have used each of these methods in our studies. Surface loading was utilized for delivery of near infrared fluorophores (NIRFs)[4] and paramagnetic Gd-chelated MRI probes,[6] core binding for delivery of NIRFs and PDT agents,[3, 5] and protein loading for redirecting LDL to receptors other than LDLR.[3] These applications will be described below after a short description of the loading procedures.

The first method, protein loading, is performed by covalent attachment of the probe to side-chains of the apo B-100. A number of investigators have attached chelating groups to lysine side chains and used these to bind ^{111}In or ^{99m}Tc for SPECT imaging[10, 11] or ^{68}Ga for PET imaging.[10] Attachment of ^{18}F-containing ligands to the lysine-ε-amino groups has also been utilized for PET imaging.[12] Radio-iodination of tyrosine side chains for SPECT detection has been attempted but leads to modification of the transport properties of the protein.[13] The advantage of protein labeling is that stable products are formed. The disadvantage is that covalent binding may modify the delivery characteristics of LDL since the receptor-binding site includes reactive lysines (see below).

The second method of LDL labeling is core labeling that introduces lipophilic probes into the core of the particle. Krieger et al.[14] demonstrated that in the presence of starch, the lipid core of LDL can be extracted with a nonpolar solvent such as heptane while keeping the phospholipid and protein "shell" intact. The lipophilic probe is introduced into the extract, the solvent evaporated *in vacuo* and the particle spontaneously reassembles with recovery of over 50% of the LDLR binding activity.

The third method of LDL labeling is surface labeling, which can be used to non-covalently bind probes to the phospholipid surface of the phospholipid monolayer. In general the probe is conjugated to one or two long hydrophobic chains and sometimes also to a cholesterol group. These entities intercalate between the phospholipid fatty acids and may even protrude into the lipid core (thereby anchoring the ligand in the particle). Very high yields have been achieved with fluorescent[4, 5] or paramagnetic[6] probes. The method is easy to implement but prone to high leakage rates. The reason for this is that transfer of the surface probe to the outer phospholipid layer of the cell plasma membrane is thermodynamically favorable. Hence, substantial delivery of probes to cells could occur by pathways not involving receptor delivery.

4.2. LDL Surface and Core Loading of NIRFs/PDT Agents

Krieger et al. used a cholesterol laurate scaffold for surface binding of the fluorophore pyrene to LDL.[15] We used the same strategy to attach tricarbocyanine[4, 16] and tris[(porphinato)zinc(II)][17] fluorophores to LDL. Figure 1A exhibits the structure of a tricarbocyanine NIR dye IRD41 (LI-COR Biotechnology Division (Lincoln, NE) conjugated to an amino group on cholesterol laurate via a thiourea linkage. Pyropheophorbide cholesterol esters (Pyro-CE) of oleic acid (Fig. 1B) were similarly prepared.[5] The tricarbocyanine conjugates are charged facilitating surface loading (although core loading is also feasible), whereas the Pyro-CE is neutral and most suitable for core loading. Pyropheophorbide is both a NIRF and a PDT agent.

Delivery of both types of LDL-NIRF conjugates to B16 melanoma and HepG2 tumor cells was confirmed by confocal microscopy.[4, 5] Intense NIR fluorescence distributed uniformly throughout the cytoplasm demonstrated internalization of the NIRFs in the cytoplasm after digestion of the particles in the lysosomes. Excess unlabeled LDL blocked delivery confirming that the dyes were delivered via LDLRs.

Reconstituted LDL (r-LDL) core labeled with Pyro-CE was injected into the tail veins of nude mice with subcutaneous B16 or HepG2 tumors. Analysis of blood specimens indicated clearance of the r-LDL in ~30 min (data not shown). Localization of the dye in the tumor was monitored by cryo-spectrophotometry using reflectance spectrophotometry by the method of Quistorff et al.[18] ("redox scanning"). Images were taken of 10 μm slices with an isotropic in-plane resolution of 100 μm. Figure 2 shows that in the B16 melanoma the dye accumulated mostly in the periphery of the tumor. Large areas of central necrosis were virtually devoid of label. In the hepatoma much more intense binding was noted in the center of the tumor, and there was much less necrosis. Scatchard analysis of binding of LDL surface labeled with the commercial visible monocarbocyanine dye DiI to isolated B16 and HepG2 cells indicated that both tumors exhibited similar numbers of LDLRs per cell, but the affinities of the HepG2 receptors were about twice those of the B16 receptors.[4] However, the distribution of Pyro-CE LDL appears to be determined primarily by the perfusion properties of these tumors, with the melanoma exhibiting a well perfused periphery and a poorly perfused central region, whereas perfusion of the hepatoma was probably more uniform.

A preliminary study with surface loaded and core loaded LDL particles was performed comparing NIR images of tumor bearing mice with HepG2 xenografts on their

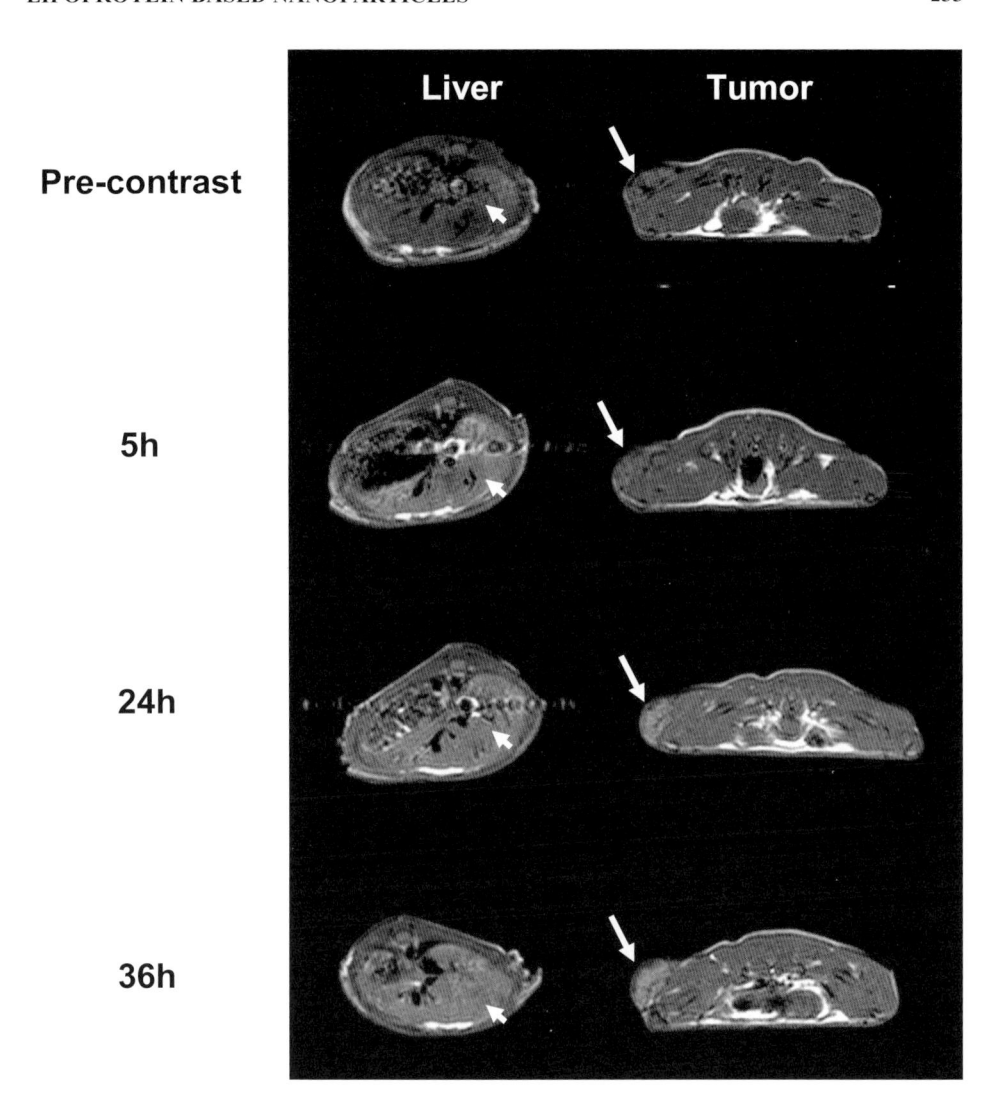

Figure 5. Comparison of T_1-weighted MRI of nude mouse with subcutaneous HepG2 tumor before injection of GdDTPA-SA-LDL (180:1; 0.04 mmol/kg) (top images), 5 hr after injection (liver 55% enhancement, tumor 9%), 24 hrs after injection (liver 25% enhancement, tumor 25%), and 36 hr after injection (liver 24% enhancement, tumor 23%).

thighs. The commercial NIR tricarbocyanine DiR was used to label the LDL. Figure 3 shows much more extensive and non-specific binding of the surface loaded dye. Therefore, whenever possible core loading should be used to label LDL.

4.3. Surface Binding of Gd-Chelates for MRI Imaging

For MRI detection, the GdDTPA bis-stearylamide (SA) complex, a bidentate phospholipid intercalator, was synthesized and conjugated to LDL by surface loading (Figure.

Table 1. Relaxivity ($mM^{-1}s^{-1}$) of Gadolinium Analogs at 4.7 T (21°C)

Parameter	Gadodiamide	Gd-DTPA-SA	Gd-DTPA-SA-LDL
Gd/molecule	1	1	180
Relaxivity/Gd	4.1	4.8	6.5
Relaxivity/molecule	4.1	4.8	1170

4).[6] In this instance surface loading was required since relaxation enhancement occurs by direct contact of the Gd with water. Between 100 and 500 GdDTPA SA complexes could be conjugated to the LDL surface. The optimum relaxivity was achieved with 180 Gd complexes per LDL. Table 1 compares the relaxivities of commercial Gd-diamide MRI contrast agent with those of the isolated GdDTPA SA and its complex with LDL (180 Gd/LDL particle). Similar relaxivities were obtained with the isolated chelates, but the complex of the nanoparticle with 180 Gds produced much higher relaxation enhancement.

Figure 5 compares the relaxation enhancement of liver and tumor in axial MR images of a mouse with a subcutaneous HepG2 hepatoma xenograft at various times after i.v. injection of the GdDTPA-SA-LDL complex. The enhancement of the tumor becomes significant 24 hr post-injection.

These studies represent initial attempts at using LDL as a vehicle for delivery of Gd. Substantially higher relaxivities can be achieved with other Gd chelates, particularly with those containing two water ligands per Gd. Further studies are being pursued to optimize the Gd enhancements induced by agents conjugated with LDL. It must be noted that these particles will be degraded upon cellular incorporation, and the identity and properties of the intracellular Gd complex are as yet undetermined.

4.4. Rerouting of LDL to Folate Receptors

High levels of LDLR in various normal tissues, particularly liver and the adrenals, limit the specificity of LDL based nanoparticles (LDLNP) for tumor cells. Since a number of much more tumor-specific receptors have been identified, we sought to develop a method to redirect the LDLRB to these alternative targets. Our strategy was to alkylate the lysine ε-amino side chains with appropriate receptor-targeting ligands. To demonstrate the principle, we chose folic acid receptors since these are over expressed in a number of tumors, particularly ovarian. Key to the success of this strategy is the lower pK_a (8.9) and, hence, higher sensitivity to alkylation of the lysine amino side chains of the LDLR-targeting region of LDL.[19] Figure 6 shows that five of the eight amino acids of

Apo B: Arg-Leu-Thr-Arg-Lys-Arg-Leu-Lys
Apo E: Arg-Lys-Leu-Arg-Lys-Arg-Leu-Arg

Figure 6. The LDLR-binding sequences of apo B100 and Apo E contains clusters of cationic amino acids.

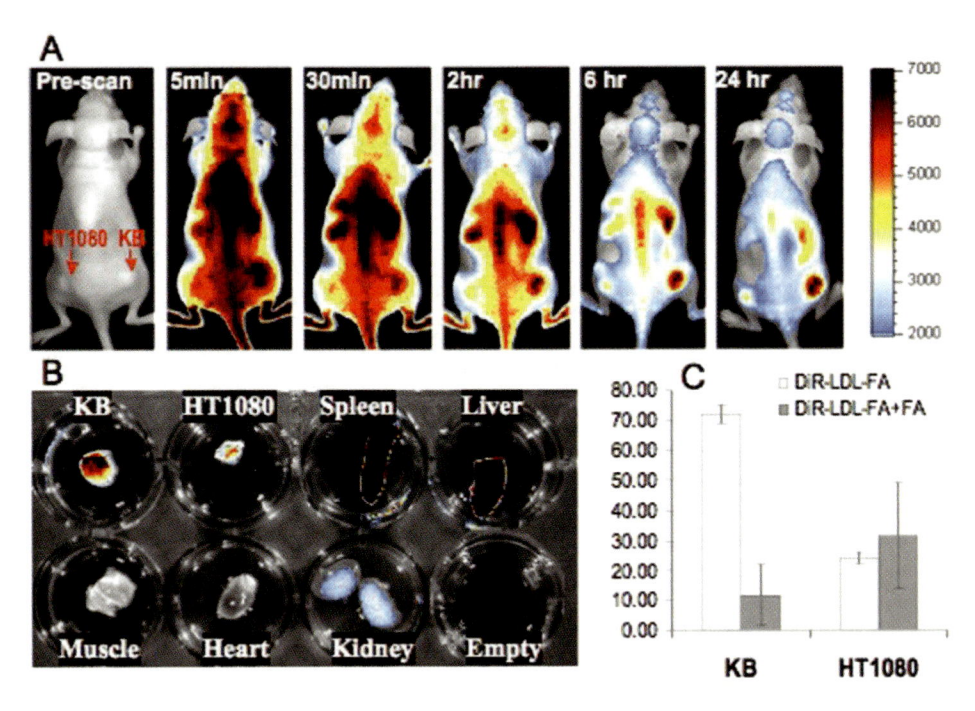

Figure 7. A) Real time in vivo fluorescence images of KB/HT1080 dual tumor mice with iv injection of DiR-LDL-FA (5.8 μ M); (B) fluorescence images of tissues and tumors excised at 24 h postinjection; (C) fluorescence readings of tumor extracts from in vivo FA inhibition assay.

has six of eight cationic amino acids at its receptor binding site. Therefore, alkylation of the lysine side chains of LDL rapidly eliminates LDLR-binding activity at the same time as the addition of folic acid ligands to these amino groups redirects the particle to folate

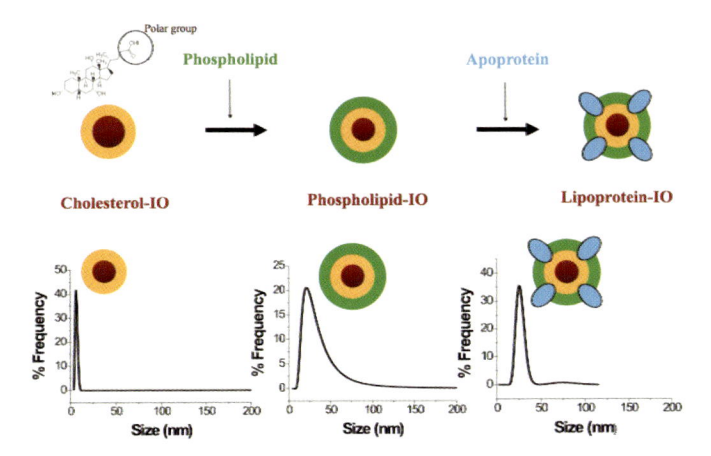

Figure 8. Scheme for preparation of semi-synthetic lipoprotein-iron oxide particles. The apoprotein can be apo A1 or apo E for targeting HDL or LDL receptors, respectively.

receptors. Alkylation of 20% of the lysine side chains totally abolishes LDLR binding.[19]

Apo B100 contains 357 lysine side chain amino groups, of which 225 are exposed and 132 buried. Among the exposed lysines, there are 53 active lysine amino groups (pK_a 8.9) with the rest exhibiting normal pK_as (10.5) and normal reactivity.[3] The presence of clusters of cationic amino acids in the backbone sequence could account for all or some of these anomalously acidic amino side chains; however, regions of low dielectric constant could also facilitate deprotonation of some of the charged ammonium groups.

We attached folate groups to 170 of or 47.6% of the lysine side chains of LDL, which totally abolished binding to LDLR.[3] To enable confocal microscopy, we surface loaded the commercial visible dye DiI (Invitrogen Corp., Carlsbad, CA) to the folate-conjugated LDL. The molar ratio of the DiL:LDL:FA (folic acid) was 55:1: ~150-200; the diameter of the particle measured by electron microscopy was 26.1±3.0 nm, which is slightly more than native LDL (see above). We also prepared LDL-FA particles core loaded with the silicon phthalocyanine (r-SiPc-BOA-LDL-FA) PDT agent with composition r-SiPc-BOA:LDL:FA = 1500:1:~150-200; 24.0±4.3 nm. The latter were adequately fluorescent for confocal microscopy.

Confocal microscopy studies of cell lines with specific types of receptors confirmed receptor mediated delivery of DiI-LDL-FA to FA receptors (FR) and absence of uptake by LDLR.[3] Briefly, nasopharyngeal carcinoma cells (FR+) internalized DiI-LDL-FA; addition of excess FA abolished uptake, but addition of excess unlabled LDL did not. Chinese hamster ovary (CHO) cells (FR-) and HT1080 cells (FR-) failed to take up DiI-LDL-FA, as did HepG2 cells (FA⁻, LDLR⁺), but the latter bound DiI-LDL. Flow cytometry demonstrated a monotonic increase in fluorescence intensity of KB cells incubated with increasing concentrations of DiI-LDL-FA; but FA competitively inhibited uptake of the DiI-LDL-FA by these cells. Similar studies were performed with r-SiPc-BOA-LDL-FA: KB (FR⁺) cells that internalized r-SiPc-BOA-LDL-FA, excess FA inhibited uptake, but excess unmodified LDL did not. CHO cells (FR⁻) did not internalize the r-SiPc-BOA-LDL-FA. HepG2 (LDLR⁺) cells did not take up r-SiPc-BOA-LDL-FA but did accumulate r-Pc-BOA-LDL.

Preliminary *in vivo* tumor localization studies were performed on mice with a subcutaneous HT1080 (FR⁻) tumor on one thigh and a KB (FR⁺) tumor on the other thigh.[20] At t=0, 0.77 μM LDL-FA surface loaded with the NIRF DiR ((ex: 748nm, em: 780nm); DiR: LDL:FA = 8:1:105) was injected i.v. into the tail vein. Xenogen images (ex. 710-760nm, em. 810-875nm) measured at various times post-injection are shown in Figure 7.[20] Five minutes after injection the dye was distributed throughout the animal and in both tumors. At 6 hr post-injection the dye had cleared from the FR⁻ HT1080 tumor but was retained in the FR⁺ KB tumor. At 24 hr fluorescent intensity had increased in the KB tumor and diminished in the abdomen. The study demonstrates that the LDL particle can diffuse through leaky blood vessels into both tumors, but is retained and accumulates in the tumor that contains the FRs. There also appears to be slow transfer of the nanoparticles between binding sites in the abdomen and in the tumor. The mechanism underlying this transfer requires further study.

4.5. Super-Paramagnetic Iron-Oxide Nanoparticles

Super-paramagnetic iron-oxide (SPIO) nanoparticles have been utilized as highly sensitive MRI markers for tracking cells and antibodies.[21-25] Weissleder's laboratory has attached dextran-coated SPIOs to transferrin and delivered these particles to tumors via transferrin-receptors.[24, 25] However, the tumors had to be transfected to prevent down-regulation of the transferrin receptors at high cellular iron concentrations. Since lipoprotein receptors are not down regulated under these conditions, we have sought to core load SPIOs into LDL. Similarly, iron oxide has been used for cells tracking, but the iron was diluted as the cells divided.[21-23] By delivering iron *in situ* in the animal to cell surface receptors on the targeted cell, we should be able to overcome these limitations. Efforts to load LDL with iron oxide have proven unsuccessful even after lipid coating of the iron-oxide particles (A. Stolpen and J.S. Leigh, personal communication). However, Choi, Zhou and Chen have developed methods for targeting SPIOs to folate receptors[26] and for incorporating SPIOs into semi-synthetic lipoproteins. The latter strategy is depicted in Figure 8. Using 5 nm SPIO clusters, lipoprotein-iron oxide particles with apo A-1 for targeting HDL receptors have been prepared; alternatively Apo E can be used for targeting LDLR or synthetic 4F, an apo A-1 mimetic amphipathic α-helical D-retropeptides can be attached to the nanoparticle.[27] These particles can also be redirected to folate or other receptors by alkylation of the lysine side chains. Such targeted particles can be used for T_2-weighted MRI and for delivery of therapeutic agents to specific tumor receptors, vascular receptors on atherosclerotic plaque, stem cells or macrophages.

4.6. Advantages and Disadvantages of Lipoprotein-Based Nanoparticles

In summary, the LBNPs constitute a highly versatile natural platform for delivery of diagnostic and therapeutic agents. In the absence of extensive chemical modification, they should be non-immunogenic, thereby avoiding a major potential limitation of synthetic nanoparticles. They are multifunctional, can be targeted to their natural receptors or modified to bind other natural receptors or perhaps even unique unnatural receptors that could be transfected into cells (N. Sachs, personal communication). They provide a highly efficient mechanism for amplification of the NIR, MRI, PET or SPECT signals detection of the targeted cells in tumors, atherosclerotic plaque or stem cells. Multivalency can be achieved by binding more than one targeting agent or therapeutic agent to the particles. In a similar manner, the platform facilitates multiple imaging modalities.

A key limitation is the existence of receptors on normal cells, such as the reticular endothelial system that leads to high background binding. This can be overcome to some extent by judicious choice of the targeting particle dimensions, by attaching polyethylene glycol groups to the surface to minimize non-specific binding or by choice of appropriate receptors, but it cannot yet be totally eliminated and remains a confounding problem in the use of this delivery system.

Many of the lipoproteins are isolated from human blood. Consequently, there is concern about introduction of pathogens that produce hepatitis or AIDS. Commercial sources are available for providing pathogen-free blood proteins, but this increases the cost of lipoproteins (CSL Behring-Biotherapies, www.cslbehring.com). Small proteins such as human Apo A-1 and Apo E can be produced in a recombinant form from bacterial sources, but the cost of these is quite high.

5. ACKNOWLEDGEMENTS

The authors are indebted to Dr. Theo van Berkel of Leiden University for providing the B16 melanoma and HepG2 hepatoma cell lines used in this study. This research was conducted with support from the following grants: N01-C037119 (GZ), R24-CA83105 (JDG), P20-CA86255 (JDG), R21/R33-CA114463 (GZ), Oncological Fdn. of Buffalo (GZ), RSNA (GZ), Bioadvance Program of Pennsylvania, University of Pennsylvania Institute of Medicine and Engineering (RZ, HC and I-WC). Imaging was performed at the Small Animal Imaging Facility (SAIF) in the Department Radiology at the University of Pennsylvania.

Declaration of Commercial Interest. Two participants in this study (JDG and GZ) are cofounders of Marillion Pharmaceuticals, Inc., established with a grant from the Bioadvance Program of Pennsylvania. Part of this research was supported by funds from this company.

6. REFERENCES

1. D. E. Vance and J. E. Vance. Biochemistry of Lipids, Lipoproteins and Membranes (Elsevier Science, Amsterdam, 2002).
2. J. M. Shaw. Lipoproteins as Carriers of Pharmacological Agents (Marcel Dekker, New York, 1991).
3. G. Zheng, J. Chen, H. Li and J. D. Glickson. Rerouting lipoprotein nanoparticles to selected alternate receptors for the targeted delivery of cancer diagnostic and therapeutic agents. Proceedings of the National Academy of Sciences of the United States of America 102, 17757-17762 (2005).
4. H. Li, Z. H. Zhang, D. Blessington, D. S. Nelson, R. Zhou, S. Lund-Katz, B. Chance, J. D. Glickson and G. Zheng. Carbocyanine labeled LDL for optical Imaging of tumors. Academic Radiology 11, 669-677 (2004).
5. G. Zheng, H. Li, M. Zhang, S. Lund-Katz, B. Chance and J. D. Glickson. Low-density lipoprotein reconstituted by pyropheophorbide cholesteryl oleate as target-specific photosensitizer. Bioconjugate Chemistry 13, 392-396 (2002).
6. I. R. Corbin, H. Li, J. Chen, S. Lund-Katz, R. Zhou, J. D. Glickson and G. Zheng. Low-density lipoprotein nanoparticles as magnetic resonance imaging contrast agents. Neoplasia 8, 488-498 (2006).
7. R. A. Firestone. Low density lipoprotein as a vehicle for targeting antitumor compounds to cancer cells. Bioconjugate Chem. 5, 105-113 (1994).
8. P. C. de Smidt and T. J. C. van Berkel. Prolonged serum half-life of antineoplastic drugs by incorportation into the low density lipoprotein. Cancer Res. 50, 7476-82 (1990).
9. P. C. N. Rensen, R. M. Schiffelers, J. Versluis, M. K. Bijsterbosch, M. E. M. J. van Kuijk-Meuwissen and T. J. C. van Berkel. Human recombinant apolipoprotein E-enriched liposomes can mimic low-density lipoproteins as carriers for the site-specific delivery of antitumor agents. Molec. Pharmacol. 52, 445-455 (1997).
10. S. M. Moerlein, A. Daugherty, B. E. Sobel and M. J. Welch. Metabolic Imaging with Gallium-68-Labeled and Indium-111-Labeled Low-Density-Lipoprotein. Journal of Nuclear Medicine 32, 300-307 (1991).
11. E. Ponty, G. Favre, R. Benaniba, A. Boneu, H. Lucot, M. Carton and G. Soula. Biodistribution Study of Tc-99m-Labeled LDL in B16-Melanoma-Bearing Mice - Visualization of a Preferential Uptake by the Tumor. International Journal of Cancer 54, 411-417 (1993).
12. J. Pietzsch, R. Bergmann, K. Rode, C. Hultsch, B. Pawelke, F. Wuest and J. van den Hoff. Fluorine-18 radiolabeling of low-density lipoproteins: a potential approach for characterization and differentiation of metabolism of native and oxidized low-density lipoproteins in vivo. Nuclear Medicine and Biology 31, 1043-1050 (2004).
13. G. Sobal, U. Resch and H. Sinzinger. Modification of low-density lipoprotein by different radioiodination methods. Nuclear Medicine and Biology 31, 381-388 (2004).
14. M. Krieger, M. S. Brown, J. R. Faust and J. L. Goldstein. Replacement of endogenous cholesteryl esters of low density lipoprotein with exogenous cholesteryl linoleate. Reconstitution of a biologically active lipoprotein particle. Journal of Biological Chemistry 253, 4093-101 (1978).
15. M. Krieger, Y. K. Ho and J. R. Falck. Reconstitution of LDL with lipophilic fluorescein derivatives: Quantitative analysis of the receptor activity of human lymphocytes. J. Receptor Res. 3, 361-75 (1983).
16. G. Zheng, H. Li, K. Yang, D. Blessington, K. Licha, S. Lund-Katz, B. Chance and J. D. Glickson. Tricarbocyanine cholesteryl laurates labeled LDL: New near infrared fluorescent probes (NIRFs) for monitoring tumors and gene therapy of Familial hypercholesterolemia. Bioorganic & Medicinal Chemistry Letters 12, 1485-1488 (2002).

17. S. P. Wu, I. Lee, P. P. Ghoroghchian, P. R. Frail, G. Zheng, J. D. Glickson and M. J. Therien. Near-infrared optical Imaging of B16 melanoma cells via low-density lipoprotein-mediated uptake and delivery of high emission dipole strength tris[(porphinato)zinc(II)] fluorophores. Bioconjugate Chemistry 16, 542-550 (2005).

18. B. Quistorff, J. C. Haselgrove and B. Chance. High resolution readout of 3-D metabolic organ structure: An automated, low-temperature redox ratio-scanning instrument. Anal. Biochem. 148, 389-400 (1985).

19. S. Lund-Katz, J. A. Ibdah, J. Y. Letizia, M. T. Thomas and M. C. Phillips. A 13C NMR characterization of lysine residues in apolipoprotein B and their role in binding to the low density lipoprotein receptor. Journal of Biological Chemistry 263, 13831-8 (1988).

20. J. Chen, I. R. Corbin, H. Li, W. G. Cao, J. D. Glickson and G. Zheng. Ligand conjugated low-density lipoprotein nanoparticles for enhanced optical cancer imaging in vivo. Journal of the American Chemical Society 129, 5798- (2007).

21. J. W. M. Bulte, A. S. Arbab, T. Douglas and J. A. Frank. in Imaging in Biological Research, Pt B 275-299 (2004).

22. K. A. Hinds, J. M. Hill, E. M. Shapiro, M. O. Laukkanen, A. C. Silva, C. A. Combs, T. R. Varney, R. S. Balaban, A. P. Koretsky and C. E. Dunbar. Highly efficient endosomal labeling of progenitor and stem cells with large magnetic particles allows magnetic resonance imaging of single cells. Blood 102, 867-872 (2003).

23. J. W. M. Bulte, I. D. Duncan and J. A. Frank. In vivo magnetic resonance tracking of magnetically labeled cells after transplantation. Journal of Cerebral Blood Flow and Metabolism 22, 899-907 (2002).

24. R. Weissleder and U. Mahmood. Molecular imaging. Radiology 219, 316-333 (2001).

25. R. Weissleder, A. Bogdanov, E. A. Neuwelt and M. Papisov. Long-circulating iron-oxides for MR-imaging. Advanced Drug Delivery Reviews 16, 321-334 (1995).

26. H. Choi, S. R. Choi, R. Zhou, H. F. Kung and I. W. Chen. Iron oxide nanoparticles as magnetic resonance contrast agent for tumor imaging via folate receptor-targeted delivery. Academic Radiology 11, 996-1004 (2004).

27. M. Navab, G. M. Anantharamaiah, S. T. Reddy, S. Hama, G. Hough, V. R. Grijalva, N. Yu, B. J. Ansell, G. Datta, D. W. Garber and A. M. Fogelman. Apolipoprotein A-I mimetic peptides. Arteriosclerosis Thrombosis and Vascular Biology 25, 1325-1331 (2005).

PROGNOSTIC POTENTIAL OF THE PRE-THERAPEUTIC TUMOR OXYGENATION STATUS

Peter Vaupel[*]

Abstract: Hypoxia, a characteristic feature of locally advanced solid tumors, has emerged as a key factor of the tumor pathophysiome, since it can promote tumor progression and resistance to therapy. Independent of established prognostic parameters, such as clinical tumor stage, histology, histological grade and nodal status, hypoxia has been identified as an adverse prognostic factor for patient outcome. Studies of pretreatment tumor hypoxia involving direct assessment (polarographic oxygen tension measurements) have suggested a poor prognosis for patients with hypoxic tumors. These investigations indicate a worse disease-free survival for patients with hypoxic cancers of the uterine cervix or soft tissue sarcomas. In head & neck cancers, the studies suggest that pretherapeutic hypoxia is prognostic for survival and local control.

1. INTRODUCTION

Hypoxia, a characteristic feature of locally advanced solid tumors, has emerged as a pivotal factor of the tumor (patho-) physiome since it is capable of promoting tumor progression and resistance to therapy[1-10]. Hypoxia represents a "Janus face" in tumor biology because (a) it is associated with restrained proliferation, differentiation, necrosis or apoptosis, but (b) contrarily it can also lead to the development of an aggressive phenotype[11]. While being independent of standard prognostic factors such as tumor stage, histology, histological grade and nodal status, hypoxia has been proposed as an adverse prognostic factor for patient outcome. Studies of tumor hypoxia involving direct assessment of the oxygenation status with polarographic O_2 microelectrodes have indicated worse disease-free survival for patients with hypoxic cervical cancers or soft tissue sarcomas. In head & neck cancers the studies suggest that hypoxia is prognostic for survival and local control[6, 12].

[*] Peter Vaupel, Institute of Physiology and Pathophysiology, University of Mainz, Duesbergweg 6, 55099 Mainz, Germany, e-mail address: vaupel@uni-mainz.de

In the following sections, information on the impact of tumor hypoxia on clinical outcome is summarized for cancers of the uterine cervix, head & neck tumors and soft tissue sarcomas. In these studies, the tumors have been treated with radiation (± chemotherapy or surgery). Due to some overlap in patients reported from the same clinical institution, a selection of relevant data has been performed so that only one relevant communication from each clinical center has been considered[13].

2. HYPOXIA IN CANCERS OF THE CERVIX AND CLINICAL OUTCOME

The first data suggesting that pretherapeutic hypoxia in primary tumors could be a prognostic factor for patient outcome was published in 1993 by Höckel et al.[14]. In a first analysis of 31 cervix cancer patients, the authors could show that patients with hypoxic tumors (median pO_2 < 10 mmHg) had a significantly lower overall and recurrence-free survival. These observations were confirmed in a later study involving 103 patients[1]. The survival differences were independent of stage, histology and grade. Differences in local control were not apparent upon multivariate analysis (see Table 1).

Differences in survival were also observed by Fyles et al.[15] in 106 patients using a cutoff pO_2 of 5 mmHg. The impact of hypoxia in this latter study, however, was observed only in node-negative patients. Again, hypoxia did not appear to be of prognostic value when local control was assessed. Two smaller studies by Knocke et al.[16] and Sundfør et al.[17] also confirmed the prognostic impact of hypoxia on disease-free and overall survival in cervical cancers. Sundfør et al.[17] demonstrated hypoxia to additionally be a prognostic factor for local control. In contrast, the prospective international multi-center study by Nordsmark et al.[18] involving 120 patients with cervical cancer yielded conflicting data with no impact of hypoxia on the outcome. The reason for these conflicting results remains unclear.

Table 1. Prognostic significance of hypoxia for irradiated cervix cancers (* multivariate analysis). n = number of patients, DFS = disease-free survival, OS = overall survival, LC = local control, PFS = progression-free survival, n.s. = not significant.

Authors	N	Oxygenation parameter	Endpoint	p*
Höckel et al.[1]	89	median pO_2 < 10 mmHg	DFS	= 0.009
			OS	= 0.004
Knocke et al.[16]	51	median pO_2 ≤ 10 mmHg	DFS	< 0.02
Sundfør et al.[17]	40	subvolume with pO_2 readings < 5 mmHg (%)	DFS	= 0.0001
			OS	= 0.0004
			LC	= 0.0006
Fyles et al.[15]	106	pO_2 readings < 5 mmHg (%)	PFS	< 0.004
Nordsmark et al.[18]	120	median pO_2 < 4 mmHg	LC, OS	n.s.

Table 2. Prognostic significance of hypoxia for irradiated head & neck tumors (primary SCCs, *multivariate analysis). For further explanations see title of Table 1.

Authors	N	Oxygenation parameter	Endpoint	p*
Nordsmark et al.[21]	67	pO$_2$ readings < 2.5 mmHg (%)	LC	= 0.01
Dunst et al.[22]	125	subvolume with pO$_2$ readings < 5 mmHg (%)	OS	= 0.001

3. HYPOXIA IN PRIMARY AND METASTATIC CANCERS OF THE HEAD AND NECK AND CLINICAL OUTCOME

The accessibility of primary and metastatic squamous cell carcinomas of the head and neck for tumor oxygenation assessment has meant that these tumors have received much attention, with a large number of studies having been documented. Already in 1988, the study of Gatenby et al.[19] using O$_2$ needle electrodes in the clinical setting demonstrated hypoxia in head & neck tumors. These authors have convincingly shown that hypoxia in metastatic lesions was associated with a poor prognosis upon radiotherapy.

Hypoxia also appears to be prognostic for outcome in head & neck cancers, with data suggesting that hypoxia is prognostic for survival and local control (see Tables 2 and 3). The international multi-center study by Nordsmark et al.[20] involving 397 patients with head & neck tumors provided further evidence that tumor hypoxia is associated with a poor prognosis in patients with advanced head & neck cancer following primary radiotherapy. In head & neck cancers, hypoxia not only predicts for survival (as is also the case in cervical cancers) but also for local control, suggesting hypoxia-induced radiation resistance as a major factor for local failure. In the study of Adam et al.[25], only a small number of patients was assessed and hypoxia did not appear to be a prognostic factor in disease-free survival and local control.

4. HYPOXIA IN SOFT TISSUE SARCOMAS AND CLINICAL OUTCOME

Studies of soft tissue sarcomas also suggest worse disease-free survival for patients with hypoxic tumors (see Table 4). In these studies, however, the small number of patients

Table 3. Prognostic significance of hypoxia for irradiated head & neck tumors (lymph node metastases, *multivariate analysis). For further explanations see title of Table 1.

Authors	N	Oxygenation parameter	Endpoint	p*
Gatenby et al.[19]	31	pO$_2$ < 5 mmHg	LC	< 0.001
Brizel et al.[23]	63	median pO$_2$ < 10 mmHg	DFS	= 0.005
			OS	= 0.02
			LC	= 0.01
Rudat et al.[24]	134	pO$_2$ readings < 2.5 mmHg (%)	OS	= 0.004
Adam et al.[25]	25	median pO$_2$ < 20 mmHg	LC, OS	n.s.

Table 4. Prognostic significance of hypoxia for irradiated soft tissue sarcomas. DF = distant failure. For further explanations see title of Table 1.

Authors	n	Oxygenation parameter	Endpoint	p
Brizel et al.[26]	22	median $pO_2 \leq 10$ mmHg	DF	$= 0.01$
Nordsmark et al.[27]	31	median $pO_2 \leq 19$ mmHg	OS	$= 0.01$

did not allow multivariate analyses. For this reason, the impact of hypoxia on local control and distant spread has, so far, not been clarified.

5. OUTLOOK AND LIMITATIONS

The oxygenation status of cancers of the uterine cervix and of the head & neck, as well as oxygenation parameters of soft tissue sarcomas, have been identified as independent, adverse prognostic factors. In cancer of the uterine cervix, tumor hypoxia has been reported to be a "new and promising factor"[28]. Thus, identification of tumor hypoxia may allow an assessment of a tumor's potential to develop an aggressive phenotype and for acquired treatment resistance, both of which lead to poor prognosis[29]. Detection of hypoxia in the clinical setting may therefore be helpful in selecting high-risk patients for individual and/or more intensive treatment schedules. Furthermore, tumor hypoxia could be exploited for therapeutic advantage using selective targeting of hypoxic tumor cells[30-32].

The interpretation of the data presented on these three tumor types is however complicated by a series of unresolved issues: (a) the selection of the optimal endpoints characterizing the oxygenation status of tumors[33], (b) the role of heterogeneity in tumor oxygenation[34], (c) the impact of heterogeneous treatment protocols[18], (d) insufficient sample sizes[18], (e) pronounced inter-institutional (inter-observer) differences for the same tumor type (f) pO_2 readings in necrotic regions, (g) laboratory-dependent calibration differences[35], and (h) inconsistent handling of negative pO_2 readings. In this context, it has to be mentioned that a combination of the O_2 microsensor technique with other existing techniques has not yet been proven to be helpful[13].

6. ACKNOWLEDGEMENT

We acknowledge grant support from the Deutsche Krebshilfe (# 106758).

7. REFERENCES

1. M. Höckel, K. Schlenger, B. Aral, M. Mitze, U. Schäfer, and P. Vaupel, Association between tumor hypoxia and malignant progression in advanced cancer of the uterine cervix, *Cancer Res.* 56, 4509-4515 (1996).
2. M. Höckel and P. Vaupel, Tumor hypoxia: Definitions and current clinical, biologic and molecular aspects, *J. Natl. Cancer Inst.* 93, 266-276 (2001).
3. P. Vaupel, O. Thews, and M. Höckel, Treatment resistance of solid tumors: Role of hypoxia and anemia, *Med. Oncol.* 18, 243-259 (2001).

4. P. Vaupel, A. Mayer, and M. Höckel, Tumor hypoxia and malignant progression, *Methods Enzymol.* 381, 335-354 (2004).

5. P. Vaupel, The role of hypoxia-induced factors in tumor progression, *Oncologist* 9 (Suppl. 5), 10-17 (2004).

6. P. Vaupel and A. Mayer, Hypoxia in cancer: significance and impact on clinical outcome, *Cancer Metastasis Rev.* 26, 225-239 (2007).

7. G.L. Semenza, Hypoxia, clonal selection, and the role of HIF-1 in tumor progression, *Crit. Rev. Biochem. Mol. Biol.* 35, 71-103 (2000).

8. G.L. Semenza, Involvement of hypoxia-inducible factor 1 in human cancer, *Intern. Med.* 41, 79-83 (2002).

9. G.L. Semenza, HIF-1 and tumor progression: Pathophysiology and therapeutics, *Trends Mol. Med.* 8, S62-S67 (2002).

10. A.L. Harris, Hypoxia – A key regulatory factor in tumour growth, *Nature Rev. Cancer* 2, 38-47 (2002).

11. P. Vaupel and A. Mayer, Effects of anaemia and hypoxia on tumour biology, in: *Anaemia in Cancer,* edited by C. Bokemeyer and H. Ludwig, 2nd Edition (Elsevier, Edinburgh, London, 2005), pp. 47-66.

12. B.J. Moeller, R.A. Richardson, and M.W. Dewhirst, Hypoxia and radiotherapy: opportunities for improved outcomes in cancer treatment, *Cancer Metastasis Rev.* 26, 241-248 (2007).

13. P. Vaupel, M. Höckel, and A. Mayer, Detection and characterization of tumor hypoxia using pO_2 histography, *Antioxid. Redox Signal.* 9, 1221-1235 (2007).

14. M. Höckel, C. Knoop, K. Schlenger, B. Vorndran, E. Baussmann, M. Mitze, P.G. Knapstein, and P. Vaupel, Intratumoral pO_2 predicts survival in advanced cancer of the uterine cervix, *Radiother. Oncol.* 26, 45-50 (1993).

15. A. Fyles, M. Milosevic, D. Hedley, M. Pintilie, W. Levin, L. Manchul, and R.P. Hill, Tumor hypoxia has independent predictor impact only in patients with node-negative cervix cancer, *J. Clin. Oncol.* 20, 680-687 (2002).

16. T.H. Knocke, H.D. Weitmann, H.J. Feldmann, E. Selzer, and R. Potter, Intratumoral pO_2-measurements as predictive assay in the treatment of carcinoma of the uterine cervix, *Radiother. Oncol.* 53, 99-104 (1999).

17. K. SundfØr, H. Lyng, C.G. Trope, and E.K. Rofstad, Treatment outcome in advanced squamous cell carcinomas of the uterine cervix: relationship to pretreatment tumor oxygenation and vascularization, *Radiother. Oncol.* 54, 101-107 (2000).

18. M. Nordsmark, J. Loncaster, C. Aquino-Parsons, S.C. Chou, V. Gebski, C. West, J.C. Lindgaard, H. Havsteen, S.E. Davidson, R. Hunter, J.A. Raleigh, and J. Overgaard, The prognostic value of pimonidazole and tumour pO_2 in human cervix carcinomas after radiation therapy: A prospective international multi-center study, *Radiother. Oncol.* 80, 123-131 (2006).

19. R.A. Gatenby, H.B. Kessler, J.S. Rosenblum, L.R. Coia, P.J. Moldofsky, W.H. Hartz, and G.J. Broder, Oxygen distribution in squamous cell carcinoma metastases and its relationship to outcome of radiation therapy, *Int. J. Radiat. Oncol. Biol. Phys.* 14, 831-838 (1988).

20. M. Nordsmark, S.M. Bentzen, V. Rudat, D. Brizel, E. Lartigau, P. Stadler, A. Becker, M. Adam, M. Molls, J. Dunst, and D.J. Terris, Prognostic value of tumor oxygenation in 397 head and neck tumors after primary radiation therapy. An international multi-center study, *Radiother. Oncol.* 77, 18-24 (2005).

21. M. Nordsmark and J. Overgaard, Tumor hypoxia is independent of hemoglobin and prognostic for loco-regional tumor control after primary radiotherapy in advanced head and neck cancer, *Acta Oncol.* 43, 396-403 (2004).

22. J. Dunst, P. Stadler, A. Becker, C. Lautenschläger, T. Pelz, G. Hänsgen, M. Molls, and T. Kuhnt, Tumor volume and tumor hypoxia in head and neck cancers: The amount of the hypoxic volume is important, *Strahlenther. Onkol.* 179, 521-526 (2003).

23. D.M. Brizel, R.K. Dodge, R.W. Clough, and M.W. Dewhirst, Oxygenation of head and neck cancer: changes during radiotherapy and impact on treatment outcome, *Radiother. Oncol.* 53, 113-117 (1999).

24. V. Rudat, P. Stadler, A. Becker, B. Vanselow, A. Dietz, M. Wannenmacher, M. Molls, J. Dunst, and H.J. Feldmann, Predictive value of the tumor oxygenation by means of pO_2 histography in patients with advanced head and neck cancer, *Strahlenther. Onkol.* 177, 462-468 (2001).

25. M.F. Adam, E.C. Gabalski, D.A. Bloch, J.W. Oehlert, J.M. Brown, A.A. Elsaid, H.A. Pinto, and D.J. Terris, Tissue oxygen distribution in head and neck cancer patients, *Head Neck* 21, 146-153 (1999).

26. D.M. Brizel, S.P. Scully, J.M. Harrelson, L.J. Layfield, J.M. Bean, L.R. Prosnitz, and M.W. Dewhirst, Tumor oxygenation predicts for the likelihood of distant metastases in human soft tissue sarcoma, *Cancer Res.* 56, 941-943 (1996).

27. M. Nordsmark, J. Alsner, J. Keller, O.S. Nielsen, O.M. Jensen, M.R. Horsman, and J. Overgaard, Hypoxia in human soft tissue sarcomas: Adverse impact on survival and no association with p53 mutations, *Brit. J. Cancer* 84, 1070-1075 (2001).

28. G. Pitson and A. Fyles, Uterine Cervix Cancer, in: *Prognostic Factors in Cancer*, edited by M.K. Gospodarowicz, D.E. Henson, R.V.P. Hutter, B. O'Sullivan, L.H. Sobin, and Ch. Wittekind, 2nd Edition (Wiley-Liss, New York, Chichester, 2001), pp. 501-514.

29. P. Okunieff, I. Ding, P. Vaupel, and M. Höckel, Evidence against hypoxia as the primary cause of tumor aggressiveness. Adv. Exp. Med. Biol. 510, 69-75 (2003).

30. J.M. Brown and A.J. Giaccia, The unique physiology of solid tumors: opportunities (and problems) for cancer therapy, *Cancer Res.* 58, 1408-1416 (1998).

31. J.M. Brown, Exploiting the hypoxic cancer cell: mechanisms and therapeutic strategies, *Mol. Med. Today* 6, 157-162 (2000).

32. S. Kizaka-Kondoh, M. Inoue, H. Harada, and M. Hiraoka, Tumor hypoxia: A target for selective cancer therapy, *Cancer Sci.* 94, 1021-1028 (2003).

33. S.M. Evans and C.J. Koch, Prognostic significance of tumor oxygenation in humans, *Cancer Lett.* 195, 1-16 (2003).

34. C. Menon and D.L. Fraker, Tumor oxygenation status as prognostic marker, *Cancer Lett.* 221, 225-235 (2005).

35. J.L. Tatum, G.J. Kelloff, R.J. Gillies, J.M. Arbeit, J.M. Brown, K.S. Chao, J.D. Chapman, W.C. Eckelman, A.W. Fyles, A.J. Giaccia, R.P. Hill, C.J. Koch, M.C. Krishna, K.A. Krohn, J.S. Lewis, R.P. Mason, G. Melillo, A.R. Padhani, G. Powis, J.G. Rajendran, R. Reba, S.P. Robinson, G.L. Semenza, H.M. Swartz, P. Vaupel, D. Yang, B. Croft, J. Hoffman, G. Liu, H. Stone, and D. Sullivan, Hypoxia: importance in tumor biology, noninvasive measurement by imaging, and value of its measurement in the management of cancer therapy, *Int. J. Radiat. Biol.* 82, 699-757 (2006).

HISTOLOGICAL BASIS OF MR/OPTICAL IMAGING OF HUMAN MELANOMA MOUSE XENOGRAFTS SPANNING A RANGE OF METASTATIC POTENTIALS

He N. Xu, Rong Zhou, [$]Shoko Nioka, [$]Britton Chance, Jerry D. Glickson, and Lin Z. Li[*]

Abstract: Predicting tumor aggressiveness will greatly facilitate cancer treatment. We have previously reported investigations utilizing various MR/optical imaging methods to differentiate human melanoma mouse xenografts spanning a range of metastatic potentials. The purpose of this study was to explore the histological basis of the previously reported imaging findings. We obtained the cryogenic tumor sections of three types of human melanoma mouse xenografts with their metastatic potentials falling in the rank order A375P<A375M<C8161. Both H&E and DAPI counter-stained TUNEL analysis showed distinct core-rim difference in aggressive tumors, while the core has apparently many viable cells forming structure of vascular-like networks and the rim appears viable-cell dense. The least aggressive ones (A375P) are relatively more homogenous without distinct core-rim difference. However, our previous study showed the core of more aggressive melanoma has higher Fp/NADH redox ratio, indicative of nutritional deprivation. Additionally, the low perfusion/blood vessel permeability measured previously by DCE-MRI indicated these cells should be under starvation presumably accompanied with more cell death. Thus, it remains an open question what the survival status of the cells in the core of more aggressive melanoma is. We are currently investigating whether these cells are in autophagic state, a possible cell survival mechanism under starvation conditions.

---

* Lin Z. Li (215-898-1805, linli@mail.mmrrcc.upenn.edu), He N. Xu, Rong Zhou, Jerry D. Glickson, Molecular Imaging Laboratory, Department of Radiology, Philadelphia, Pennsylvania 19104. Shoko Nioka, Britton Chance, Department of Biochemistry & Biophysics, Johnson Research Foundation, University of Pennsylvania, Philadelphia, Pennsylvania 19104.

P. Liss et al. (eds.), *Oxygen Transport to Tissue XXX*, DOI 10.1007/978-0-387-85998-9_37, **247**
© Springer Science+Business Media, LLC 2009

1. INTRODUCTION

Predicting tumor metastatic potentials has high clinical significance since aggressive tumors in general exhibit faster growth progression and higher metastasis than less aggressive ones. Our long term goal is to identify imaging markers of metastatic potentials (or aggressiveness) in mouse models of human melanoma and eventually translate these markers into clinical applications. We have established mouse xenografts of five melanoma cell lines (A375P, A375M, A375P5, A375P10, and C8161), whose metastatic potentials in mouse models and/or invasive potentials *in vitro* have been investigated.[1-6] We have reported in previous ISOTT meetings that DCE-MRI and $T_{1\rho}$-MRI can differentiate the most aggressive melanoma xenografts from the least aggressive ones,[7] and that low temperature NADH/Fp fluorescence imaging or redox scanning can distinguish the aggressiveness of five melanoma cell lines (reported at ISOTT 2006 Conference). The results from the three methods all showed distinct difference between tumor core and rim in more aggressive melanomas. For example, DCE-MRI indicated that the aggressive C8161 tumors have a well-perfused rim and poorly-perfused core; $T_{1\rho}$-weighted MRI gave smaller $T_{1\rho}$ value in the areas near the necrotic center of the aggressive C8161 tumors; the redox scanning results also showed the least aggressive tumor A375P is largely reduced with low redox ratios and the other four types of more aggressive tumors have oxidized core with high redox ratios and reduced rim with low redox ratios. The mean redox ratios in the more oxidized core highly correlate with the invasive potentials of the five lines measured in Boyden chamber (R^2=0.93). This seems to suggest that it is the core that accounts for the differences in the metastatic potential. As part of our research efforts to understand the results obtained from the MR/optical imaging methods, we further performed histological analysis of frozen tumor sections. In this paper, we report our preliminary research findings and some insights obtained from H&E and DAPI counter-stained TUNEL assay analysis of three types of human melanoma mouse xenografts with their metastatic potentials falling in the rank order A375P<A375M <C8161.

2. METHODS

2.1. Animal Models

Three human melanoma cell lines, A375P, A375M, and C8161 were cultured in 10% FBS-added RPMI 1640 medium at 37°C under 5% CO_2. The metastatic potentials of A375P, A375M, C8161 have been evaluated by counting the number of metastasis in distant organs in mice inoculated with human melanoma cell lines,[1-3] with C8161 being the most aggressive and A375P being the least aggressive. The invasive potentials of the three cell lines were also measured by the *in vitro* Membrane Invasion Culture System (MICS),[5,6] falling in the rank order A375P(3%)<A375M(7%) <C8161(13.5%). The propagated cells were collected when they reach 75% confluence and subcutaneously inoculated into the athymic nude mice and grew into tumors in about 4-8 weeks. Two size groups of tumor were chosen for our study: smaller than 7 mm and bigger than 10 mm in diameter. In total, we have studied 13 tumors.

2.2. Histological Analysis of Tumor Tissue Sections by H&E and DAPI Counter-stained TUNEL Assay

The tumors were excised and immediately placed in OCT medium and stored at -20°C. The OCT-embedded frozen-tissue samples were then sliced 10μm thick at various depths of the tumor, with the slice plane in parallel with the mouse body surface. The frozen sections were stained by H&E and DAPI Counter-stained TUNEL Assay for histological examination. Adjacent sections were used for both H&E and TUNEL analysis, respectively.

H&E-stained tumor section slides were examined under microscope and photo images were taken with a bright light microscope at 40X. In addition, several dozens of images were photo-stitched using Photoshop® software to obtain the image of a whole tumor section for each type of tumor.

Terminal deoxynucleotidyl transferase (TdT)-mediated dUTP-biotin nick end labeling (TUNEL) assay was performed using In Situ Cell Death Detection Kit, Fluorescein (Roche Applied Science) as per the manufacturer's instructions. Cryo-perserved melanoma tissue sections treated with DNase (DNase I recombinant, Grade I) prior to the TdT reaction served as the positive control. Sections processed in the absence of TdT served as the negative control. All sections were counter-stained with DAPI contained in the mounting medium (Vectashield®) and protected by cover-slip and edge-sealed with nail polish. The TUNEL assayed slides were stored in the dark at 4°C before and after images were taken. Fluorescent photo images were acquired with a NIKON 600 upright microscope equipped with a digital camera. In addition, several dozens of adjacent images of both DAPI and FITC were photo-stitched using Photoshop® software to obtain the image of a whole tumor section for each type of tumor.

3. RESULTS AND DISCUSSION

3.1. H&E stained sections

The more aggressive melanomas showed high histological heterogeneity in terms of the distinct core and rim difference in both size groups. Distinct histological difference from core and rim of the more aggressive A375M and C8161 tumors was readily seen with the rim being more bluish and the core more reddish as shown in Figure 1(a). In addition, the histology of the core appears more like fishnet constituted of viable cells. Similar structure is also often observed in the metastatic melanoma patients and is called vascular-like network[8] that is rich in extracellular matrix.[9, 10] Most of the cells were morphologically different from the necrotic ones but also different from the ones in the rim. Their intact nuclei indicate that they were not completely dead yet. The rim was viable-cell dense.

No distinct difference from core and rim was observed from the least aggressive A375P tumors (Figure 1(b)) although very few small areas of morphologically different cells can be found sporadically in the bigger size group, indicating the histological homogeneity relative to the more aggressive ones.

3.2. DAPI counter-stained TUNEL assay results:

Correlating well with the histological result from H&E, results from TUNEL assay, in terms of core and rim difference, also showed high heterogeneity for the more aggressive A375M and C8161 melanomas (Figure 1(c) & (g)); while the least aggressive A375P tumors did not have core and rim difference (Figure 1(e)).

Vascular-like networks in the core of the more aggressive tumors can be seen readily. DAPI images clearly show many intact nuclei constituting the histological networks regardless whether the tumors appear to have visible ulceration or not (note: all tissue sections were carefully sliced to have excluded any ulceration). This is consistent with the observations from the H&E analysis. The FITC images give more interesting results. For the bigger aggressive tumors, very few fluorophore colocalizations in the overlay image of DAPI and FITC images were found, suggesting very few TUNEL positive cells. The large area having bright fluorescence that shows no color colocalization with DAPI staining in the FITC images of the more aggressive tumors is presumably background of autofluorescence. For the smaller aggressive tumors with visible ulceration, relatively more TUNEL positive cells were found in the tissue section near the ulcerated area compared to the bigger ones, indicating some of the core cells were apoptotic. The rim of the more aggressive tumors was viable-cell dense and no apoptotic cells were found regardless of the tumor size.

Also consistent with the H&E result, the least aggressive A375P tumors were viable-cell dense, showing no distinct core and rim difference. The bigger one has <10% of apoptotic cells in sporadic areas. Almost no TUNEL-positive cells were found in the smaller ones.

3.3. Discussion

Our previous DCE-MRI study indicated limited blood/nutrient supply in C8161 tumor core, which suggests the core of the aggressive melanoma should be accompanied by increased cell death. We also observed previously that the core of more aggressive melanoma has higher Fp/NADH redox ratio. High Fp/NADH ratios have been shown to indicate high oxidative metabolism in mitochondria.[11, 12] High Fp/NADH ratios were also seen in heart tissues under ischemia, indicating that cells were under starvation and cell death was occurring.[13] Hence, high Fp/NADH ratios indicate hypermetabolic activity or starvation/cell death. However, in our histological study, we found very few apoptotic cells in the core of the bigger aggressive ones and low percentage of apoptotic cells in the core of the smaller aggressive ones. Meanwhile we observed a large number of intact nuclei from DAPI images of the core. DAPI (4'-6-Diamidino-2-phenylindole) is known to form fluorescent complexes with natural double-stranded DNA. The intactness of the nuclei seen in our DAPI images suggests these cells were still viable. From the corresponding H&E images, we observed that these cells maintained their integrity although they were morphologically different from the ones in the rim, which also suggests high likelihood of their viable status. Thus, it remains an open question what the survival status of the cells in the core of more aggressive melanoma is.

Cancer cells may undergo apoptosis when under metabolic stress or treatment. However, study shows that the apoptosis-defective cells evade immediate death fate by entering into autophagic state.[14] An autophagic cell lives by eating its own intracellular macromolecules, or eating neighboring cells (cannibalism).[15, 16] Autophagy can be a cell-survival mechanism,[17, 18] and may be linked with poor treatment prognosis in melanoma clinic (personal communication with Ravi Amaravadi at the Department of Medicine, University of Pennsylvania).

Our prior experience indicates that melanoma growth rates seem to positively correlate with their invasive potentials. For example, the tumor volumes measured at day 33 (N=3~4 for each cell line) correlate with the invasive potentials of the 5 cell lines with a correlation coefficient of $R^2=0.9662$ (unpublished data). Whether the imaging/histological findings reflect the difference in tumor invasiveness or growth rate or both is still a question for future investigation.

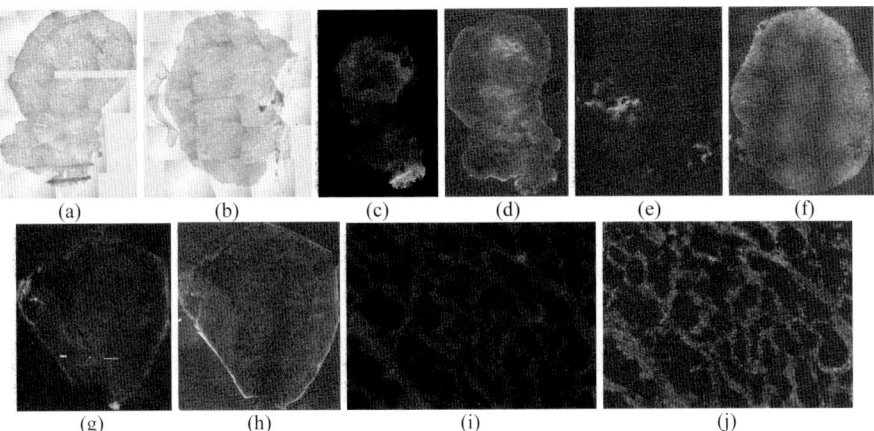

Figure 1. (a)-(h) Photo-stitched images of a tumor section and (i)-(j) expanded view of a core area of the most aggressive C8161 showing few apoptotic cells (20X). (a) & (b) H&E staining image (40X) of tumor sections of A375M (a) and A375P (b); FITC (c) and DAPI (d) of A375M and FITC (e) and DAPI (f) of A375P; FITC (g) and DAPI (h) of C8161; (i) FITC of C8161, and (j) DAPI of C8161. The diameter of the tumors shown here is greater than 11 mm. The strong fluorescence in Figure 1(c) is not specific FITC-signal for apoptosis, rather it maybe tissue autofluorescence.

4. CLOSING REMARKS

We have reported our histological findings from H&E staining and TUNEL assay on frozen tissue sections of human melanoma xenografts in athymic nude mice. In summary, for more aggressive melanomas, we found 1) tumor core and rim is distinctly different; 2) vascular-like networks constituted by viable cells dominate the core; 3) bigger tumors have very few apoptotic cells; 4) the core of the smaller tumors without visible necrotic center have more apoptotic cells than that of the bigger ones; 5) the smaller tumors with necrotic center still have many viable cells in the core; 6) tumor rim is viable-cell dense. For the least aggressive melanomas, we found that they appear to be relatively more homogenous, viable-cell dense. No distinct core-rim difference was observed. For bigger tumors, the less aggressive ones seem to have more apoptotic cells than the aggressive. This was not the case for the smaller ones. We may need to study more tumors to verify our preliminary results. It may be useful to have more quantitative measurements of the amount of apoptosis and necrosis in different types of tumors.

Since our previous study indicates that the tumor core seems to hold the answer for tumor aggressiveness, and in our preliminary histological study we did not find many apoptotic cells in the core but rather viable cells that are morphologically different from those in the rim were present, it is necessary to understand their survival status. We are currently investigating whether these cells are in autophagic state, a possible cell survival mechanism under starvation conditions.

5. ACKNOWLEDGEMENTS

We thank Drs. Mary J.C. Hendrix and Elisabeth Seftor (Children's Memorial Research Center, Northwestern University, Evanston, IL 60208) for providing the melanoma cell lines and their in vitro invasiveness data for this research. This research was supported by NIH PO1-CA56690 (PI: Dennis Leeper, Project Leader: Jerry D. Glickson), and two Pilot Grants awarded to Lin Z. Li from NIH SPORE of Skin Cancer at Wistar Institute (PI: Meenhard Herlyn) and Penn Network of Translational Research in Optical Imaging at the University of Pennsylvania (PI: Wafik El-Deiry).

6. REFERENCES

1. M. J. Hendrix, E. A. Seftor, Y. W. Chu, R. E. Seftor, R. B. Nagle, K. M. McDaniel, S. P. Leong, K. H. Yohem, A. M. Leibovitz, F. L. Meyskens, Jr. and et al., Coexpression of vimentin and keratins by human melanoma tumor cells: correlation with invasive and metastatic potential, *Journal of the National Cancer Institute* **84**(3), 165-174 (1992).
2. J. M. Kozlowski, I. R. Hart, I. J. Fidler and N. Hanna, A human melanoma line heterogeneous with respect to metastatic capacity in athymic nude mice, *Journal of the National Cancer Institute* **72**(4), 913-917 (1984).
3. D. R. Welch, J. E. Bisi, B. E. Miller, D. Conaway, E. A. Seftor, K. H. Yohem, L. B. Gilmore, R. E. Seftor, M. Nakajima and M. J. Hendrix, Characterization of a highly invasive and spontaneously metastatic human malignant melanoma cell line, *International Journal of Cancer* **47**(2), 227-237 (1991).
4. M. J. Hendrix, E. A. Seftor, Y. W. Chu, K. T. Trevor and R. E. Seftor, Role of intermediate filaments in migration, invasion and metastasis, *Cancer Metastasis Rev* **15**(4), 507-525 (1996).
5. E. A. Seftor, R. E. Seftor and M. J. Hendrix, Selection of invasive and metastatic subpopulations from a heterogeneous human melanoma cell line, *Biotechniques* **9**(3), 324-331 (1990).
6. M. J. Hendrix, E. A. Seftor, R. E. Seftor and I. J. Fidler, A simple quantitative assay for studying the invasive potential of high and low human metastatic variants, *Cancer Letters* **38**, 137-147 (1987).
7. L. Z. Li, R. Zhou, T. Zhong, L. Moon, E. J. Kim, H. Qiao, S. Pickup, M. J. Hendrix, D. Leeper, B. Chance and J. D. Glickson, Predicting melanoma metastatic potential by optical and magnetic resonance imaging, *Advances in Experimental Medicine and Biology* **599**,67-78 (2007).
8. A. J. Maniotis, R. Folberg, A. Hess, E. A. Seftor, L. M. Gardner, J. Pe'er, J. M. Trent, P. S. Meltzer and M. J. Hendrix, Vascular channel formation by human melanoma cells in vivo and in vitro: vasculogenic mimicry, *American Journal of Pathology* **155**(3), 739-752 (1999).
9. M. J. Hendrix, E. A. Seftor, A. R. Hess and R. E. Seftor, Vasculogenic mimicry and tumour-cell plasticity: lessons from melanoma, *Nature Reviews. Cancer* **3**(6), 411-421 (2003).
10. M. J. Hendrix, E. A. Seftor, D. A. Kirschmann, V. Quaranta and R. E. Seftor, Remodeling of the microenvironment by aggressive melanoma tumor cells, *Annals of the New York Academy of Sciences* **995**,151-161 (2003).
11. B. Chance, Flavoproteins of mitochondrial fatty acid oxidation in *Flavins and Flavoproteins*, edited by E. C. Slater(Elsevier, 1966), pp498-510.
12. B. Chance and H. Baltscheffsky, Respiratory Enzymes in Oxidative Phosphorylation. VII. Binding of IntramitochondrialReduced Pyridine Nucleotide, *The Journal of Biological Chemistry* **233**(3), 736-739 (1958).
13. M. Ranji, S. Kanemoto, M. Matsubara, M. A. Grosso, J. H. Gorman, 3rd, R. C. Gorman, D. L. Jaggard and B. Chance, Fluorescence spectroscopy and imaging of myocardial apoptosis, *Journal of Biomedical Optics* **11**(6), 064036 (2006).

14. K. Degenhardt, R. Mathew, B. Beaudoin, K. Bray, D. Anderson, G. Chen, C. Mukherjee, Y. Shi, C. Gelinas, Y. Fan, D. A. Nelson, S. Jin and E. White, Autophagy promotes tumor cell survival and restricts necrosis, inflammation, and tumorigenesis, *Cancer Cell* **10**(1), 51-64 (2006).
15. V. Karantza-Wadsworth, S. Patel, O. Kravchuk, G. Chen, R. Mathew, S. Jin and E. White, Autophagy mitigates metabolic stress and genome damage in mammary tumorigenesis, *Genes & Development* **21**(13), 1621-1635 (2007).
16. J. J. Lum, D. E. Bauer, M. Kong, M. H. Harris, C. Li, T. Lindsten and C. B. Thompson, Growth factor regulation of autophagy and cell survival in the absence of apoptosis, *Cell* **120**(2), 237-248 (2005).
17. A. L. Edinger and C. B. Thompson, Death by design: apoptosis, necrosis and autophagy, *Current Opinion in Cell Biology* **16**(6), 663-669 (2004).
18. J. Marx, Autophagy: Is It Cancer's Friend or Foe? *Science* **312**(5777), 1160-1161 (2006).

THE RELATIONSHIP BETWEEN VASCULAR OXYGEN DISTRIBUTION AND TISSUE OXYGENATION

Alexandru Daşu and Iuliana Toma-Daşu[*]

Abstract: Tumour oxygenation could be investigated through several methods that use various measuring principles and can therefore highlight its different aspects. The results have to be subsequently correlated, but this might not be straightforward due to intrinsic limitations of the measurement methods. This study describes an analysis of the relationship between vascular and tissue oxygenations that may help the interpretation of results. Simulations have been performed with a mathematical model that calculates the tissue oxygenation for complex vascular arrangements by taking into consideration the oxygen diffusion into the tissue and its consumption at the cells. The results showed that while vascular and tissue oxygenations are deterministically related, the relationship between them is not unequivocal and this could lead to uncertainties when attempting to correlate them. However, theoretical simulation could bridge the gap between the results obtained with various methods.

1. INTRODUCTION

Tissue oxygenation is one of the important prognostic factors for the outcome of radiotherapy since the presence of tumour hypoxia worsens the response to treatment through the radioresistance it confers to the affected cells or through the selection of a more aggressive malignant phenotype[1-2]. Several methods have therefore been proposed to determine the extent of tumour hypoxia. The methods use various measuring principles ranging from electrochemical measurements with a polarographic electrode[1,3] or the metabolisation of bioreductive drugs labelled with fluorescence markers or with radioactive isotopes up to complex magnetic resonance measurements, and they provide different sensitivities and discrimination powers. Thus, some methods use nitroimidazole compounds that are reduced in hypoxic conditions and become trapped intracellularly to map the distribution of hypoxia in tumours. When used in combination with fluorescence techniques[4], the method provides good spatial resolution down to the cellular level, but its major disadvantage is that the tumour has to be excised before analysis which limits its applicability mostly to experimental settings. For clinical applications, radiolabelled

[*] Alexandru Daşu, Norrland University Hospital, Department of Radiation Physics, 901 85 Umeå, Sweden. Iuliana Toma-Daşu, Stockholm University and Karolinska Institutet, Department of Medical Radiation Physics, 171 76 Stockholm, Sweden.

P. Liss et al. (eds.), *Oxygen Transport to Tissue XXX*, DOI 10.1007/978-0-387-85998-9_38,
© Springer Science+Business Media, LLC 2009

compounds are imaged with PET[5-6], but the major limitation is the spatial resolution provided by the method that involves averaging over rather large volumes. Other methods study the oxygen-dependent quenching of the phosphorescence of a hydrophilic agent present in the vasculature, thus yielding only the vascular oxygenation[7-8].

Correlating the results from different methods is not a straightforward task as each method investigates a different facet of the tissue oxygenation. Furthermore, the limitations in sensitivity or resolution provided by each method may also be reflected into the final results and may thus interfere with the interpretation of results. However, theoretical modelling may be used to investigate the efficiency of various measuring methods and to provide a better understanding of the potential difficulties that may appear for comparisons of results from various measurement techniques. Thus, this study describes a theoretical analysis of the relationship between vascular and tissue oxygenations that will facilitate the quantification of the results from techniques investigating these aspects of tumour oxygenation.

2. MATERIALS AND METHODS

The simulations were performed using a previously developed model for tissue oxygenation[9-10] that has been adapted for distributions of vascular oxygenations. It uses Monte Carlo algorithms to position blood vessels and to assign the oxygen content to each of them according to biologically relevant distributions. The distributions of intervascular distances were in agreement with measurements of vascular corrosion casts[11]. Similarly, the distributions of vascular oxygenations were in agreement with experimental phosphorescence studies[7-8]. Thus, it was assumed that vascular oxygenations follow log-normal distributions centred on oxygen tensions in the range 30-60 mmHg that covers an interval which might be encountered for clinical tumours as their vasculature is thought to have a venous origin with poor oxygen content[12]. Tissue oxygenations were calculated numerically from the diffusion equation (equation 1).

$$ -\nabla(D\nabla p) + q(p) = 0 \qquad (1) $$

where D is the oxygen diffusion coefficient, p is the local oxygen tension and $q(p)$ is the local consumption of oxygen dependent on concentration.

The simplified expression in equation 1 was derived assuming that all the blood vessels in the tissue are perfused and that the oxygen concentration in each vessel does not vary in time[9]. This is a reasonable simplification since for functional vessels the blood flow prevents the oxygen source from being exhausted. Furthermore, the presence of non-perfused vessels could be simulated by removing some of the vessels in the tissue as proposed by Daşu and co-workers[9] which is equivalent to using a different distribution of intervascular distances that takes into consideration only the functional vessels. The relevance of the model for practical applications has been verified by comparing the numerical results obtained with the analytical calculations and the experimental observations of Thomlinson and Gray[13].

Tissue oxygenations were calculated for various distributions of vascular oxygenations as well as for different distributions of intervascular distances. The results were compared in terms of median oxygenations and hypoxic fractions defined as fractions of values below certain thresholds considered relevant for radiobiological hypoxia.

3. RESULTS AND DISCUSSION

Figure 1 presents the result of a typical simulation of tissue oxygenation starting from a given vascular distribution centred on 30 mmHg. It shows that consumption in the intervascular space considerably lowers the tissue oxygen content in comparison to the vascular one. Similar results were obtained for all the distributions taken into consideration, illustrating that this is not an artefact due to poor vascular oxygenations. The difference between the two distributions means that one cannot be used as a substitute for the other in quantitative assessments of tumour response to radiation. However, this does not exclude the possibility to establish qualitative relationships between the two distributions and in fact this could explain the correlations found experimentally such as that by Ziemer and co-workers[8]. The question however is whether the features of the two distributions can be linked through simple relationships that could subsequently be used to convert the results of one measurement technique when the other one is not available. Features of interest in this case could be for example the fractions of values below certain thresholds (the hypoxic fractions) or the median values of the two distributions.

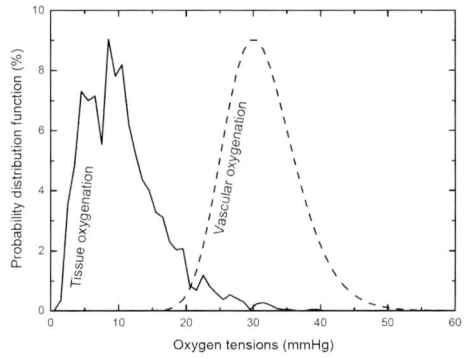

Figure 1. Tissue oxygenation (solid line) calculated with the diffusion equation for a vascular structure with the mean intervascular distance of 100 μm and relative standard deviation of 0.05 and a vascular distribution centred on 30 mmHg (dashed line). Consumption of oxygen decreases the oxygen content in the tissue.

Figure 2 shows the relationship between hypoxic fractions in the vascular and tissue oxygenation for various combinations of intervascular distributions and vascular oxygen content. The two hypoxic thresholds illustrated, 2.5 and 5 mmHg, are considered relevant for defining clinically the amount of radiobiological hypoxia. Indeed it has been suggested that patients could be ranked according to the amount of tumour hypoxia in order to identify the group that might benefit from aggressive treatment approaches. The success of any such strategy however depends strongly on the correct identification of the hypoxic fraction in individual patients. From this point of view, figure 2 shows that for a rather broad range of cases, the vascular distributions fail to show any hypoxia, even though the corresponding tissue oxygenations contain considerable hypoxic fractions. This means that patient ranking solely on the hypoxic fraction of the vascular oxygenation might overlook many patients that would otherwise benefit from dedicated treatment approaches.

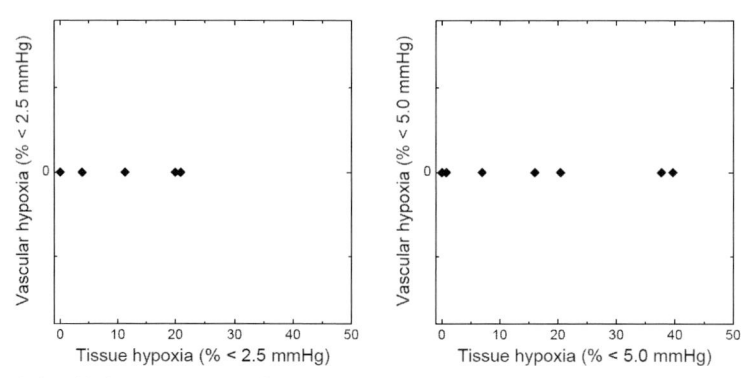

Figure 2. Relationship between hypoxic fractions determined from the vascular and tissue oxygenations for a broad range of distributions.

Vascular oxygen content is however one of the major determinants for tissue hypoxia and therefore correlations could be sought between other parameters such as the mean or the median value for each distribution. Figure 3 shows the relationship between the median values for the vascular and tissue oxygenations for a broad range of distributions covering various cases that could be encountered clinically for tumour vasculature. It therefore appears that the relationship between tissue and vascular oxygenations is not unequivocal as tissue oxygenation depends not only on the oxygen content of the blood vessels but also on the vascular structure. Thus for the same median vascular oxygenations, several tissue oxygenations are possible if the disposition of blood vessels changes both in terms of mean value and variance. Indeed broader distributions of intervascular distances indicate the presence of microregions with sparse vasculature where the fine equilibrium between supply and consumption is very much perturbed and cannot be balanced by the supply from regions with dense vasculature. This leads to an overall lower tissue oxygenation than for cases with relatively narrow intervascular distributions. Furthermore, tissues with sparse vasculature might have poor tissue oxygenation even for good vascular oxygenation as their oxygen supply is severely disrupted. These differences were indeed mentioned as potential limitations for attempts to characterise tissue oxygenation solely from vascular oxygenation[8].

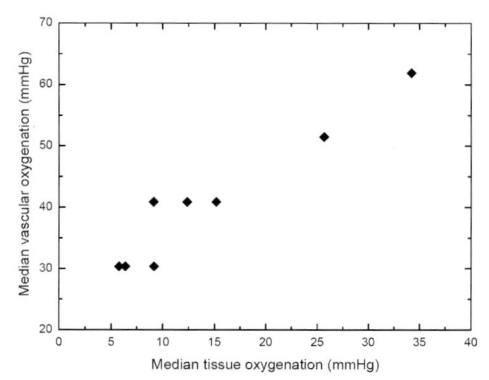

Figure 3. Relationship between the median values of a broad range of vascular and tissue oxygenations.

Further discordances may arise when attempting to correlate results from measurement processes with different thresholds as these values may not always be related in a linear manner. Such difficulties were indeed mentioned to appear for attempts to correlate measurements of nitroimidazole binding with tumour response as the former process has a higher threshold than the range of tensions which confer considerable radioresistance[14].

Many such differences could however be resolved through theoretical analyses starting from the basic physical processes. Theoretical simulation has the advantage that it is not affected by measuring artefacts that may appear for other methods and therefore the interpretation of its results is less subject to confounding factors. This is why it could be used to analyse the efficiency of various measuring methods[15-16] which would ultimately lead to a better understanding of the advantages and limitations for each of them. The limitations of theoretical modelling are given by the assumptions made about the model used and the accuracy of its parameters, but the method has been successfully used for decades to characterise the oxygenation of many tissue types[9,13,17-19].

The results of the simulations performed in this study have shown that theoretical modelling could bridge the gap between measurements of vascular oxygenation and the characterisation of tissue oxygenation. These findings could therefore have the potential to ease the quantitative interpretation of the results and the significance of comparisons between results from various measurement methods.

4. CONCLUSIONS

Theoretical modelling has shown that while vascular and tissue oxygenations are deterministically correlated, the relationship between them is not unequivocal. This could lead to possible divergences between various measurements methods especially if they have different threshold sensitivities. These findings could provide a better understanding of the measurement processes which could ultimately be reflected into further improvement of the interpretation of the results.

5. ACKNOWLEDGEMENTS

This project was partly supported by grants from the Cancer Research Foundation in Northern Sweden.

6. REFERENCES

1. M. Nordsmark, M. Overgaard, and J. Overgaard, Pretreatment oxygenation predicts radiation response in advanced squamous cell carcinoma of the head and neck, *Radiother. Oncol.* **41,** 31-39 (1996).
2. M. Höckel and P. Vaupel, Biological consequences of tumor hypoxia, *Semin. Oncol.* **28,** 36-41 (2001).
3. T. H. Knocke, H.-D. Weitmann, H.-J. Feldmann, E. Selzer, and R. Pötter, Intratumoral pO_2-measurements as predictive assay in the treatment of carcinoma of the uterine cervix, *Radiother. Oncol.* **53,** 99-104 (1999).
4. K. I. Wijffels, J. H. Kaanders, P. F. Rijken, et al, Vascular architecture and hypoxic profiles in human head and neck squamous cell carcinomas, *Br. J. Cancer* **83,** 674-683 (2000).
5. W. J. Koh, K. S. Bergman, J. S. Rasey, et al, Evaluation of oxygenation status during fractionated radiotherapy in human nonsmall cell lung cancers using [F-18]fluoromisonidazole positron emission tomography, *Int. J. Radiat. Oncol. Biol. Phys.* **33,** 391-398 (1995).
6. N. Lawrentschuk, A. M. Poon, S. S. Foo, et al, Assessing regional hypoxia in human renal tumours using ^{18}F-fluoromisonidazole positron emission tomography, *BJU. Int.* **96,** 540-546 (2005).

7. D. F. Wilson, S. A. Vinogradov, B. W. Dugan, D. Biruski, L. Waldron, and S. A. Evans, Measurement of tumor oxygenation using new frequency domain phosphorometers, *Comp Biochem. Physiol A Mol. Integr. Physiol* **132,** 153-159 (2002).

8. L. S. Ziemer, W. M. Lee, S. A. Vinogradov, C. Sehgal, and D. F. Wilson, Oxygen distribution in murine tumors: characterization using oxygen-dependent quenching of phosphorescence, *J. Appl. Physiol* **98,** 1503-1510 (2005).

9. A. Daşu, I. Toma-Daşu, and M. Karlsson, Theoretical simulation of tumour oxygenation and results from acute and chronic hypoxia, *Phys. Med. Biol.* **48,** 2829-2842 (2003).

10. A. Daşu and I. Toma-Daşu, Theoretical simulation of tumour oxygenation--practical applications, *Adv. Exp. Med. Biol.* **578,** 357-362 (2006).

11. M. A. Konerding, W. Malkusch, B. Klapthor, et al, Evidence for characteristic vascular patterns in solid tumours: quantitative studies using corrosion casts, *Br. J. Cancer* **80,** 724-732 (1999).

12. P. Vaupel, F. Kallinowski, and P. Okunieff, Blood flow, oxygen and nutrient supply, and metabolic microenvironment of human tumors: a review, *Cancer Res.* **49,** 6449-6465 (1989).

13. R. H. Thomlinson and L. H. Gray, The histological structure of some human lung cancers and the possible implications for radiotherapy, *Br. J. Cancer* **9,** 539-549 (1955).

14. A. Yaromina, D. Zips, H. D. Thames, et al, Pimonidazole labelling and response to fractionated irradiation of five human squamous cell carcinoma (hSCC) lines in nude mice: the need for a multivariate approach in biomarker studies, *Radiother. Oncol.* **81,** 122-129 (2006).

15. I. Toma-Daşu, A. Daşu, and M. Karlsson, The relationship between temporal variation of hypoxia, polarographic measurements and predictions of tumour response to radiation, *Phys. Med. Biol.* **49,** 4463-4475 (2004).

16. I. Toma-Daşu, A. Daşu, and M. Karlsson, Conversion of polarographic electrode measurements--a computer based approach, *Phys. Med. Biol.* **50,** 4581-4591 (2005).

17. A. Krogh, The number and distribution of capillaries in muscles with calculations of the oxygen pressure head necessary for supplying the tissue, *J. Physiol.* **52,** 409-415 (1919).

18. A. V. Hill, The diffusion of oxygen and lactic acid through tissues, *Proc. Roy. Soc. B* **104,** 39-96 (1928).

19. I. F. Tannock, Oxygen diffusion and the distribution of cellular radiosensitivity in tumours, *Br. J. Radiol.* **45,** 515-524 (1972).

INTRAOPERATIVE MEASUREMENT OF COLONIC OXYGENATION DURING BOWEL RESECTION

Daya B. Singh[#], Gerard Stansby[Φ], Iain Bain[#] and David K. Harrison[#]

Abstract: Recently lightguide spectrophotometry (LGS) has been investigated for assessing bowel mucosal oxygenation and may prove helpful in the diagnosis of bowel ischaemia. This pilot study explores the use of LGS and laser Doppler flowmetry (LDF) to measure SO_2 and perfusion in the bowel during key stages of colon surgery. SO_2 and perfusion in the mucosal and serosal layers of the rectum, sigmoid and descending colon were measured in 7 patients by LGS (Whitland Research, UK) and LDF (Moor Instruments, UK) respectively at four stages (baseline, after mobilisation of the sigmoid, after ligation of the inferior mesenteric artery (IMA) and after complete devascularisation of the sigmoid). The sigmoid mucosal SO_2 and LDF values were significantly lower than the baseline after the ligation of IMA and devascularisation. Mean (SD) baseline sigmoid mucosal SO_2 (73%) decreased to 55% after ligation of IMA and to 39% after complete devascularisation. The sigmoid serosal SO_2 did not show any change after ligation of IMA and showed only 7% decrease after devascularisation. There was no difference in baseline SO_2 and LDF values in different parts of the bowel but the mean mucosal baseline SO_2 (75%) was significantly lower than that in the serosa (87%). In conclusion, mucosal SO_2 measurements can accurately diagnose bowel ischaemia but serosal SO_2 does not reflect mucosal ischaemia.

1. INTRODUCTION

Diagnosis of bowel ischaemia (BI) poses significant difficulty for the clinicians due to its non specific presentations and lack of a simple diagnostic test. Colonoscopic appearances in BI are very non specific and mucosal biopsy is necessary to confirm the diagnosis. Mesenteric angiography is considered as the gold standard investigation for BI. This is an invasive investigation with significant morbidity. The interpretation of

---

[#] Department of Medical Physics, University Hospital of North Durham, Durham, DH1 5TW, UK.

[Φ] Northern Vascular Unit, Freeman Hospital, Newcastle upon Tyne, NE7 7DN, UK

P. Liss et al. (eds.), *Oxygen Transport to Tissue XXX*, DOI 10.1007/978-0-387-85998-9_39,
© Springer Science+Business Media, LLC 2009

angiography findings can be difficult and inconclusive in the presence of widespread atherosclerosis and partial blocks[1].

Laser Doppler flowmetry has been used in a number of clinical studies to measure tissue blood flow[2-4]. Although surface laser Doppler flowmetry is non-invasive, there are considerable problems with reproducibility, making it difficult to use in the clinical setting. Furthermore, tissue blood flow is not necessarily an accurate measure of tissue functional activity. Direct measurement of tissue oxygen levels or intramural pH levels may be more sensitive and less prone to variations. Indeed, a comparison of intramural pH (measured in an animal study using tonometry) and laser Doppler flowmetry showed that intramural pH was indeed a more sensitive and better quantitative method. Intramural pH is clearly related directly to tissue oxygen levels[5]. Tissue oxygen levels have been measured in the past in human colon using a Clark type electrode, and have been shown to be a predictor of colonic anastomotic healing[6].

Optical measurement of tissue oxygen saturation has recently been described. Harrison et al[7] used lightguide spectrophotometry (LGS) to measure skin tissue oxygen saturation in the lower limb of patients with peripheral vascular disease and has been used to define the level of amputation in these patients. LGS has been used recently to measure colonic mucosal oxygenation[8]. This technique gives a direct measurement of tissue oxygen levels in the colon mucosa and can be performed through a colonoscope/sigmoidoscope without significant risk to the patients. The relative concentration of oxy-haemoglobin is estimated from the attenuation of visible light (500-620nm) by the tissue. Optical fibres deliver light to the surface of the tissue and detect that which is returned. Computer software analyses the resulting attenuation spectrum to obtain a value of haemoglobin oxygen saturation.

This study reports preliminary results of measuring SO_2 and laser Doppler flow in the colon during different stages of surgery including after complete devascularisation.

2. AIM

The aim of this study was to explore the use of LGS and laser Doppler flowmetry (LDF) to measure SO_2 and perfusion in the bowel during key stages of colon surgery including devascularised colon.

3. MATERIALS

3.1 Lightguide Spectrophotometer

The lightguide spectrophotometer used in this study was a Whitland Research RM200 SO_2 monitor (Whitland Research, Whitland, UK). The visible wavelength used in this instrument is 500-586nm with a resolution of 3nm. Briefly, light is shone into the tissue and the light scattered back from the tissue is returned to the instrument. The machine has an inbuilt reference spectrum for different values of SO_2 measured from haemoglobin solutions: one fully oxygenated and one fully deoxygenated. The algorithm used here incorporates the Kubelka–Munk theory[9] of optical scattering in a non-homogeneous medium. Oxy and deoxygenated haemoglobin references are combined in different proportions iteratively until the best fit with the processed measured spectrum is achieved. The proportion with the best fit gives the SO_2. There is logarithmic correction for interference from skin pigment melanin.

3.2 Laser Doppler

A Moor Instruments DRT4 was used for Laser Doppler flux measurements (Moor Instruments Ltd, Axminster, UK).

3.3 Probes

Probes used for LGS
1. Surface probe: The surface probe was a standard surface probe by Whitland Research (Figure 1, a). This was used for measuring serosal SO_2 and comprises 500 micron transmitting and receiving fibre.
2. Catheter probe: 2.1mm diameter end delivery (Moor Instruments DP6a). This probe had a length of 4m so that it could pass through the colonoscope for the measurements (Figure 1, b)

Probes used for Laser Doppler
1. Surface probe: This was a DP1T-V2 probe (Figure 1, c) by Moor Instruments, UK. It was used for measuring serosal blood flow.
2. Catheter probe: This was the same probe as for the LGS (Moor Instruments DP6a) and was used for measuring mucosal blood flow.

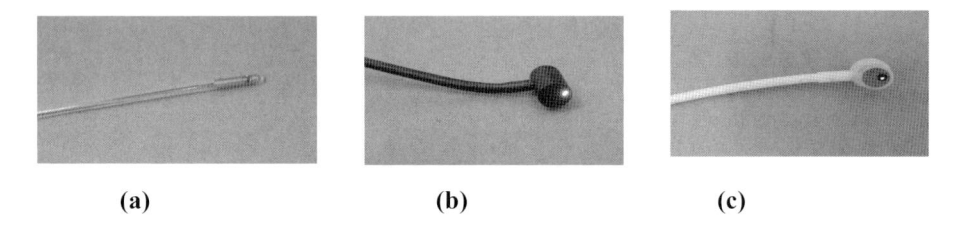

(a)   **(b)**   **(c)**

Figure 1. The probes: (a) Catheter probe, (b) LGS surface probe, and (c) Laser Doppler surface probe

4. METHODS

Ethical approval for this study was obtained from the Local Research Ethics Committee, Durham, UK. Seven patients, undergoing elective left-sided bowel resections (left hemicolectomies, sigmoid colectomies or anterior resections), were recruited into this study following informed consent. Patients with obstructive bowel lesions or those undergoing emergency surgery were excluded from the study. All the patients had standard general anaesthesia along with epidural analgesia and were haemodynamically monitored during the surgery. SO_2 and LDF were measured on the serosal and mucosal surface of each of the rectum, sigmoid colon and descending colon at four different stages of surgery. The first set of measurements was taken just after opening the abdomen; this was considered as baseline. The second set of measurements was taken after mobilisation of the sigmoid colon from the surrounding structures without interfering with its blood supply. The third set of measurements was taken after ligation of the inferior mesenteric artery (IMA). This would serve as a partial ischaemic model for the sigmoid. The fourth and final set of measurements was taken after complete devascularisation of the sigmoid colon. The third and the fourth sets of measurements were taken within a period of 5 – 15 minutes from the interventions (i.e ligation of the

IMA and complete devascularisation. The SO_2 and Laser Doppler signals usually reach a stable state within 5 minutes.   For the serosal measurements, surface probes were used. The probes were contained in a sterile transparent plastic bag which is normally used for holding the camera cable during laparoscopic surgery. The mucosal measurements were taken through a colonoscope by inserting the catheter probe through the biopsy channel of the colonoscope. Five values were recorded for each of the SO_2 and flux measurements at a particular site in the bowel and the mean values taken for analysis.

Statistical analysis of the data was done using SPSS® version 13.0. SO_2 values were recorded as percentage and flux was recorded as perfusion units and the mean values were used for analysis. Due to the small number, nonparametric tests were used to compare the measurements. Kruskal-Wallis and Mann Whitney U tests were used and P<0.05 was considered significant.

5. RESULTS

The total number of patients included in this preliminary results was seven (n=7) with five males (Table 1)

Table 1. Patient demographic data

Median age (range) in years	65 (41-71)
Sex ratio (M:F)	5:2
Body mass index, mean (± sd)	25.6 (± 2.4)
Median operation time in minutes (range)	140 (120-160)
Smoker : non smoker	1:6
Indications of surgery :	
Cancer	5
Benign conditions	2

The serosal and mucosal SO_2 in different parts of the colon at the four stages of surgery are presented in table 2. There were no significant differences in baseline SO_2 between the rectum, sigmoid and descending colon. The mean (± sd) baseline mucosal SO_2 in the colon was 75% (± 4.6) whereas the serosal SO_2 was 87% (± 5.3). The mean baseline serosal SO_2 (87% ± 5.3)) were significantly higher than the mucosal baseline (75% + 4.6) values (P < 0.001). The mean sigmoid mucosal SO_2 decreased after ligation of IMA from a baseline value of 73% (± 4.9) to 55% (± 12.2). After complete devascularisation, the mean sigmoid mucosal SO_2 was reduced to 39% (± 12.2). There were no significant changes in the sigmoid serosal SO_2 after mobilisation and ligation of IMA. The mean sigmoid serosal SO_2 was reduced to 79% (± 10) from a baseline value of 86% (± 7.3) after complete devascularisation.

The baseline serosal and mucosal LDF measurements (table 3) in the different parts of the bowel were comparable. The mean sigmoid mucosal flux decreased to 140 from a baseline of 216 after ligation of IMA and it furthered decreased to 57 after complete devascularisation. Similar trend was seen on the serosal flux measurements.

Table 2. The mean mucosal and serosal SO_2 (±sd) in the rectum, sigmoid, and descending colon at different stages of surgery.

	Baseline		Mobilisation		Ligation of IMA		Devascualarisation of sigmoid	
	Mucosal % (sd)	Serosal % (sd)	Mucosal % (sd)	Serosal % (sd)	Mucosal % (sd)	Serosal % (sd)	Mucosal % (sd)	Serosal % (sd)
Rectum	75 (±3.9)	88 (±4.8)	76 (±7.7)	89 (±4.8)	72 (±11.7)	83 (±4.3)	73 (±10.8)	87 (±6.8)
Sigmoid	73 (±4.9)	86 (±7.3)	73 (±6.3)	87 (±9)	55 (±12.2)	86 (±5.6)	39 (±12.2)	79 (±10)
Descending colon	76 (±5.2)	86 (±3.7)	77 (± 7)	88 (±4.9)	66 (±14.3)	86(±10.7)	74 (±14)	90 (±6.8)

Table 3. The mean mucosal and serosal flux (± sd) in the rectum, sigmoid, and descending colon at different stages of surgery.

	Baseline		Mobilisation		Ligation of IMA		Devascualarisation of sigmoid	
	Mucosal	Serosal	Mucosal	Serosal	Mucosal	Serosal	Mucosal	Serosal
Rectum	265 (±79)	241 (±87)	248 (±42)	281 (±80)	224 (±66)	181 (±66)	224 (±93)	195 (±39)
Sigmoid	216 (±60)	241 (±117)	231 (±47)	227 (±82)	140 (±78)	143 (±123)	57 (±21)	46 (±16)
Descending Colon	250 (±115)	287 (±111)	222 (±69)	262 (±54)	218 (±78)	270 (±116)	209 (±99)	308(±127)

6. DISCUSSION

The sigmoid mucosal SO_2 measurements showed a significant reduction from the baseline following the ligation of IMA and after devascularisation. Mobilisation of the sigmoid had no effect on the mucosal SO_2. The sigmoid colon is primarily supplied by the IMA but it also receives its blood supply from other collateral blood vessels. The degree of sigmoid ischaemia following ligation of IMA depends on the state of the collaterals and this was represented by the range of SO_2 values (41% – 74%) found in this study and the mean decrease after ligation of IMA was by 25%. The mean mucosal SO_2 in the completely devascularised sigmoid was 39% (± 12.2). This was a mean decrease by 46% from the baseline. Friedland et al,[8] in their animal model, demonstrated a decrease of 40% or more in the mucosal SO_2 following complete ligation of the blood supply to the colon and this is comparable with the results in this study.

The serosal SO_2 in the sigmoid was not affected by mobilisation and ligation of IMA. After complete devascularisation, the sigmoid serosal SO_2 decreased to 79%, which was 7% lower than the baseline. This difference was statistically significant but all the values in this group were above the anticipated ischaemic range described in the literature.[8] This makes us speculate that the serosal SO_2 measurements do not represent mucosal SO_2 and also fail to pick up ischaemia. Failure of the serosal SO_2 to fall to a lower level following ligation of IMA and devascularisation may be due to slow utilisation of oxygen in the serosa and/or the SO_2 may have been maintained by diffusion of atmospheric oxygen into the serosa. Smoking has the potential to affect SO_2 measurements in the intestine by two different mechanisms, firstly by directly affecting the circulation and secondly by a possible interference in the measurements by Spectrophotometry due to the presence of pathologic forms of haemoglobin. In this study it was difficult to comment on the effects of smoking on the SO_2 values as there was just one smoker in the group. Interestingly, this patient had the lowest mucosal SO_2 values in the descending colon and the rectum after ligation of the IMA (43% and 55% respectively).

The mucosal and serosal flux in the sigmoid showed significant decrease after ligation of IMA and devascularisation but the values showed high variability. The serosal flux measurements were able to pick up decreased blood flow in the sigmoid following ligation of IMA and devascularisation.

7. CONCLUSIONS

The results show that mucosal SO_2 measurements can accurately diagnose bowel ischaemia but serosal SO_2 does not reflect mucosal ischaemia. Doppler flux measurements could be useful in detecting ischaemia from the serosal surface of the bowel.

8. REFERENCES

1. L. J. Brandt and S. J. Boley, Colonic ischemia, *Surg Clin North Am* **72**(1): 203-29 (1992).
2. A. Vignali, L. Gianotti, M. Braga, G. Radaelli, L. Malvezzi and V. Di Carlo. Altered microperfusion at the rectal stump is predictive for rectal anastomotic leak. *Dis Colon Rectum* **43**, 76-82 (2000).
3. M. Sailer, E. S. Debus, K. H. Fuchs, J. Beyerlein and A. Thiede. Comparison of anastomotic microcirculation in coloanal J-pouches versus straight and side-to-end coloanal reconstruction: an experimental study in the pig. *Int. J. Colorectal Dis.* **15**, 114-117 (2000).
4. O. Hallbook, K. Johansson, and R. Sjodahl, Laser Doppler blood flow measurement in rectal resection for carcinoma--comparison between the straight and colonic J pouch reconstruction. *Br. J. Surg.* **83**, 389-392 (1996).
5. A. Senagore, J. W. Milsom, R. K. Walshaw, R Dunstan, W. P. Mazier, I. H. Chaudry. Intramural pH: a quantitative measurement for predicting colorectal anastomotic healing. *Dis Colon Rectum*. 33 (3):175-9 (1990)
6. W. G. Sheridan, R. H. Lowndes, H. L. Young Tissue oxygen tension as a predictor of colonic anastomotic healing. *Dis Colon Rectum* 30(11):867-71 (1987)
7. D. K. Harrison, P. T. McCollum, D. J. Newton, P. Hickman and A. S. Jain, Amputation level assessment using lightguide spectrophotometry. *Prosthet. Orthot. Int.* **19** 139-147 (1995).
8. S. Friedland, D. Benaron, I. Parachikov and R. Soetikno, Measurement of mucosal capillary hemoglobin oxygen saturation in the colon by reflectance spectrophotometry. *Gastrointest. Endosc.* **57** 492-497 (2003).
9. P. Kubelka, and F. Munk. Ein Beitrag zur Optik der Farbanstriche. *Zeitschrift fur Technische Physik* **12**: 593-601 (1931).

QUANTIFYING TUMOUR HYPOXIA BY PET IMAGING - A THEORETICAL ANALYSIS

Iuliana Toma-Daşu, Alexandru Daşu and Anders Brahme[*]

Abstract: Information on tumour oxygenation could be obtained from various imaging methods, but the success of incorporating it into treatment planning depends on the accuracy of quantifying it. This study presents a theoretical analysis of the efficiency of measuring tumour hypoxia by PET imaging. Tissue oxygenations were calculated for ranges of biologically relevant physiological parameters and were then used to simulate PET images for markers with different uptake characteristics. The resulting images were used to calculate dose distributions that could lead to predefined tumour control levels. The results have shown that quantification of tumour hypoxia with PET may lead to different values according to the tracer used and the tumour site investigated. This would in turn be reflected into the dose distributions recommended by the optimisation algorithms. However, irrespective of marker-specific differences, focusing the radiation dose to the hypoxic areas appears to reduce the average tumour dose needed to achieve a certain control level.

1. INTRODUCTION

Clinical and experimental observations have shown that the presence of hypoxia decreases the tumour response to treatment through the radioresistance it confers to the cells and the induction of genetic mutations that may lead to malignant progression[1-2]. In the light of these observations the treatment outcome can be improved by effective therapy of the hypoxic regions in tumours.

Information on tumour hypoxia in individual patients could be obtained from various imaging modalities and could in principle be used for more efficient optimisation of radiation treatments[3-4]. Positron Emission Tomography (PET) imaging nitroimidazole compounds that are metabolised in hypoxic regions and become trapped intracellularly is one of the clinically available methods. This method has the advantage that it could provide information both on the severity and on the spatial distribution of the hypoxic regions[5-6]. Furthermore, it is almost non-invasive and it could be relatively easily used in

---

* Iuliana Toma-Daşu, Stockholm University and Karolinska Institutet, Department of Medical Radiation Physics, 171 76 Stockholm, Sweden. Alexandru Daşu, Norrland University Hospital, Department of Radiation Physics, 901 85 Umeå, Sweden. Anders Brahme, Karolinska Institutet, Department of Medical Radiation Physics, 171 76 Stockholm, Sweden.

P. Liss et al. (eds.), *Oxygen Transport to Tissue XXX*, DOI 10.1007/978-0-387-85998-9_40,
© Springer Science+Business Media, LLC 2009

the clinic. Indeed, clinical studies have shown good correlations between hypoxia measured with PET and treatment outcome, indicating the relevance of such measurements for radiotherapy[7]. However, the success of incorporating PET information into the biological optimisation of treatment planning depends on the accuracy of the quantification of images and knowledge about the associated radiation resistance[8-9]. From this point of view, several aspects have to be taken into consideration and be analysed with respect to the potential implications they may have for the optimisation process. Thus, PET imaging involves averaging the signal over large volumes which results in limitations with respect to the spatial resolution of the analysed images. Furthermore, the various markers that are now available for clinical imaging of tumour hypoxia may differ in their uptake characteristics and discrimination power which are ultimately related to the achieved image contrast[6,10]. Last, but not least, the heterogeneity of the hypoxic regions as well as their temporal variation is another aspect that has to be taken into account when quantifying tumour hypoxia from PET imaging. It is the aim of the present study to investigate the impact of these factors on the optimisation process and ultimately on the predictions for treatment outcome.

2. MATERIALS AND METHODS

The analysis was performed on tumour models with morphologically realistic subregions with heterogeneous oxygenations. The relevant oxygen distributions for the tumour regions were calculated with a previously developed model that simulates the oxygen diffusion from complex vascular structures and its consumption at the cells[11-12]. The selective binding of several hypoxic tracers was subsequently simulated for the modelled tumours taking into account the binding characteristics derived from published experimental data[6,10]. PET images were eventually obtained by convoluting the resulting maps with an averaging mask that simulates the effects of the finite resolution of the imaging technique due to photon scatter and attenuation. The resulting images were then used as input for further simulations of the radiation treatment and for the calculation of optimal dose distributions that could lead to predefined tumour control levels. Comparisons were carried out between the results from several targeting strategies using different dose distributions designed to counteract the effects of the hypoxic regions in the tumours.

The tumour control probability was described by a Poisson model taking into account the initial number of clonogenic cells and the cell survival at the end of the radiation treatment. Cell survival in each point in the tumour was calculated with the linear quadratic (LQ) model[13-15] with the generic parameters $SF_2=0.5$ and $\alpha/\beta=10$ Gy^{16} for the fully oxic cells. For cells with less oxygenation, it was assumed that the radiation sensitivity varies with the local oxygen tension (or pO_2) through the equation proposed by Alper and Howard-Flanders[17].

3. RESULTS AND DISCUSSION

Figure 1 shows a tumour oxygenation pattern generated according to the methods described above and the simulated uptake of two hypoxic markers, FMISO and CuATSM. The distribution of each marker mirrors the distribution of oxygen in tumours according to its binding characteristics. Thus, marker concentration increases in the hypoxic regions of the tumours. However, the appearance of the hypoxic regions in relation to the better oxygenated ones is not the same for the two markers due to their different binding characteristics. Thus, FMISO is very little uptaken in regions with intermediate oxygenation which provides good discriminating power for the hypoxic regions, in spite of its poor overall binding. CuATSM on the other hand has considerable binding in tissues with intermediate oxygenation which counteracts the high binding in the very hypoxic regions by leading to a poor ratio of concentrations in the two regions.

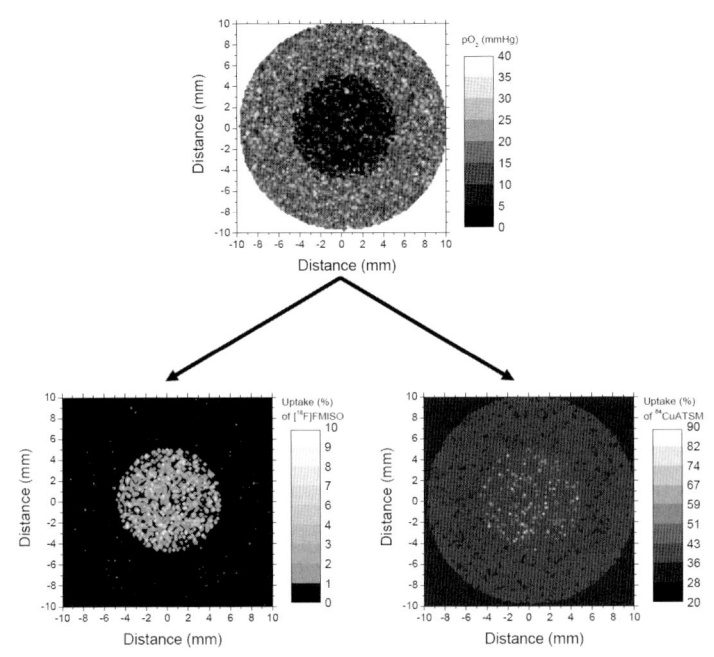

Figure 1. The heterogeneous tumour oxygenation model (upper panel) and the simulated uptake of two hypoxic tracers, FMISO and CuATSM, respectively (lower panels).

The unique relationship between marker concentration and the local oxygen tension could be used to calculate the predicted marker uptake in the tumour as well as to determine the local oxygenation from the measured marker uptake. An important issue for PET imaging is that the measured marker uptake is not identical to the real concentration as the measured value is influenced by other factors such as the averaging over relatively large volumes. The impact of these differences was investigated in a theoretical setting that is not influenced by the limitations that may be associated with practical measurement techniques.

Figure 2 shows in the left panel the simulated image of the FMISO distribution in figure 1 taking into consideration the spatial averaging and in the right panel the tissue oxygenation derived from the image. A comparison of the right panel of Figure 2 with the upper panel of Figure 1 shows the influence of the averaging of the imaging method. The derived tissue oxygenation corresponds to the real one, but it lacks much of the fine resolution of the former.

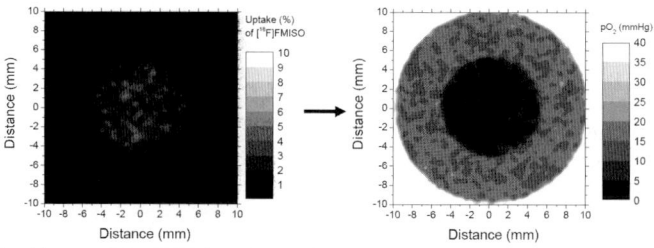

Figure 2. Simulated image of the FMISO marker (left panel) and the derived tissue oxygenation (right panel).

The combined effect of the binding characteristics of the marker and the averaging of the imaging method are illustrated in Figure 3. It shows the differences between the histogram distributions of the real oxygenation and those determined with different hypoxic markers. While the median values of the measured distributions are rather similar, there are considerable differences between the hypoxic fractions predicted from measurements with different markers. Thus, in the example given, the real hypoxic fraction (defined as the fraction of values below 2.5 mmHg) is 8%, while the values determined by the simulated measurements are 9% for FMISO and 14% for CuATSM. It thus appears that the poor selectivity of the latter marker, combined with spatial averaging leads to an overestimation of tissue hypoxia. Such differences may in fact be the cause for uncertainties when attempting to correlate measurement results with clinical outcome, as observed by Yaromina and co-workers[18].

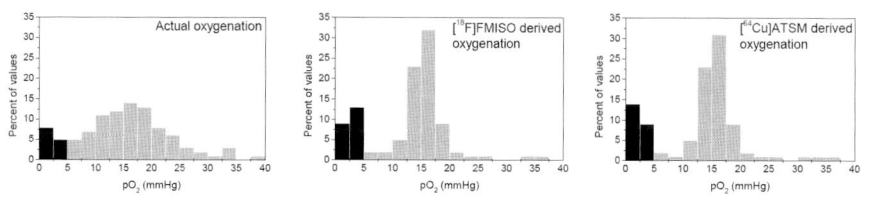

Figure 3. The real tissue oxygen distribution and derived distributions from measurements with different hypoxic markers.

The derived tissue oxygenations from PET measurements (such as that in the right panel of Figure 2) could also be used as input for treatment optimisation strategies which calculate optimum dose distributions that might counteract the adverse effects of tumour hypoxia.

Figure 4 shows two dose distributions calculated under the assumption that tumour hypoxia follows the distributions in the images obtained either with FMISO or with CuATSM. It was further assumed that the tumour contains 10^8 cells at the beginning of the treatment and that local temporal changes in pO_2 occur throughout the treatment without altering the overall oxygenation. The target control level was 90%. The overestimation of the amount of tumour hypoxia measured with CuATSM is translated into an increase of the dose needed for the central region of the tumour. Thus, the dose distribution calculated from the simulated FMISO image has 87 Gy to the central region of the tumour and 66 Gy to the outer ring, while the simulated CuATSM image leads to 93 Gy and 66 Gy respectively.

However, the common point of the two distributions is that they propose a redistribution of the dose, so that the central part of the tumour with increased hypoxia would require considerably higher dose than the outer ring with better oxygenation. This redistribution of the dose is very important as a uniform dose to the same target that would lead to the same control level of 90% is in excess of 80 Gy. A uniform increase in dose to the whole target would inevitably lead to an increase in the dose to the healthy tissues surrounding the tumour for most irradiation techniques available today which in turn would increase the rate of complications associated with the treatment. However, a higher dose to the hypoxic regions is expected to avoid these problems as the dose to the tumour regions neighbouring normal tissues is rather low.

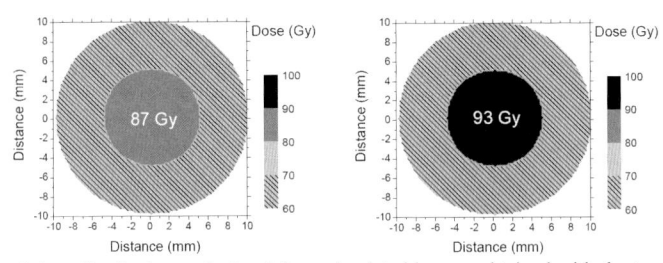

Figure 4. Targeted dose distributions calculated from simulated images obtained with the two hypoxic tracers, FMISO (left panel) and CuATSM (right panel).

This study has therefore shown that it is possible to use PET images to objectively improve the treatment of the hypoxic regions in a tumour. A simple method has also been proposed to transform the measured activities into radiation doses needed to achieve a prescribed tumour control level. This is a significant improvement over the empirical approaches to prescribe doses depending only on the appearance of functional images. Further developments of the method will take into consideration the theoretical approaches previously proposed by Brahme and co-workers[3-4,19] for the calculation of optimal dose distributions for the eradication of heterogeneous tumours taking also into account the uncertainties of the radiobiological response parameters[20].

4. CONCLUSIONS

The results of this study have shown that different markers yield different images both for the apparent oxygen distribution and for the magnitude of the activity relative to the oxic tissues. Quantification of tumour hypoxia may therefore result in different values according to the tracer used and the tumour site investigated. If not corrected, these differences may pose some problems when interpreting PET images of tumour hypoxia. Therefore these marker-specific aspects would have to be carefully taken into consideration when physiological information is included into the treatment planning process. However, irrespective of these differences, focusing the radiation dose to the hypoxic areas in the tumour has the potential to reduce the average tumour dose and indirectly the dose to the normal tissues adjacent to the tumour.

5. REFERENCES

1. M. Nordsmark, M. Overgaard, and J. Overgaard, Pretreatment oxygenation predicts radiation response in advanced squamous cell carcinoma of the head and neck, *Radiother. Oncol.* **41,** 31-39 (1996).
2. M. Höckel and P. Vaupel, Biological consequences of tumor hypoxia, *Semin. Oncol.* **28,** 36-41 (2001).
3. A. Brahme, Biologically optimized 3-dimensional in vivo predictive assay-based radiation therapy using positron emission tomography-computerized tomography imaging, *Acta Oncol.* **42,** 123-136 (2003).
4. A. Brahme, Fractionated and biologically optimized IMRT using in vivo predictive assay based radiation therapy (BIOART), *Proceedings of the Fifth International Symposium on Target Volume Definition in Radiation Oncology* (Limburg, 2005).
5. J. S. Rasey, W. J. Koh, M. L. Evans, L. M. Peterson, T. K. Lewellen, M. M. Graham, and K. A. Krohn, Quantifying regional hypoxia in human tumors with positron emission tomography of [18F]fluoromisonidazole: a pretherapy study of 37 patients, *Int. J. Radiat. Oncol. Biol. Phys.* **36,** 417-428 (1996).
6. J. S. Lewis, D. W. McCarthy, T. J. McCarthy, Y. Fujibayashi, and M. J. Welch, Evaluation of 64Cu-ATSM in vitro and in vivo in a hypoxic tumor model, *J. Nucl. Med.* **40,** 177-183 (1999).
7. F. Dehdashti, P. W. Grigsby, M. A. Mintun, J. S. Lewis, B. A. Siegel, and M. J. Welch, Assessing tumor hypoxia in cervical cancer by positron emission tomography with 60Cu-ATSM: relationship to therapeutic response-a preliminary report, *Int. J. Radiat. Oncol. Biol. Phys.* **55,** 1233-1238 (2003).
8. B. K. Lind and A. Brahme, The radiation response of heterogeneous tumors, *Physica Medica* (in press).
9. I. Toma-Daşu, Theoretical modelling of tumour oxygenation and influences on treatment outcome, PhD Thesis, Umeå University (2004).
10. J. S. Rasey, P. D. Hofstrand, L. K. Chin, and T. J. Tewson, Characterization of [18F]fluoroetanidazole, a new radiopharmaceutical for detecting tumor hypoxia, *J. Nucl. Med.* **40,** 1072-1079 (1999).
11. A. Daşu, I. Toma-Daşu, and M. Karlsson, Theoretical simulation of tumour oxygenation and results from acute and chronic hypoxia, *Phys. Med. Biol.* **48,** 2829-2842 (2003).
12. A. Daşu and I. Toma-Daşu, Theoretical simulation of tumour oxygenation--practical applications, *Adv. Exp. Med. Biol.* **578,** 357-362 (2006).
13. B. G. Douglas and J. F. Fowler, Fractionation schedules and a quadratic dose-effect relationship, *Br. J. Radiol.* **48,** 502-504 (1975).
14. G. W. Barendsen, Dose fractionation, dose rate and iso-effect relationships for normal tissue responses, *Int. J. Radiat. Oncol. Biol. Phys.* **8,** 1981-1997 (1982).
15. J. F. Fowler, The linear-quadratic formula and progress in fractionated radiotherapy, *Br. J. Radiol.* **62,** 679-694 (1989).
16. H. D. Thames, S. M. Bentzen, I. Turesson, M. Overgaard, and W. Van den Bogaert, Time-dose factors in radiotherapy: a review of the human data, *Radiother. Oncol.* **19,** 219-235 (1990).
17. T. Alper and P. Howard-Flanders, Role of oxygen in modifying the radiosensitivity of E. Coli B., *Nature* **178,** 978-979 (1956).
18. A. Yaromina, D. Zips, H. D. Thames, W. Eicheler, M. Krause, A. Rosner, M. Haase, C. Petersen, J. A. Raleigh, V. Quennet, S. Walenta, W. Mueller-Klieser, and M. Baumann, Pimonidazole labelling and response to fractionated irradiation of five human squamous cell carcinoma (hSCC) lines in nude mice: the need for a multivariate approach in biomarker studies, *Radiother. Oncol.* **81,** 122-129 (2006).
19. A. Brahme and A. K. Agren, Optimal dose distribution for eradication of heterogeneous tumours, *Acta Oncol.* **26,** 377-385 (1987).
20. G. Kåver, B. K. Lind, J. Löf, A. Liander, and A. Brahme, Stochastic optimization of intensity modulated radiotherapy to account for uncertainties in patient sensitivity, *Phys. Med. Biol.* **44,** 2955-2969 (1999).

NEW MEASUREMENTS FOR ASSESSMENT OF IMPAIRED CEREBRAL AUTOREGULATION USING NEAR-INFRARED SPECTROSCOPY

Dominique De Smet, Joke Vanderhaegen, Gunnar Naulaers, and Sabine Van Huffel[*]

Abstract: Some preterm infants have poor cerebral autoregulation. The concordance between cerebral intravascular oxygenation (HbD), computed as the difference between oxygenated (HbO$_2$) and deoxygenated (Hb) haemoglobin, and mean arterial blood pressure (MABP) reflects impaired autoregulation. As HbD is not an absolute value, we developed mathematics to prove that the cerebral tissue oxygenation (TOI), an absolute signal computed as the ratio of HbO$_2$ to total haemoglobin (Hb+HbO$_2$), may replace HbD. In the meantime, we attempt to theoretically predict the true level of autoregulation of a patient by defining a critical percentage of the signal recording time (CPRT). 20 preterm infants with need for intensive care were studied in the first days of life. HbD and TOI were obtained with the NIRO-300 (© Hamamatsu, Japan). Invasive MABP was measured continuously. All mathematics showed a strong similarity between HbD and TOI.

1. INTRODUCTION

The most important forms of brain injury in premature infants are partly caused by disturbances in cerebral autoregulation, the mechanism which limits cerebral blood flow (CBF) variation over a range of cerebral perfusion pressures. Autoregulation may be absent not only in sick preterm infants, but also in those that are clinically well, and there is also evidence that CBF is independent of MABP over a wide pressure range in preterm infants, suggesting that autoregulation may actually be effective in the immature brain.

[*] Dominique De Smet (Dominique.DeSmet@esat.kuleuven.be, Kasteelpark Arenberg no. 10 box 2446, B-3001 Heverlee, Belgium, tel. +3216321067, fax. +3216321986) and Sabine Van Huffel, SCD, Dept. of Electrical Engineering, KULeuven, Belgium. Joke Vanderhaegen and Gunnar Naulaers, Neonatal Intensive Care Unit, University Hospitals Gasthuisberg, KULeuven, Belgium.

P. Liss et al. (eds.), *Oxygen Transport to Tissue XXX*, DOI 10.1007/978-0-387-85998-9_41,   **273**
© Springer Science+Business Media, LLC 2009

As HbD reflects CBF, autoregulation can be measured by studying the concordance between HbD and MABP, assuming no changes in oxygen consumption, in oxygen saturation and in blood volume.[1] As HbD is unfortunately not an absolute value, we focused our attention on TOI, another absolute-valued signal. We explored the relationship between HbD and TOI to prove that TOI can also be used to assess impaired cerebral autoregulation in neonates. We developed for this purpose three similarity measurements based either on a correlation coefficient between two curves, or on a multiple linear regression using least squares. However, there are some other differences between the HbD and TOI signal. In terms of the physiology the TOI reflects oxygen saturation mostly in the cerebral venous compartment. Assuming constant cerebral metabolic rate for oxygen, changes in cerebral venous oxygen saturation will parallel variations of CBF according to the Fick principle. On the other hand the HbD signal reflects the mismatch between HbO_2 and Hb. In terms of the near-infrared algorithm the TOI calculation is based on spatially resolved spectroscopy (SRS) whereas the measurement of HbD is based on the modified Beer-Lambert law assuming a path length factor. The two methodologies also happen to have different depth resolutions.

The direct concordance (DiCo) between HbD and MABP has in the literature been studied, in the field of near-infrared spectroscopy, mostly by means of the correlation (COR) and coherence (COH) coefficients, the latter one measuring the degree of linear dependence between the frequency spectra of two signals. In 2000, Tsuji et al.[1] applied COH to detect impaired cerebral autoregulation from continuous measurements. In 2001, Morren et al.[4] pursued Tsuji's work using the same methodology. Recently, Soul et al.[5] further optimized COH for continuous MABP and HbD. The partial coherence (PCOH) coefficient also measures the degree of linear dependence -in the frequency domain- between two signals, but after having eliminated in a least squares sense the contribution of other influencing signals.[2]

The above defined DiCo scores are not handy enough for the physician to handle with. Because of time variability resulting from the continuous nature of the measurements, the DiCo score value describing the infant's cerebral autoregulation differs from one time instant to another one during the recording. Therefore, we define here a measurement that synthesizes the level of autoregulation of a patient for the whole recording time. During the last years, some authors derived mathematics to synthesize in a single parameter the level of autoregulation of a patient. So did Tsuji et al.[1] when computing the mean COH of each patient for fixed periods of time. The higher this mean DiCo score, the worse the cerebral autoregulation of the patient. Last year, Soul et al.[5] proposed a pressure-passive index (PPI): dividing first the signal recordings into consecutive epochs of constant duration,[4] they computed the percentage of epochs with significant low-frequency COH between MABP and HbD. If the mean COH was higher than a certain threshold value, then score 1 was assigned to the considered epoch, otherwise it received score 0. In this paper, we defined a similar parameter called CPRT, also based on the previously presented DiCo scores.

2. DATASETS

From 20 premature infants with need for intensive care, MABP, SaO_2 (arterial oxygen saturation), HbD, and TOI were measured simultaneously at the University Hospitals Leuven in the first days of life. The babies were characterized by a mean PMA of 28.7 weeks (24-39) and a weight ranging from 570 to 1470g. The recording time ranges from 1h30 to 23h35. The data were recorded at a sampling frequency of 100Hz by a data acquisition system Codas (CODAS©, Dataq Instruments, USA) and stored on a PC. MABP (assessed by intravascular catheterization) and SaO_2 (assessed by pulse oximetry) are analog and were digitized afterwards by the CODAS-system. The NIRO-300 signals are digital with a sampling rate of 6Hz. They were converted to analog signals with a sample-and-hold function before their introduction in the CODAS-system. From this 100Hz data, the average values for non-overlapping 5-seconds intervals were calculated (0.2Hz). This sampling frequency is still high enough, since major physiological importance can be attributed to the frequency band of 0-0.01Hz (i.e. changes occurring over several minutes).[1] The differential path length taking into account the scattering of the infrared light into the brain was set at 1.39 as commonly accepted in the literature, and encoded into the PC as a constant value.

3. METHODS

3.1. Direct Concordance Methods

We applied the COR, COH and PCOH DiCo methods. We described them previously.[3] COR was scaled such that its value matches the interval [0,1] instead of [-1,1]. In this way the COR score may be more easily compared with the ones obtained with the other methods. Before computing COR, the signals were low pass filtered to the frequency range of 0–0.01Hz.[1] We based our work on computing COH using the Welch's method as Morren et al.[4] described it. The average of COH over the frequency band of interest (0–0.01Hz) was used as DiCo score for the considered signal duration.[1,3-5] Since the concordance between the signals might vary as a function of time, a sliding window approach was used (the DiCo scores were calculated over 30-minute epochs).[4] Before applying COH and COR, we first filtered all signals to keep only the recording intervals during which the variation in SaO_2 around its mean did not exceed 5%.[3] In this way the condition of constant SaO_2 was satisfied. This procedure was not followed for PCOH, as SaO_2 was the signal of which we took off the influence on MABP and HbD/TOI. For further information on the operating conditions, we refer the reader to the literature.[3, 4] It is important to recognize here the potential limitations of the use of correlation and transfer function analyses to investigate moment-to-moment autoregulatory mechanisms, as these approaches assume a linear and stationary relationship between the signals HbD/MABP or TOI/MABP. This approach can produce misleading results in a system with non-linear and non-stationary properties.

3.2. The CPRT: A Synthesized Measurement of Cerebral Autoregulation

The acronym CPRT stands for critical percentage of the recording time, i.e. the percentage of the recording time during which the DiCo score is greater than a fixed value, called critical score value (CSV). In this work the CSV will be set by default at half the maximum DiCo score value (i.e. 0.5). In contrast to the PPI, the CPRT is not computed from boolean scores over epochs -which means a loss of accuracy-, but from the DiCo scores. As seen previously, the DiCo scores are proportional to the autoregulation impairment. The proposed definition is moreover not limited to COH, but is applicable to all existing methods. We define the CPRT as

$$CPRT = \frac{\sum_i t_i}{t_{total}} \times 100$$

where the t_i's represent the durations of the parts of the DiCo score curve lying above the CSV, and t_{total} is the total recording time.

3.3. Similarity Measurements

The first similarity measurement computes the correlation between the interpolated DiCo score curves (CBS) related to both signals. These are the curves of the time-varying DiCo scores[3, 4] computed from MABP and HbD, and from MABP and TOI. As the CBS holds only for one infant, it has to be computed for all infants, afterwards a mean of the obtained CBSs (mCBS) is computed for each DiCo method.

The second one is based on the CPRT definition, and is done at the patient group level from the CPRTs. Figure 1 illustrates the CPRTs for all patients and for each DiCo method. For each DiCo method, we then computed a correlation coefficient between the interpolated curves related to signal sets MABP/HbD and MABP/TOI (CB-CPRT).

The last way to look at the similarity between HbD and TOI was achieved by computing a multiple linear regression, using least squares, from the mean DiCo scores of all patients (see Fig. 2). On the x-axis and y-axis, we considered the mean DiCo scores related to signal set MABP/HbD, and MABP/TOI, respectively. We used two parameters to assess the regression efficiency. The first is the mean of the absolute values of the regression residuals r_i defined as

$$|\bar{r}| = \frac{\sum_{i=1}^{n} |r_i|}{n}$$

where n is the number of patients. The second parameter is the regression mean squared error (MSE), which is an estimator of the variance of the random disturbances.

4. EXPERIMENTAL RESULTS AND DISCUSSION

The mCBS is equal to 0.76 for COH, 0.63 for COR, and 0.70 for PCOH. The CB-CPRT is equal to 0.81 for COH, 0.71 for COR, and 0.78 for PCOH (see Fig. 1). A larger follow-up study is however needed to evaluate statistically the CPRT efficiency with respect to the infant clinical outcomes. As a scaled correlation coefficient belonging to interval [0,1] was considered to compute mCBS and CB-CPRT, the obtained results are very positive.

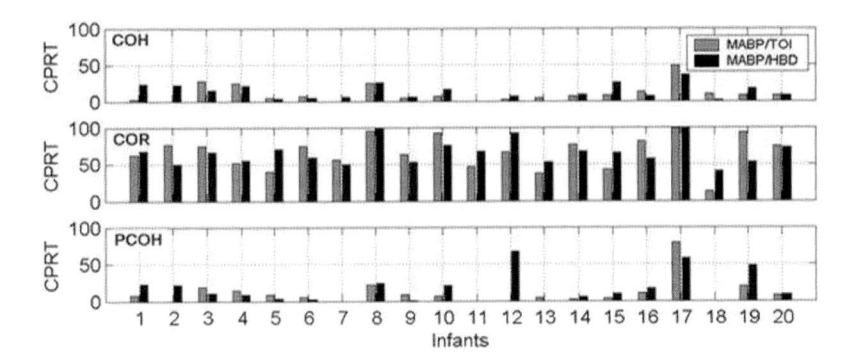

Figure 1. Illustration of the CPRTs related to each infant and for each score. For each DiCo method, a correlation coefficient (the CB-CPRT) was then computed from the interpolated CPRT curves related to both signal sets: it is equal to 0.81 for COH, 0.71 for COR, and 0.78 for PCOH.

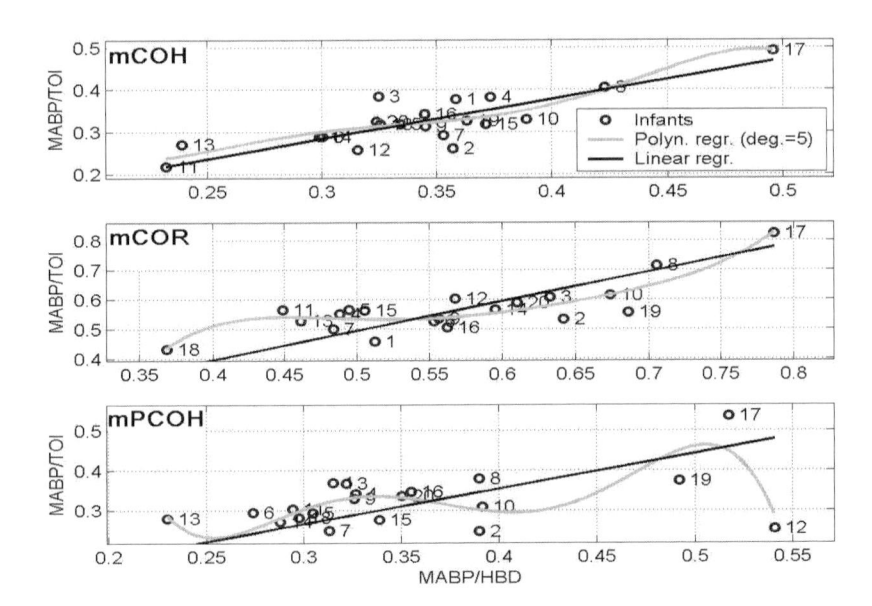

Figure 2. Multiple linear regression using least squares, achieved from the mean -over time- DiCo scores (mCOH, mCOR and mPCOH) of all patients. On the x-axis we considered the mean DiCo scores related to signal set MABP/HbD, and following the y-axis the ones of signal set MABP/TOI. The mean of the absolute values of the residuals is 0.027 for COH (corresponds to 2.7% of the full interval width [0,1] in which the DiCo scores belong), 0.043 for COR (4.3%), and 0.041 (4.1%) for PCOH. The MSE is 0.0013 for COH, 0.0031 for COR, and 0.0038 for PCOH. The figure also shows a polynomial regression of degree 5 of the data.

The mean of the absolute values of the residuals r_i is 0.027 (corresponds to 2.7% of the full interval width [0,1] in which the DiCo scores take their values) for COH, 0.043 (4.3%) for COR, and 0.041 (4.1%) for PCOH (see Fig. 2). The MSE is 0.0013 for COH, 0.0031 for COR, and 0.0038 for PCOH. The linear regression shows thus that TOI is similar to HbD as the regression parameters are low. Please see Table 1 for all results.

Table 1. Similarity measurements for proving that the behavior of TOI is similar to the one of HbD regarding cerebral autoregulation. They are based either on a correlation coefficient between two curves (CB-CPRT and mean CBS), or on a multiple linear regression using least squares (MSE and mean of |r|).

	mCBS	CB-CPRT	Mean of |r|	MSE
COH	0.76	0.81	0.027	0.0013
COR	0.63	0.71	0.043	0.0031
PCOH	0.70	0.78	0.041	0.0038

5.   CONCLUSION

First, the measurements presented to study the similarities between signal sets MABP/HbD and MABP/TOI show very positive results, which brings us to claim that TOI may be used for the calculation of cerebral autoregulation in neonates. Secondly, a new synthesized measurement of cerebral autoregulation -the CPRT- was presented. In opposition to previously defined measurements, it is applicable to all existing DiCo methods, and generates a measurement of the autoregulation impairment proportional to the time-varying DiCo score between the signals of interest.

6.   ACKNOWLEDGEMENTS

The research was supported by Research Council KUL: GOA AMBioRICS, CoE EF/05/006, by FWO projects G.0519.06 (Noninvasive brain oxygenation), and G.0341.07 (Data fusion), by Belgian Federal Science Policy Office IUAP P5/22.

7.   REFERENCES

1. M. Tsuji, J. Saul, A. du Plessis, E. Eichenwald, J. Sobh, R. Crocker; and J. Volpe, Cerebral intravascular oxygenation correlates with mean arterial pressure in critically ill premature infants, Pediatrics 106 (4), 625-632 (2000).
2. J. Leuridan, and B. Rost, Multiple input estimation of frequency response functions: diagnostic techniques for the excitation, ASME Paper Number 85-DET-107, 5 pages, (1985).
3. D. De Smet, S. Van Huffel, J. Vanderhaegen, G. Naulaers, and E. Dempsey, Detection of autoregulation in the brain of premature infants, EMBS/BSMBEC, Proc. of IEEE/EMBS Annual Benelux Symposium and of Belgian Day on Biomedical Engineering, 215-218 (2006).
4. G. Morren, P. Lemmerling, S. Van Huffel, G. Naulaers, H. Devlieger, and P. Casaer, Detection of autoregulation in the brain of premature infants using a novel subspace-based technique, EMBS, Proc. of the 23rd Annual Intern. Conf. of the IEEE (2), 2064-2067 (2001).
5. J. Soul, P. Hammer, M. Tsuji, J. Saul, H. Bassan, C. Limperopoulos, D. Disalvo, M. Moore, P. Akins, S. Ringer, J. Volpe, F. Trachtenberg, and A. du Plessis, Fluctuating pressure-passivity is common in the cerebral circulation of sick premature infants, Pediatric Research 61 (4), 467-473 (2007).

SINGLE SHOT $T_{1\rho}$ MAGNETIC RESONANCE IMAGING OF METABOLICALLY GENERATED WATER *IN VIVO*

Eric A. Mellon[*], R. Shashank Beesam*, Mallikarjunarao Kasam[1], James E. Baumgardner[2], Arijitt Borthakur*, Walter R. Witschey Jr. *, Ravinder Reddy*

Abstract: The use of Oxygen-17 MRI provides great promise for the clinically-useful quantification of metabolism. To bring techniques based on ^{17}O closer to clinical application, we demonstrate imaging of metabolically generated $H_2^{17}O$ in pigs after $^{17}O_2$ delivery with increased temporal resolution $T_{1\rho}$-weighted imaging and precision delivery of $^{17}O_2$ gas. The kinetics of the appearance of $H_2^{17}O$ in pig brains are displayed with one to two minutes of $^{17}O_2$ delivery, the shortest delivery times reported in the literature. It is also shown that $H_2^{17}O$ concentrations can be quantified with single shot $T_{1\rho}$ imaging based on a balanced steady state free precession readout, and that with this strategy pausing to reduce T_1 saturation increases sensitivity to $H_2^{17}O$ over acquisition in the steady state. Several additional considerations with this sequence, which can be generalized to any pre-encoding cluster, such as energy deposition are considered.

1. INTRODUCTION

A demand for sequences that rapidly obtain $T_{1\rho}$-weighted images has been created by the use of the non-radioactive, naturally abundant T_2 and $T_{1\rho}$ shortening agent $H_2^{17}O$ *in vivo* to measure the cerebral metabolic rate of oxygen consumption $(CMRO_2)$[1]. The molecular form, $^{17}O_2$ gas, can be inhaled by the subject but is invisible to MRI due to its paramagnetic nature causing extremely fast relaxation. However, during cellular respiration $^{17}O_2$ is reduced to the detectable form $H_2^{17}O$ water. By modeling or measuring the kinetics of $^{17}O_2$ delivery and $H_2^{17}O$ recirculation, it is possible to quantify $CMRO_2$.

---

[*] Department of Radiology, MMRRCC University of Pennsylvania, Philadelphia, Pennsylvania, 19104-6100
[1] Department of Radiology, University of Miami, Florida, 33136
[2] Department of Anesthesiology, University of Pennsylvania, Philadelphia, Pennsylvania 19104

P. Liss et al. (eds.), *Oxygen Transport to Tissue XXX*, DOI 10.1007/978-0-387-85998-9_42,
© Springer Science+Business Media, LLC 2009

The detection of $H_2{}^{17}O$ is typically achieved by direct[2] or indirect[3-5] measurements. Direct techniques obtain signal from the ^{17}O nucleus by exciting and receiving at the ^{17}O Larmor frequency. These techniques are very difficult to perform in vivo in large animals due to the low gyromagnetic ratio of ^{17}O, low sensitivity of the ^{17}O nucleus, and short relaxation times. As a result, ^{17}O measurements of $CMRO_2$ are improved by increasing the magnetic field, typically to field strengths far above those used clinically, and using very small coils to increase signal that can not give whole brain coverage in large animals.

By contrast, the indirect method is based on the enhancement of T_2 or $T_{1\rho}$ relaxation of protons in bulk water by ^{17}O, where protons experience spin-spin (J-) coupling to ^{17}O and exchange at a rate on the order of 1 kHz between $H_2{}^{17}O$ and $H_2{}^{16}O$. Because the J-coupling interaction that leads to relaxation enhancement is a field strength independent interaction, lower field strengths make detection of $H_2{}^{17}O$ easier *in vivo*. That is, higher starting T_2 values for tissue make the absolute T_2 reduction by $H_2{}^{17}O$ greater at lower field strengths. Also, lower tissue T_1 values allow for faster averaging of the $H_2{}^{17}O$ effect. Further, diminished susceptibility reduces the competing signal effects of cerebral blood and sequence artifacts. As a result, indirect techniques hold a great deal of promise at today's clinically useful field strengths (typically 1.5 Tesla). The goal of this work is to improve upon the pulse sequence groundwork for future measurements of $CMRO_2$ in large animals and humans on clinical scanners.

Nevertheless, ^{17}O is rarely used as a metabolic tracer due to numerous difficulties. For example, the relaxation enhancement due ^{17}O is subtle, which leads to long times to echo in conventional imaging sequences to develop contrast and resultant $T_2{}^*$ effects in gradient echo-based sequences that lead to unstable artifacts that often change throughout the study. The approach used here attempts low amplitude spin locking, where field inhomogeneities are eliminated but contrast based on $H_2{}^{17}O$ remains virtually the same. A further benefit is that with high amplitude spin locking, $H_2{}^{17}O$ contrast is mitigated, allowing one to quantify the amount of $H_2{}^{17}O$[6].

The pulse sequence of choice for $T_{1\rho}$-based $H_2{}^{17}O$ studies has been $T_{1\rho}$ preparation of turbo spin echo. Increasing turbo factors to reduce time per image has led to unacceptable k-space blurring *in vivo* and as such it has taken multiple repetitions in order to fill all of k-space. These long times per image on the order of 10s/image have proven unacceptable when trying to measure the kinetics of $^{17}O_2$ metabolism. This is because we propose to measure $CMRO_2$ based off of the initial dip in signal before the metabolically generated water has had time to recirculate, a time on the order of 30-60 seconds. Also, in attempting to define the recirculation of the tracer in large animals, we propose to analyze arterial blood signal to detect $H_2{}^{17}O$ "wash-in".

Still, many complexities have appeared in the attempt to reduce this time per image. For example, implied in each $T_{1\rho}$-based indirect model of $H_2{}^{17}O$ measurement is the assumption that T_2 does not change while images are being generated and therefore the contrast used to calculate $H_2{}^{17}O$ is entirely based on the pre-encoding cluster. This assumption produces slight error for sequences with short TE since the T_2 and $T_{1\rho}$ both change as $H_2{}^{17}O$ is added. However, most standard single shot sequences use linear phase encoding, acquiring several lines of k-space before the center line of k-space where most contrast information is acquired. Minimum TE starts at 20ms for Echo Planar Imaging (EPI)[7] and increases for other single shot sequences, creating increasing amounts of error. Centric encoding of k-space, provided with balanced steady state free precession readouts[8], with the associated short effective TE greatly diminishes this problem.

Here we report single shot $T_{1\rho}$ imaging with readout based on $T_{1\rho}$-prepared balanced Steady State Free Precession (bSSFP). We use pigs as a model for their similar lung capacities and circulations to humans. While the detection of H$_2$^{17}O with a steady state sequence has been reported[9], we show theoretically H$_2$^{17}O sensitivity is improved by a sequence with a T$_{1\rho}$ or T$_2$ pre-encoding cluster followed by signal acquisition without the steady state directly after pre-encoding. The efficacy is demonstrated by phantom experiments and *in vivo*.

2. MATERIALS AND METHODS

Images were acquired from a Siemens 1.5T Sonata MR scanner. Simulations were performed with custom-written software in Matlab.

2.1. Pulse Sequence Design

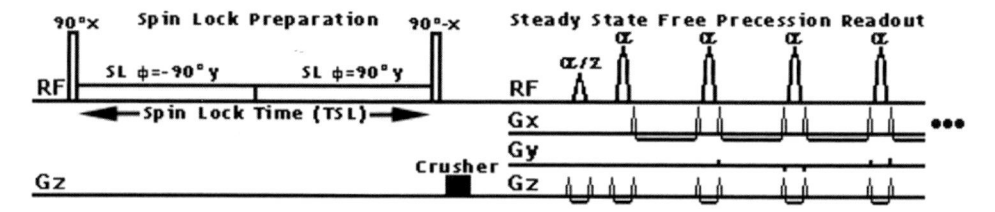

Figure 1. The T$_{1\rho}$-prepared, center-out sampled bSSFP sequence uses a spin-lock cluster to generate T$_{1\rho}$ contrast. The preparation cluster includes a 90 degree pulse to put magnetization in the transverse plane, followed by spin lock (SL) pulses of spin lock time (TSL) duration and with the provided phase alternation, and finally a 90 degree pulse to place the magnetization back in the longitudinal plane. After this cluster, a crusher removes unwanted magnetization in the transverse plane. The bSSFP readout uses a $\alpha/2$ RF pulse and a train of $+/-\alpha$ pulses with fully balanced gradients. K-space sampling begins at the center to maximize the effects of the preparation cluster. The image is taken in a single shot and a delay before the next repetition allows for T$_1$ recovery.

The T$_{1\rho}$ pre-encoded bSSFP pulse sequence consists of a T$_{1\rho}$ preparatory cluster followed by a centrically-encoded 2D bSSFP readout without dummy scans (Fig. 1).

2.2. Estimation of T$_{1\rho}$ in 17-Oxygen Water Phantoms

The ability of the sequence to detect changes in H$_2$^{17}O concentration was measured by imaging six 15mL conical tubes filled with 1X Phosphate Buffered Saline doped with H$_2$^{17}O (Isotec, Miamisberg, OH) in steps of 5mM from 20mM (natural abundance) to 45mM. These tubes were sealed and placed into a 9cm jar filled with water. A receive only loop coil was placed around the jar and used in combination with the body transmit coil. Cross-sectional images of the tubes were acquired with a spin lock amplitude of 100Hz and spin lock times of 200,400,600,800, and 1000msec. Imaging parameters are: FoV 90mm^2, ST 5mm, TR/TE 10.4/5.2ms, Matrix 128^2, BW 130Hz/Px, α=180°.

The concentrations of the phantoms were determined experimentally according to the following equation:

$$f(p) = f_n + \left[\ln\left(\frac{S_{T_{1\rho},NA}}{S_{PD,NA}}\right) - \ln\left(\frac{S_{T_{1\rho},^{17}O}}{S_{PD,^{17}O}}\right) \right] \frac{1}{R_{1\rho} \bullet TSL} \qquad \text{Eq. (1)}$$

Where f_n is the natural abundance $[H_2{}^{17}O]$, S is the signal from a pixel or ROI with natural abundance (NA) $[H_2{}^{17}O]$ or $[H_2{}^{17}O]$ added (^{17}O) taken as a proton density weighted image to account for coil sensitivity or when $T_{1\rho}$ pre-encoded ($T_{1\rho}$). $R_{1\rho}$ is the relaxivity due to $H_2{}^{17}O$ and TSL is the spin lock time.

2.3 Animal imaging

The Institutional Animal Care and Use Committee of the University of Pennsylvania approved all animal experiments. A live pig was placed supine inside a 1.5T Siemens Sonata clinical scanner on a vendor supplied surface coil. Once placed inside, the intubation tube was connected to a custom precision delivery breathing circuit (manuscript in review) that drove the pig's breathing at 6 breaths/min at a tidal volume of 18mL times the weight in kg. A standard T1-weighted localizer sequence was run to find a suitable coronal image of the middle portion of the pig brain that included cortex, brainstem, and ventricle. A T1-weighted IR-prepared GRE sequence (MP-RAGE) was performed to obtain anatomical images of the pig brain and the desired slice. Sequence parameters for the $T_{1\rho}$-prepared bSSFP are as follows: TR/TE 9.7/4.8ms, ST 5 or 10mm, FoV 200mm^2, 128^2 matrix, BW 130Hz/Px, $\alpha=180°$, SL Amp 100Hz, fat saturation on, 1.6 second readout time, 2 second delay between acquisition, time per image 3.6/sec.

3. RESULTS AND DISCUSSION

3.1 Phantom Studies

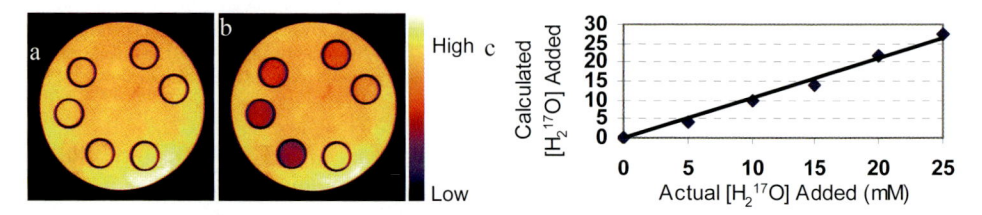

Figure 2. Quantification of $[H_2{}^{17}O]$ is performed in phantoms and quantified using the relaxivity of $H_2{}^{17}O$ in PBS. Above are cross-sectional images of 15mL conical tubes filled with phosphate buffered saline and doped with increasing amounts of $H_2{}^{17}O$ to concentrations from 20mM (natural abundance) to 45mM (25mM added). The left image **(a)** shows the ordering of the phantoms (labeled by concentration) in a proton density image that uses only two 90 degree pulses and no spin locking as a preparation cluster. The image **(b)** shows the contrast developed with 100Hz spin locking. The relaxivity of these phantoms taken from a series of $T_{1\rho}$-weighted images (not shown) is applied to equation 1 along with the signal intensities from the above images to calculate ^{17}O concentrations **(c)**.

Table 1. Shown is a comparison of bSSFP for detecting a T_2 reducing agent without a preparation cluster in the steady state or with a preparation cluster not in the steady state.

	Steady State	Non-Steady State
Equation	$M_{SS} = M_0 \dfrac{\sin(\alpha)}{1+\cos(\alpha)+(1-\cos(\alpha))(T_1/T_2)}$	$M = M_0 \exp\left(\dfrac{-t}{T_2 \text{ or } T_{1\rho}(\omega)}\right)$
Contrast	$C_{SS} = M_{SS}(T_{2,1}) - M_{SS}(T_{2,2})$	$C = M(T_{2,1}) - M(T_{2,2})$
Optimal α	60°	180°
Contrast	.0059M$_0$	.0223M$_0$
Scan Time	1.28s	3.6s
CNR Efficiency	1	2.25
SNR Efficiency	1	1.66

A demonstration of the $T_{1\rho}$-prepared bSSFP sequence shows that measurements of $H_2{}^{17}O$ concentrations can be made with single shot imaging very rapidly (Fig. 2). To show that high SNR and contrast are maintained by using a sequence with a pre-encoding cluster versus detecting pure T_2 changes in the steady state, Bloch equation simulations were performed to show that detecting in the steady state gives less contrast and CNR than methods not in the steady state. In essence, by filling the center of k-space first with a high flip angle, contrast is much higher than when obtaining T_2 contrast in the steady state. The simulated differences are summarized in Table 1. Despite the possibility of generating images faster, steady state acquisition offers less SNR and CNR efficiency.

Table 2. Because SAR is a concern in fast imaging sequences, shown is how contrast and SAR reduce with decreases in Flip Angle. CNR Efficiency vs. Steady State compares the CNR from a non-steady state acquisition with the given flip angle against CNR from a steady state acquisition with optimum flip angle (~60°) for CNR in the steady state.

Flip Angle (α)	Contrast vs. 180°	CNR Efficiency vs. Steady State	SAR
180°	100%	121%	100%
150°	96%	116%	69%
120°	86%	104%	44%

3.2 Animal Studies

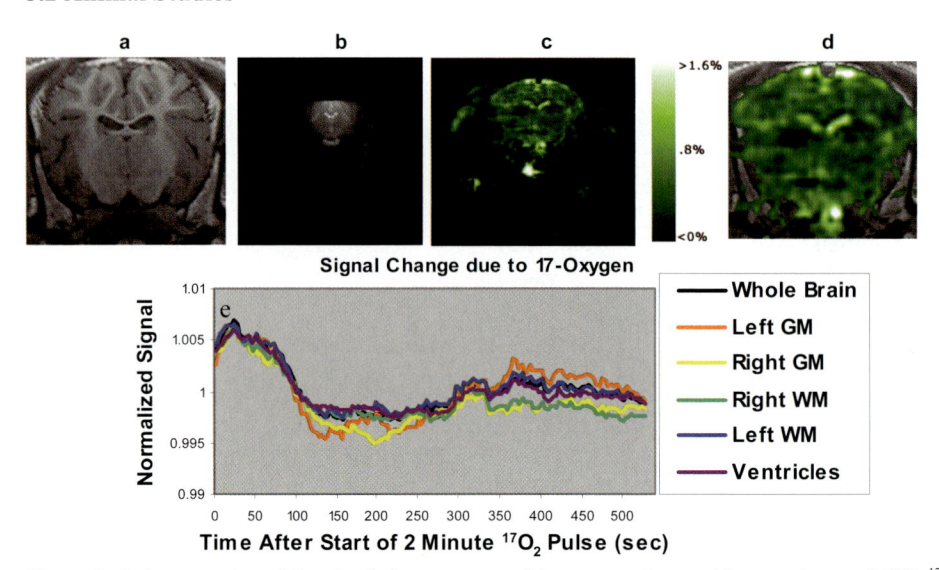

Figure 3. A demonstration of the signal change generated by an experiment with two minutes of 20% $^{17}O_2$ (70% enriched) delivery mixed with 80% nitrogen. **(a)** A $T_{1\rho}$ weighted image of the imaging slice is presented. **(b)** shows an example $T_{1\rho}$-prepared bSSFP image. A series of 376 of these images were obtained (3.6sec/image). After 10 minutes imaging with room air, 2 minutes of 70% $^{17}O_2$ (NUKEM Group, Germany) was delivered and room air resumed. To obtain the difference map in **(c)**, 32 images after $^{17}O_2$ delivery (150sec-262 sec) were subtracted from 32 baseline images. The color bar describes the signal decrease due to $H_2^{17}O$. **(d)** Difference map overlaid on the image from (a). **(e)** moving average signal traces are plotted over time from the beginning of $^{17}O_2$ delivery (time 0).

To demonstrate detection of $H_2^{17}O$ *in vivo*, $^{17}O_2$ gas was delivered to a pig during imaging (Figure 3). Signal decreases demonstrate the detection of $^{17}O_2$ conversion to $H_2^{17}O$. To demonstrate pulsed delivery of $^{17}O_2$ gas, 1 breath of 40% gas was delivered to three pigs either as one breath or diluted with nitrogen to simulate room air. A consistent signal drop is observed after each delivery. Because only one breath of gas is used to show change, this represents the smallest volume delivery of gas published.

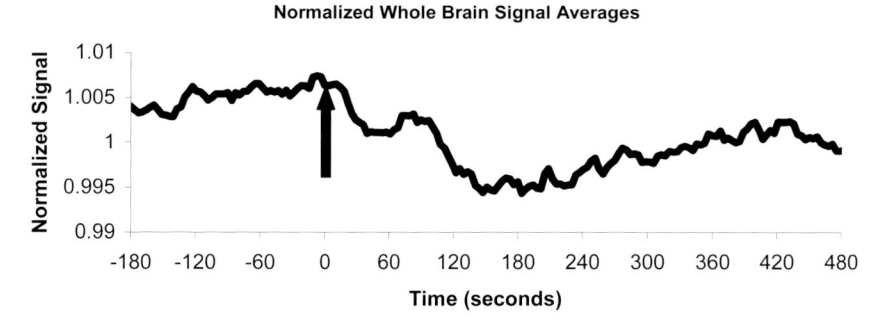

Figure 4. Three experiments were performed with 40% ^{17}O and their results are pooled. These were performed with TSL=200ms to develop a higher percent signal change, but with SNR about 40:1 as opposed to 150:1 at TSL=75ms in Figure 3. The whole brain signal trace is an average of these three experiments, though ROIs from different regions show similar kinetics. Two pigs were imaged with a protocol of 5 minutes of 100% O_2, 2 breaths of 40% ^{17}O enriched O_2, then 5 minutes of 100% O_2. The third pig was imaged with 5 minutes of room air then 2 breaths of 20% O_2 (40% enriched $^{17}O_2$) mixed with 80% N_2. The start of $^{17}O_2$ is indicated by the arrow. Because the normalized signal trace from the third experiment is so similar, it has been pooled. Signal changes are statistically significant when compared to baseline (data not shown).

In conclusion, it is hoped that this manuscript provides many of the technical details of and proof of concept for improvements in indirect detection of $H_2{}^{17}O$ by $T_{1\rho}$ on clinical scanners. The ultimate goal of this work is the clinical *in vivo* imaging of metabolism. However, to go from detection of $H_2{}^{17}O$ to computation of $CMRO_2$ is not a straightforward task and requires many additional considerations. These considerations will be the focus of a future manuscript. Still, the following provides a brief discussion of these issues.

In order to compute the $CMRO_2$, several parameters require modeling or measurement. First, the delivery rate and uptake of $^{17}O_2$ must be known. With the precision-delivery breathing circuit a step change of gas concentration is provided to the airway, and the rate of uptake through the lungs and tissue transport can be modeled. The rate of production of $H_2{}^{17}O$ is related to signal change. However, a parameter requiring consideration is the arterial input function (AIF) from recirculation of produced $H_2{}^{17}O$.

There are two ways of measuring the AIF. The first method assumes that it will take some time for $H_2{}^{17}O$ to recirculate through the body after production in the mitochondria. For example, at least 60 seconds elapse before $H_2{}^{17}O$ "wash-in" after the start of $^{17}O_2$ delivery in a cat[10]. That gives a window with which to compute $CMRO_2$ while ignoring recirculation. In the three experiments that were used to prepare Figure 4, a small hump was observed, always at around 90 seconds. It could be speculated that after this point wash-in due to recirculation begins, however this effect is not seen in every experiment.

A second method involves the imaging of an artery during $H_2{}^{17}O$ delivery. To this end, Zhang et al demonstrated an implantable coil method for small animals using direct ^{17}O measurements[11]. They showed with this invasive technique that there was little to no time for recirculation in that model; however, the recirculation time for a very small animal like a rat is thought to be much shorter than for a large animal like a pig. Experiments are underway with this bright blood, fast imaging sequence to address both possible methods for measuring AIF. For example, in a large animal the carotid arteries can be imaged with a surface coil over time to look at changes in $H_2{}^{17}O$ concentration.

4. ACKNOWLEDGEMENTS

This work was supported in part by NIH R01EB004349, RR02305 and DANA foundation grants. The authors wish to thank Letitia Cheatham and Marion Knaus for their assistance in animal handling. We also wish to thank Kiarash Emami, John Woodburn-McDuffy, and Vahid Vahdat for their assistance with gas handling as well as Sridhar Charagundla, Dharmesh Tailor, and Mark A. Elliot for helpful discussions.

5. REFERENCES

1. D. R. Tailor, J. E. Baumgardner, R. R. Regatte, J. S. Leigh, R. Reddy, Proton MRI of metabolically produced H2 17O using an efficient 17O2 delivery system, *Neuroimage* **22**(2), 611-618 (2004).
2. X. H. Zhu, N. Zhang, Y. Zhang, X. Zhang, K. Ugurbil, W. Chen, In vivo 17O NMR approaches for brain study at high field, *NMR Biomed* **18**(2), 83-103 (2005).
3. A. J. de Crespigny, H. E. D'Arceuil, T. Engelhorn, M. E. Moseley, MRI of focal cerebral ischemia using (17)O-labeled water, *Magn Reson Med* **43**(6), 876-883 (2000).
4. I. Ronen, H. Merkle, K. Ugurbil, G. Navon, Imaging of H217O distribution in the brain of a live rat by using proton-detected 17O MRI, *Proc Natl Acad Sci U S A* **95**(22), 12934-12939 (1998).
5. T. Arai, S. Nakao, S. Morikawa, T. Inubushi, T. Yokoi, K. Shimizu, K. Mori, Measurement of local cerebral blood flow by magnetic resonance imaging: in vivo autoradiographic strategy using 17O-labeled water, *Brain Res Bull* **45**(5), 451-456 (1998).
6. R. Reddy, A. H. Stolpen, J. S. Leigh, Detection of 17O by proton T1 rho dispersion imaging, *J Magn Reson B* **108**(3), 276-279 (1995).
7. A. Borthakur, J. Hulvershorn, E. Gualtieri, A. J. Wheaton, S. Charagundla, M. A. Elliott, R. Reddy, A pulse sequence for rapid in vivo spin-locked MRI, *J Magn Reson Imaging* **23**(4), 591-596 (2006).
8. K. Scheffler, S. Lehnhardt, Principles and applications of balanced SSFP techniques, *Eur Radiol* **13**(11), 2409-2418 (2003).
9. A. L. Hopkins, E. M. Haacke, J. Tkach, R. G. Barr, C. B. Bratton, Improved sensitivity of proton MR to oxygen-17 as a contrast agent using fast imaging: detection in brain, *Magn Reson Med* **7**(2), 222-229 (1988).
10. J. Pekar, L. Ligeti, Z. Ruttner, R. C. Lyon, T. M. Sinnwell, P. van Gelderen, D. Fiat, C. T. Moonen, A. C. McLaughlin, In vivo measurement of cerebral oxygen consumption and blood flow using 17O magnetic resonance imaging, *Magn Reson Med* **21**(2), 313-319 (1991).
11. X. Zhang, X. H. Zhu, R. Tian, Y. Zhang, H. Merkle, W. Chen, Measurement of arterial input function of 17O water tracer in rat carotid artery by using a region-defined (REDE) implanted vascular RF coil, *Magma* **16**(2), 77-85 (2003).

BRAIN TISSUE OXYGEN CONSUMPTION AND SUPPLY INDUCED BY NEURAL ACTIVATION:

Determined under suppressed hemodynamic response conditions in the anesthetized rat cerebral cortex

Kazuto Masamoto[*], Alberto Vazquez, Ping Wang, and Seong-Gi Kim

Abstract: The dynamic changes in cerebral metabolic rate of oxygen ($CMRO_2$) and oxygen supply during brain functions have not been well-characterized. To examine this issue, experiments with electrophysiology, oxygen microelectrode and laser-Doppler flowmetry were performed in the anesthetized rat somatosensory cortex. During neural activation, brain tissue partial pressure of oxygen (Po_2) and local cerebral blood flow (CBF) were similarly increased. To separate the Po_2 changes originating from the increase in $CMRO_2$ and the increase in oxygen supply, the same experiments were repeated under a vasodilator-induced hypotension condition in which evoked CBF change was minimal. In this condition, evoked Po_2 monotonically decreased, indicating an increase in $CMRO_2$. Then, $CMRO_2$ was determined at resting as well as activation periods using a dynamic oxygen exchange model. Our results indicated that the changes in $CMRO_2$ were linearly related with the summation of evoked field potentials and further showed that the oxygen supply in the normal condition was about 2.5 times larger than the demand. However, this oxygen oversupply was not explainable by the change in CBF alone, but at least partly by the increase in oxygenation levels at pre-capillary arterioles (e.g., 82% to 90% O_2 saturation level) when local neural activity was evoked.

1. INTRODUCTION

Hemodynamic-based brain imaging techniques, such as blood oxygen level dependent functional magnetic resonance imaging (BOLD fMRI) and intrinsic changes in

[*] Kazuto Masamoto, National Institute of Radiological Sciences, Molecular Imaging Center, 4-9-1 Anagawa, Inage-ku, Chiba, 263-8555, Japan, Phone: +81-43-206-3425, Fax: +81-43-206-0818, E-mail: masamoto@nirs.go.jp, Alberto Vazquez, Ping Wang, Seong-Gi Kim, Department of Radiology, University of Pittsburgh, Pittsburgh, Pennsylvania 15203, USA.

P. Liss et al. (eds.), *Oxygen Transport to Tissue XXX*, DOI 10.1007/978-0-387-85998-9_43,
© Springer Science+Business Media, LLC 2009

optical imaging signal (OIS), rely on the difference between oxygen demand and supply. Thus, it is important to determine the quantitative relation between the activity-dependent cerebral metabolic rate of oxygen ($CMRO_2$) and oxygen supply induced by local changes in cerebral blood flow (CBF). However, the dynamic relation between $CMRO_2$ and CBF remains uncharacterized because of the difficulty to separate signal changes induced by demand and supply. To detect imaging signals only induced by the increased $CMRO_2$, our group developed a hypotension model using systemic administration of a vasodilator drug, in which evoked cerebrovascular response is minimal, while the evoked neural response is intact[1, 2]. In this condition, the cortical blood oxygenation level monotonically decreased during evoked neural activity, indicating that the neural activity-induced change in $CMRO_2$ is not negligible. However, quantitative relationship between neural activity and $CMRO_2$ values has been not determined.

In the present paper, direct measurements of neural activity, brain tissue oxygen tension, and local CBF were first performed in the anesthetized rat somatosensory cortex under normal and hypotension conditions. Then, these data were used to determine activity-induced changes in $CMRO_2$ and oxygen supply based on a dynamic oxygen exchange model[3]. The first objective of this work was to quantify the absolute changes in $CMRO_2$ induced by neural activity. The magnitude of partial pressure of oxygen (Po_2) change induced by neural stimulation under CBF-suppressed conditions was used to directly represent the $CMRO_2$ change. The quantitative relationship between $CMRO_2$ changes and evoked field potential (FP) was then determined. The second objective was to determine the activity-induced changes in tissue oxygen supply induced by CBF increase, assuming that the change in $CMRO_2$ was consistent between normal and suppressed CBF response conditions.

2. EXPERIMENT: NEURAL ACTIVITY, PO2, AND CBF MEASUREMENTS

Five male Sprague-Dawley rats (400 to 560 g) were anesthetized with isoflurane. Endotracheal intubation was performed to allow for mechanical ventilation, and the left skull was thinned. Arterial blood pressure, heart rate, body temperature, and respiratory parameters were monitored throughout all experiments. During the experiments, inspired gas was maintained with a mixture of air and O_2 (30 to 35% total O_2) and the end-tidal isoflurane concentration was adjusted to $1.3 \pm 0.1\%$. At first, intrinsic optical imaging with an irradiation of 620-nm filtered light was performed for localization of the forepaw area in the primary somatosensory cortex. Electrical forepaw stimulation (1.0 ms pulse width, 1.2 mA current) was applied with 3-sec duration at a rate of 3 Hz for twenty runs with an inter-run time of 6 sec. The activation focus (~0.5 mm in a diameter) was determined around the largest decrease in light reflectance in the map[4].

Simultaneous measurements of FP, Po_2, and CBF were then performed at the activation foci, while electrical forepaw stimulation (1.0 ms pulse width, 1.2 mA current) was applied with 10-sec duration at a rate of 6 Hz for fifteen runs with an inter-run time of 80 sec. To avoid biasing due to different sampling positions, we performed simultaneous measurements of FP and Po_2 with an APOX double electrode micro-sensor (30-μm tip diameter; Unisense A/S) at a depth of about 0.3 mm from the cortical surface. The CBF around the electrode recording site was measured with laser-Doppler flowmetry (LDF). A needle-type LDF probe was placed on the surface of the thinned skull covering the activation focus[4]. The FP, Po_2 and CBF time-series were recorded on a PC at a rate of

1 kHz. Finally, the same experiments were repeated under the suppressed CBF response condition with systemic administration of sodium-nitroprusside (sNP), a rapid-acting vasodilator[1, 2]. The sNP (0.5 mg/ml in saline) was continuously injected and the recordings were started about 2 min after induction.

All recording runs were averaged for each condition (control and sNP-induced) based on the onset of stimulation. The mean baseline period was defined as the mean of the data obtained over the 3 sec prior to the stimulation onset, and the mean evoked response was defined as the mean of the data obtained 3 sec around the peak of activated changes. The sum of the evoked FP (ΣFP) was calculated as the sum of the amplitude of all evoked FPs measured as the difference between the positive and negative peaks. All population data are reported as mean $\pm$ SD ($N = 5$).

3. THEORY: DYNAMIC OXYGEN EXCHANGE MODEL

A two-compartment oxygen exchange model that consists of a single capillary and a well-mixed tissue compartment was used[3]. The governing equations were derived from mass conservation and Fick's diffusion law, and are shown below:

$$V_c \frac{dC_c(t)}{dt} = 2CBF(t)\big(C_a - C_c(t)\big) - PS\big(C_p(t) - C_t(t)\big) \qquad \text{(Eq. 1)}$$

$$V_t \frac{dC_t(t)}{dt} = PS\big(C_p(t) - C_t(t)\big) - CMRO_2(t) \qquad \text{(Eq. 2)}$$

where C_c, C_p and C_t are the oxygen concentration in capillary, plasma and tissue, respectively. The total capillary oxygen concentration (C_c) was determined by considering both plasma (C_p) and hemoglobin-bound oxygen using the Hill equation[3]. The Hill coefficient was set to 2.73, the oxygen solubility coefficient to 0.00139 mmol/L/mmHg and the P_{50} was set to 38 mmHg. The hemoglobin concentration was assumed to be equivalent to the values obtained from systemic arterial blood. The mean tissue oxygen concentration (C_t) was assumed to be represented by the measured Po_2 data. The volumes of capillary (V_c) and tissue (V_t) were set to 1 mL/100 g and 97 mL/100 g, respectively. The permeability (P) of oxygen across the capillary wall was used based on the reported values (78 to 115 x 10^{-4} cm/s) in the literature[5, 6]. The capillary surface area per unit volume of tissue (S) was also selected from the literature (87 to 147 mm^2/g)[7-9]. Multiplying these values, the estimated range of PS was 4000 to 10000 ml O_2/100 g/min. In the present study, we used PS = 7000 ml O_2/100 g/min based on the mean of this estimation, unless specifically noted.

For the calculation of $CMRO_2$, no variations in capillary recruitment or perfusion heterogeneity were assumed across control and sNP conditions as well as baseline and activation periods. Also, no volume change was assumed for any conditions. The baseline CBF level under control conditions was assumed to be 150 ml/100 g/min, previously measured in our laboratory using MRI[10]. The CBF values for the activation periods and sNP conditions were then corrected by taking into account the relative change in LDF data obtained in each condition in individual subjects. The effect of changes in the arterial oxygen saturation (SpO_2) and PS versus $CMRO_2$ was investigated according to their respective ranges (SpO$_2$ = 82 to 95% and PS = 4000 to 10000 ml O$_2$/100 g/min) reported in the literature. The quantitative CMRO$_2$ values reported here were calculated with SpO$_2$ = 95% and PS = 7000 ml O$_2$/100 g/min.

Oxygen supply produced by changes in CBF was estimated by subtracting the Po$_2$ data obtained under sNP conditions (i.e., *demand*) from that obtained under control conditions (i.e., *demand and supply*). This represents the "*experimentally determined supply*". Then, the oxygen exchange model was used to calculate the oxygen supply component induced by CBF change represented by the LDF data ("*theoretically determined supply*"). The baseline CMRO$_2$ was fixed to 7.1 ml O$_2$/100 g/min, and various arterial saturation levels were individually tested. Estimates of the dynamic change in oxygen supply were then generated and compared to the experimentally determined supply Po$_2$ change. Lastly, to obtain the required Po$_2$ supply, arterial oxygen concentration was dynamically changed with a baseline SpO$_2$ of 82%.

4. RESULTS

The population data showed that the 54 ± 27% increase in evoked CBF observed under control conditions was mostly suppressed under sNP (6 ± 3% at peak). There were no significant differences in evoked ΣFP under control and sNP conditions (59 ± 22 mV and 51 ± 22 mV, respectively). The baseline LDF signal was 36 ± 18% higher under sNP than that under control conditions. Accordingly, the baseline Po$_2$ increased with sNP; from 33 ± 17 mmHg to 42 ± 13 mmHg under control and sNP, respectively. The Po$_2$ increased with evoked stimulation under control conditions (10 ± 4 mmHg), whereas the Po$_2$ monotonically decreased (-4.6 ± 2.2 mmHg) over the entire period of stimulation under sNP conditions. The individual subject's data obtained under sNP conditions are reported in Table 1.

Baseline CMRO$_2$ was varied between 3 and 9 ml O$_2$/100 g/min as a function of the parameters SpO$_2$ and PS. When SpO$_2$ and PS were set to 95% and 7000 ml O$_2$/100 g/min, respectively, the baseline CMRO$_2$ under sNP conditions was calculated to be 6.8 ± 2.2 ml O$_2$/100 g/min, which was slightly lower than that calculated for the control condition (7.1 ± 2.2 ml O$_2$/100 g/min). CMRO$_2$ for activation periods was calculated to be 7.4 ± 2.2 ml O$_2$/100 g/min, which accounted for 0.6 ± 0.3 ml O$_2$/100 g/min increment or 10 ± 6% increase relative to respective baseline values. The individual subject's CMRO$_2$ values calculated with SpO$_2$ of 95% and PS of 7000 ml O$_2$/100 g/min are reported in Table 1. A strong linear correlation between the increment and the relative change in CMRO$_2$ and ΣFP were observed across the subjects (Table 1). The slope of the best fitted line was calculated to be 0.012 ml O$_2$/100 g/min/ mV.

The experimentally estimated supply of oxygen was about 2.5 times greater than the oxygen demand. To meet this oversupply of oxygen, various parameters were varied. When various constant arterial oxygen levels were used with typical values for the other model parameters (PS of 7000 ml O$_2$/100 g/min), the simulated Po$_2$ changes were always less than the experimentally determined supply Po$_2$ change. This strongly indicates that the change in CBF alone may not be sufficient to satisfy the "*necessary supply of oxygen*". If the arteriolar oxygenation level was increased from 82% SpO$_2$ at baseline to 90% SpO$_2$ at activation periods, the simulated increase in oxygen supply matched the experimentally determined supply Po$_2$ change.

Table 1. Baseline and stimulation data of Po_2 and estimated $CMRO_2$ under sNP condition in individual subjects.

animal	$\sum FP$ (mV)	$Po_{2\_base.}$ (mmHg)	$Po_{2\_stim.}$ (mmHg)	$CMRO_{2\_base.}$ (ml /100 g/min)	$CMRO_{2\_stim.}$ (ml /100 g/min)	$CMRO_{2\_change}$ (ml /100 g/min)
#1	83	52	46	4.6	5.4	0.8
#2	59	25	19	9.3	10.3	1.1
#3	22	32	31	8.9	9.2	0.2
#4	48	53	48	5.7	6.4	0.6
#5	42	50	48	5.3	5.7	0.4
Mean	51	42	38	6.8	7.4	0.6
SD	22	13	13	2.2	2.2	0.3

5. DISCUSSION

The variability in the $CMRO_2$ calculation was investigated as a function of the arterial oxygen input and the permeability-surface area product according to the respective ranges reported in the literature. Since all parameters affecting the $CMRO_2$ calculation were not measured in the present study, these additional variables could also cause systematic errors. These unknown parameters were treated as follows. First, the present study used the mean CBF previously measured in our laboratory in similar experiments but in a different group of animals[10]. However, technical differences with respect to the sampling volume, i.e., using perfusion MRI vs. Po_2 microelectrode or LDF probe, possibly generated errors in the true CBF around the Po_2 measurement location. Second, we fixed the hemoglobin concentration as measured from the systemic arterial blood. However, the hematocrit in the brain microcirculation is known to be generally lower than in the systemic arteries. This would lead to the overestimation of $CMRO_2$ under baseline conditions. Third, the longitudinal profile of oxygen concentration in the capillary was assumed to be a linear function along a direction of the flow. Because higher flow rates generate a shallower gradient, the error introduced by this linear assumption was estimated to be less than 10% for our experiment conditions. The result that activity-induced change in $CMRO_2$ closely correlates with evoked FP activity (Table 1) indicates that neural activity-induced $CMRO_2$ changes can be used as a reliable index of the evoked neural activity. Therefore, $CMRO_2$-sensitive imaging methods offer the advantage of more directly inferring neural functions compared to the overall hemodynamic response. This would be advantageous in the assessment of neural function in populations with compromised vascular function due to disease.

The oxygen concentration in blood has a longitudinal gradient from upstream large arteries to principal veins[11]. In the normoxic brain cortex, it has been shown that the oxygen loss from the first order arterioles to the arterial end accounted for 11% SpO_2 drop, whereas the capillary compartments accounted for 23% of the SpO_2 drop[12]. Taking into account the significant loss of oxygen from arterioles, the present work draws the conclusion that the increase in arterial end oxygenation level may greatly contribute to the venous oxygenation changes during activity-induced CBF changes. To examine another possibility for achieving the required supply, we evaluated the changes in permeability (P). During increased brain activity, the permeability may vary with changes in CBF, depending on the diffusivity of oxygen. However, the model predicted that the permeability needed to double at peak (data not shown), which is unlikely under physiologic conditions. Lastly, it is also possible that changes in capillary volume, affinity of hemoglobin for oxygen, and/or functional recruitment contribute to the increases in tissue P_{O2} and we are currently investing this possibility. These results support our conclusion that tissue supply of oxygen significantly relies on the change in arteriolar end oxygenation levels. Possible factors that could contribute to modulation of the arterial oxygenation level are arterio-venous shunts or arteriolar wall energetic demands[13-15].

6. ACKNOWLEDGEMENTS

This study was supported by the National Institutes of Health (EB003375, NS044589) and JSPS.

7. REFERENCES

1. T. Nagaoka, F. Zhao, P. Wang, N. Harel, R. P. Kennan, S. Ogawa, and S. G. Kim, Increases in oxygen consumption without cerebral blood volume change during visual stimulation under hypotension condition, *J Cereb Blood Flow Metab* **26**, 1043-1051 (2006).
2. M. Fukuda, P. Wang, C. H. Moon, M. Tanifuji, and S. G. Kim, Spatial specificity of the enhanced dip inherently induced by prolonged oxygen consumption in cat visual cortex: implication for columnar resolution functional MRI, *Neuroimage* **30**, 70-87 (2006).
3. R. Valabrègue, A. Aubert, J. Burger, J. Bittoun, and R. Costalat, Relation between cerebral blood flow and metabolism explained by a model of oxygen exchange, *J Cereb Blood Flow Metab* **23**, 536-545 (2003).
4. K. Masamoto, T. Kim, M. Fukuda, P. Wang, and S. G. Kim, Relationship between neural, vascular, and BOLD signals in isoflurane-anesthetized rat somatosensory cortex, *Cereb Cortex* **17**, 942-950 (2007).
5. I. G. Kassissia, C. A. Goresky, C. P. Rose, A. J. Schwab, A. Simard, P. M. Huet, and G. G. Bach, Tracer oxygen distribution is barrier-limited in the cerebral microcirculation, *Circ Res* **77**, 1201-1211 (1995).
6. C. Y. Liu, S. G. Eskin, and J. D. Hellums, The oxygen permeability of cultured endothelial cell monolayers, *Adv Exp Med Biol* **345**, 723-730 (1994).
7. T. Bär, The vascular system of the cerebral cortex, *Adv Anat Embryol Cell Biol* **59**, 1-62 (1980).
8. P. S. Manoonkitiwongsa, P. J. McMillan, R. L. Schultz, C. Jackson-Friedman, and P. D. Lyden, A simple stereologic method for analysis of cerebral cortical microvessels using image analysis, *Brain Res Brain Res Protoc* **8**, 45-57 (2001).
9. G. Pawlik, A. Rackl, and R. J. Bing, Quantitative capillary topography and blood flow in the cerebral cortex of cats: an in vivo microscopic study, *Brain Res* **208**, 35-58 (1981).
10. T. Kim, K. S. Hendrich, K. Masamoto, and S. G. Kim, Arterial versus total blood volume changes during neural activity-induced cerebral blood flow change: implication for BOLD fMRI, *J Cereb Blood Flow Metab* **27**, 1235-1247 (2007).
11. B. R. Duling, and R. M. Berne, Longitudinal gradients in periarteriolar oxygen tension. A possible mechanism for the participation of oxygen in local regulation of blood flow, *Circ Res* **27**, 669-678 (1970).
12. E. Vovenko, Distribution of oxygen tension on the surface of arterioles, capillaries and venules of brain cortex and in tissue in normoxia: an experimental study on rats, *Pflugers Arch* **437**, 617-623 (1999).
13. L. Kuo, and R. N. Pittman, Effect of hemodilution on oxygen transport in arteriolar networks of hamster striated muscle, *Am J Physiol* **254**, H331-339 (1988).
14. M. Shibata, K. Qin, S. Ichioka, and A. Kamiya, Vascular wall energetics in arterioles during nitric oxide-dependent and -independent vasodilation, *J Appl Physiol* **100**, 1793-1798 (2006).
15. A. G. Tsai, B. Friesenecker, M. C. Mazzoni, H. Kerger, D. G. Buerk, P. C. Johnson, and M. Intaglietta, Microvascular and tissue oxygen gradients in the rat mesentery, *Proc Natl Acad Sci U S A* **95**, 6590-6595 (1998).

BRAIN OXYGEN BALANCE UNDER VARIOUS EXPERIMENTAL PATHOPHYSIOLOGYCAL CONDITIONS

Michal Schechter[1], Judith Sonn, and Avraham Mayevsky

Abstract: Normally, brain tissue copes with negative oxygen balance by increasing cerebral blood flow (CBF). We examined the effects of increasing oxygen demand, by inducing spreading depression (SD) under various oxygen balance states, on brain O_2 balance. The Tissue Vitality Monitoring System was used, which enables real time simultaneous *in vivo* monitoring of CBF, mitochondrial NADH and tissue HbO_2 from the same region of the cerebral cortex. SD was induced during normoxia, hypoxia, hyperoxia, ischemia, and in normal and ischemic brain after systemic epinephrine administration. Under normoxia, hyperoxia and ischemia & epinephrine, the compensation of energy demand induced by SD, was carried out by increasing CBF. The higher oxygen delivery under hyperoxia and epinephrine did not change the pattern of recovery from SD as compared to normoxia, whereas in the ischemic and hypoxic brain, the recovery from SD was prolonged, indicating a lake in oxygen delivery. Epinephrine infusion in the ischemic rat, decreased oxyhemoglobin utilization during SD, indicating that tissue oxygen balance improves even under higher oxygen demand induced by SD.

1. INTRODUCTION

The brain is completely dependent upon continuous supply of oxygen via the autoregulation of CBF. During pathological conditions, where oxygen supply is disturbed, such as ischemia or hypoxia, any increase in oxygen demand will cause an oxygen deficiency that will terminate in severe functional disorders and irreversible brain damage [1]. Elevating tissue oxygen demand by inducing spreading depression, leads to a higher demand for oxygen that can also lead to oxygen deficiency depending on the tissue oxygen balance conditions [2]. In our previous studies, we showed that under normal oxygen supply, an increase in oxygen demand by inducing SD is compensated by an increase in CBF and NADH oxygenation [3]. Furthermore, we have found that hypoxia and ischemia induce variable responses to SD compared to normoxia, controlled by the

[1] M. Schechter, The Mina & Everard Goodman Faculty of Life Sciences and The Leslie and Susan Gonda Multidisciplinary Brain Research Center, Bar-Ilan University, Ramat-Gan, Israel, 52900.

P. Liss et al. (eds.), *Oxygen Transport to Tissue XXX*, DOI 10.1007/978-0-387-85998-9_44,
© Springer Science+Business Media, LLC 2009

limited blood supply and low blood oxygen concentration. Under these conditions, no oxidation waves were observed in NADH [3]. On the other hand, a question remained regarding the SD response under positive oxygen balance. Oxygen availability in the tissue was elevated by hyperoxia and systemic perfusion of epinephrine. In this study, an additional monitored parameter, oxyhemoglobin, provides a better insight into the oxygen utilization under these physiological conditions. Norepinephrine is a potent vasopressor that indirectly increases arterial blood supply [4]. Hence, it was used in systemic perfusion to elevate blood flow and improve oxygen tissue supply. In our previous works, we monitored tissue blood flow, NADH fluorescence and extracellular ions concentrations that helped to evaluate brain tissue function. In the present study, we used a unique multiparametric monitoring system (TVMS) that includes tissue HbO_2 measurement, providing a better insight into brain energy metabolism under various pathophysiological conditions. The aim of the present study was to examine the interrelation between hemodynamic events and energy metabolism under negative versus positive oxygen balance conditions. Here, we measured, for the first time, the oxyhemoglobin level in parallel to CBF, NADH and blood volume, which all together can lead to a better understanding of how the tissue copes with high oxygen demand under different oxygen balance conditions.

2. MATERIALS AND METHODS

2.1 The Tissue Vitality Monitoring System (TVMS)

The TVMS includes two devices for real time, simultaneous monitoring: a time-sharing fluorometer-reflectometer (TSFR) for mitochondrial NADH redox state and microcirculatory hemoglobin oxygen saturation HbO_2 measurement, combined with a laser Doppler flowmeter (LDF) for CBF monitoring [5, 6] enabling simultaneous and relative measurements of hemodynamic and metabolic activity of the rat cortex (3mm in diameter and about 1 mm in depth) in vivo. The measurement of HbO_2, reflectance at 366 nm and NADH (450 nm) were performed by the same excitation and emission fibers. The TSFR system includes a rotating wheel with 8 specific filters at the appropriate wavelengths (366 nm, 450 nm for NADH, and 585 nm, 577 nm for oxy-hemoglobin measurement) – 4 filters for excitation light and 4 filters for emitted light. The wheel rotates at about 2400 rpm, which is a very high speed with respect to the kinetics of physiological changes, thus NADH and oxyhemoglobin are simultaneously monitored. The CBF was measured by 3 optical fibers located in the center of the time sharing bundle of fibers. More details of the monitoring techniques can be found in our previous studies [6, 7].

2.2. Animal Preparation

All experiments were performed in accordance with the Guidelines of the Animal Care Committee of Bar-Ilan University. Wistar male rats (250 – 300 g.) were anesthetized with an IP injection (0.3 ml/100gw) of Equithesin [3]. In order to keep the animal under continuous anesthesia and painless during the experiment, an additional dose of 0.1 ml Equithesin/100g body weight was injected every 30 min. The rat was placed on a warming platform to keep the body temperature at 37°C. Additional details of animal preparation can be found in a previous publication [3]. The monitoring probe (6 mm diameter) and a push-pull cannula (2mm diameter), for SD waves induction (by flushing the brain with 0.5 M-1M KCl solution), were fixated to the cerebral surface using dental acrylic cement. Once the second wave developed, the brain surface was flushed with saline solution to terminate the SD after the third wave. The normal response to anoxia (100% N_2) was checked by the NADH increase and HbO_2 decrease at the point when spontaneous breathing stopped. Thereafter, anoxia was stopped and the rats were ventilated until spontaneous respiration started. The animal was sacrificed by inhalation of 100% N_2 until death.

2.2.1. Experimental Protocols

The rats were allowed to recover from surgery for 30 minutes until steady state conditions before the experiments started. Then, SD control waves were induced under normoxic conditions (spontaneous air breathing, 21% O_2).

a. **Hypoxia**: 30 minutes after the last SD control wave, hypoxia was induced by inhalation of 6%O_2+94%N_2 gas mixture for 20 minutes. One minute after hypoxia induction, the cortex was flushed with 0.5M KCl solution until SD waves started. The SD waves were stopped by saline solution, and 30 minutes from the last wave, the rats inhaled pure nitrogen until death. The hypoxia group contained 7 rats.

b. **Hyperoxia**: 30 minutes after the last control SD wave during normoxia, hyperoxia was induced by 100% O_2 inhalation for 20 minutes. One minute after hyperoxia started, SD waves were induced. Three consecutive waves were recorded and SD propagation was stopped by saline solution. 30 minutes later, the rat inhaled 100% N_2 until death. The hyperoxia group contained 6 rats.

c. **Partial Ischemia**: 24 hours before the experiment, the two carotid arteries were exposed and ligated using a surgical suture. This group contained 7 rats.

d. **Epinephrine**: In these experiments (n=8), in addition to the brain surgery, the animals underwent femoral artery and vein cannulation, for blood pressure monitoring and epinephrine infusion, respectively. SD waves were induced 30 minutes after surgery, during normoxia and under steady state conditions. One hour after the last control SD wave, a 20 minutes infusion of 0.5µg/kg/min epinephrine started. Five minutes later, during epinephrine steady state conditions, SD waves were induced. SD was stopped by saline solution, and 30 minutes from the last wave, the rat inhaled 100% N_2 until death. The same procedure was performed for the ischemic group where the control SD waves were initiated on the ischemic group before and during epinephrine infusion.

2.3 Data collection and processing

After connecting the brain to the monitoring system, all signals were set electronically to read 100% of the normoxic brain and measured the percent change of the signal [6, 7]. The differences between SD under hypoxia and hyperoxia as compared to SD under normoxia, as well as the differences between SD under ischemia as compared to SD under ischemia & epinephrine infusion were tested.

3. RESULTS

Anoxia (100% N_2 inhalation for about 20 seconds) was induced in order to find the maximal reduced NADH and minimal HbO_2 levels. The effect of anoxia on the measured parameters under normoxic and partial ischemic conditions is shown in Figure 1. Anoxia caused a decrease of (60%, 62%) in CBF, in reflectance (23%, 12%) and in HbO_2 (15%, 15%), and an increase in NADH (34%, 31%) in the normoxic and ischemic brains, respectively. Returning to spontaneous air breathing, in normoxic and ischemic brains, a large increase of (224%, 60%) in CBF which caused an increase in HbO_2 and a decrease in NADH to pre-anoxic levels. It can be seen that in returning to air breathing, the normoxic brain showed a two times larger increase in CBF, showing that the normal brain has a larger possibility for hyperemia as compared to the ischemic one.

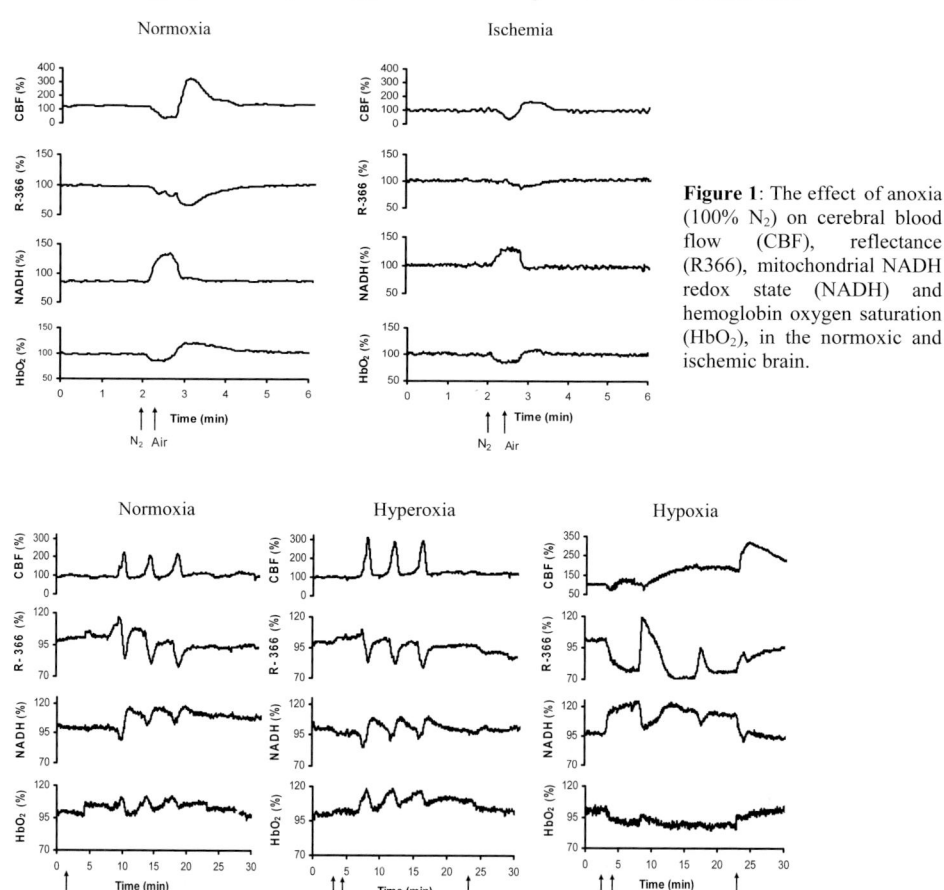

Figure 1: The effect of anoxia (100% N_2) on cerebral blood flow (CBF), reflectance (R366), mitochondrial NADH redox state (NADH) and hemoglobin oxygen saturation (HbO_2), in the normoxic and ischemic brain.

Figure 2: A typical brain response of the measured parameters to SD under normoxia, hyperoxia (Hyper.) and hypoxia. Abbreviations are as described in Fig. 1

Figure 2 shows typical responses to SD in the normoxic, hyperoxic and hypoxic brain. During normoxia, SD caused an increase in CBF, an increase-decrease in reflectance during the first wave and opposite changes in CBF during the two following waves, oxidation cycles in NADH and an increase in HbO_2.

SD during hyperoxia showed the same pattern as normoxia except that the increase in CBF was about two times higher, the oxidation cycles in NADH increased and the elevation in HbO_2 augmented as well. In contrast, during hypoxia, CBF decreased, the reflectance increased showing a reduction in blood volume, NADH augmented (reduction cycles) and HbO_2 decreased. The effect of SD on the ischemic brain and on the ischemic brain during epinephrine infusion is shown in Figure 3. CBF declined during SD in the ischemic brain, the reflectance increased, NADH increased and HbO_2 decreased showing a decline in oxygen delivery during SD. Epinephrine induction increased CBF during SD and partially diminished the reduction cycles in NADH, showing an improvement in oxygen delivery (a decrease in NADH) during the increase in oxygen consumption by SD. Furthermore, SD waves during ischemia were prolonged as compared to SD waves during ischemia & epinephrine, showing an improvement in the oxygen balance of the ischemic brain after epinephrine infusion.

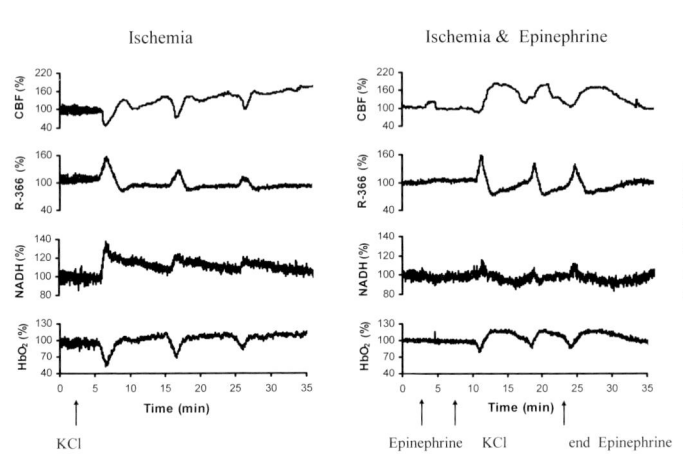

Figure 3: Typical effect of SD on the ischemic brain and ischemic brain during epinephrine infusion. Abbreviations are as described in Fig. 1

4. DISCUSSION

This study examined the relationship between CBF, tissue reflectance, HbO_2 and mitochondrial NADH redox state under negative and positive oxygen balance conditions. During anoxia (Fig. 1), in normal and partially ischemic brain, the increase in NADH and the decrease in CBF were accompanied by a decline in HbO_2. The high correlation between CBF and NADH was shown in our previous reports demonstrating the dependence of the brain upon continuous supply of oxygen [7]. The parallel decrease in HbO_2 shown in this study indicates that, under anoxia, over-extraction of oxygen from HbO_2 can occur. In returning to spontaneous air breathing, CBF increased, due to the rise in blood pressure, leading to an increase in HbO_2 and NADH oxidation. The lower reaction of the partially ischemic brain to anoxia, as compared to normoxia, can be related to the low basic levels of tissue oxygenation with reference to the decline in oxygen and blood supply [3]. It is well documented that, during the SD wave, a large increase in O_2 consumption occurs due to the stimulation of ion pumps [3, 8]. When SD is induced under relatively normal conditions, a complete coupling occurs between the increase in O_2 consumption and the large increase in CBF (Fig. 2), elevating tissue pO_2 [3].

By using oxygen electrodes, Back and coworkers found that, during SD, there is a large increase in O_2 concentration [9], that can explain the oxidation of NADH and the simultaneous increase in HbO_2. SD, under hyperoxia, increased CBF, HbO_2 and decreased NADH (Fig. 2) but no significant differences were found when compared to the normoxic SD. Presumably, the increase in O_2 availability changes tissue responses to the large O_2 demand, since there is no need to import more O_2 by increasing the blood flow. We can argue that the vasodilatation mechanism during SD acts without sensitivity to the O_2 dissolved in the blood. We can also conclude that there is no additional extraction of oxygen from HbO_2 during SD in hyperoxia and that the dissolved O_2 is the source for energy provision, resembling normoxia [10]. The fact that no changes were observed in ion levels initiating the SD wave [3, 9], supports our claim that, during SD, dilator factors are released with no sensitivity to the positive O_2 balance during hyperoxia. On the other hand, brain responses to the increase in energy demand induced by SD are altered when O_2 supply is limited (partial ischemia, hypoxia). The combination of NADH, CBF and HbO_2, used in the present study, indicates that the reversed NADH response to SD (an increase instead of a decrease) is due to the limited ability of CBF to compensate for the large oxygen shortage (Figs. 2, 3), as discussed in our previous publication [3]. Here we demonstrate, for the first time, that there is also a reversed HbO_2 response (Figs. 2, 3). Namely, the high energy demand under negative O_2 balance causes additional HbO_2 extraction. Ischemia induction simulates the conditions relevant to critical clinical situations, such as brain stroke or heart attack [7]. In these situations, the decrease in O_2 supply is the primer factor initiating the pathological state, making the tissue energy balance negative. Although the systemic administration of epinephrine produces a significant increase in CBF and oxidation of NADH in the normal brain compared to ischemia (data not shown), opposite reactions were observed during SD. Epinephrine improved the blood flow and increased CBF, in parallel to the oxidation of NADH and to the increase in HbO_2. Hence, our results demonstrate that epinephrine can help ischemic brain tissue cope with high O_2 demand situations. In summary, the following conclusions can be drawn from this work: 1. There is an added value for combining several parameters to receive a better tissue diagnosis. 2. Although more research is needed, there is significant evidence that epinephrine can be used in clinical situations, such as brain stroke or heart attack, to prevent pathological brain damage.

5. ACKNOWLEDGMENTS

This research was supported by the Israel Science Foundation, Grant No. 358/04

6. REFERENCES

1. R. Cooper, H. J. Crow, W. Greywalter, and A. L. Winter, Regional control of cerebral vascular reactivity and oxygen supply in man, *Brain Res.* **3**, 174-179 (1966)
2. T. Takano, G. F. Tian, W. Peng, N. Lou, D. Lovatt, A. J. Hansen, K. A. Kasischke, and M. Nedergaard, Cortical spreading depression causes and coincides with tissue hypoxia, *Nat. Neurosci.* **10**, 754-762 (2007)
3. J. Sonn and A. Mayevsky, Effects of brain oxygenation on metabolic, hemodynamic, ionic and electrical responses to spreading depression in the rat, *Brain Res.* **882**, 212-216 (2000)
4. A. Chieregato, A. Tanfani, C. Compagnone, R. Pascarella, L. Targa, and E. Fainardi, Cerebral blood flow in traumatic contusions is predominantly reduced after an induced acute elevation of cerebral perfusion pressure, *Neurosurgery.* **60**, 115-2 (2007)

5. H. Kutai-Asis, E. Barbiro-Michaely, A. Deutsch, and A. Mayevsky, Fiber optic based multiparametric spectroscopy *in vivo*: Toward a new quantitative tissue vitality index, *SPIE Proc.* **6083**, 608310-608310-10 (2006)

6. E. Meirovithz, J. Sonn, and A. Mayevsky, Effect of hyperbaric oxygenation on brain hemodynamics, hemoglobin oxygenation and mitochondrial NADH, *Brain Res. Rev.* **54**, 294-304 (2007)

7. A. Mayevsky and G. Rogatsky, Mitochondrial function *in vivo* evaluated by NADH fluorescence: From animal models to human studies, *Am. J. Physiol Cell Physiol.* **.**, (2007)

8. A. J. Hansen, B. Quistorff, and A. Gjedde, Relationship between local changes in cortical blood flow and extracellular K^+ during spreading depression, *Acta Physiol. Scand.* **109**, 1-6 (1980)

9. T. Back, K. Kohno, and K. A. Hossmann, Cortical negative DC deflections following middle cerebral artery occlusion and KCl-induced spreading depression: Effect on blood flow, tissue oxygenation and electroencephalogram, *J. CBF Metab.* **14**, 12-19 (1994)

10. A. Mayevsky and H. R. Weiss, Cerebral blood flow and oxygen consumption in cortical spreading depression, *J. CBF Metab.* **11**, 829-836 (1991)

KETONES SUPPRESS BRAIN GLUCOSE CONSUMPTION

Joseph C. LaManna[1], Nicolas Salem[2], Michelle Puchowicz[1], Bernadette Erokwu[1], Smruta Koppaka[2], Chris Flask[3], Zhenghong Lee[3]

Abstract: The brain is dependent on glucose as a primary energy substrate, but is capable of utilizing ketones such as β-hydroxybutyrate (βHB) and acetoacetate (AcAc), as occurs with fasting, prolonged starvation or chronic feeding of a high fat/low carbohydrate diet (ketogenic diet). In this study, the local cerebral metabolic rate of glucose consumption (CMRglu; μM/min/100g) was calculated in the cortex and cerebellum of control and ketotic rats using Patlak analysis. Rats were imaged on a rodent PET scanner and MRI was performed on a 7-Tesla Bruker scanner for registration with the PET images. Plasma glucose and βHB concentrations were measured and 90-minute dynamic PET scans were started simultaneously with bolus injection of 2-Deoxy-2[^{18}F]Fluoro-D-Glucose (FDG). The blood radioactivity concentration was automatically sampled from the tail vein for 3 min following injection and manual periodic blood samples were taken. The calculated local CMRGlu decreased with increasing plasma BHB concentration in the cerebellum (CMRGlu = -4.07*[BHB] + 61.4, r^2 = 0.3) and in the frontal cortex (CMRGlu = -3.93*[BHB] + 42.7, r^2 = 0.5). These data indicate that, under conditions of ketosis, glucose consumption is decreased in the cortex and cerebellum by about 10% per each mM of plasma ketone bodies.

1. INTRODUCTION

Neurodegeneration after oxidative stress limits the recovery of tissue response and appears to be caused by impaired glycolysis. If indeed there is a defect in glucose metabolism it might be beneficial to supplement energy metabolism with an alternate substrate. It was suggested that brain can supplement glucose as the principal energy

---

[1] Department of Anatomy, Case Western Reserve University, Cleveland, Ohio 44106
[2] Department of Biomedical Engineering, Case Western Reserve University
[3] Department of Radiology, Case Western Reserve University

P. Liss et al. (eds.), *Oxygen Transport to Tissue XXX*, DOI 10.1007/978-0-387-85998-9_45,
© Springer Science+Business Media, LLC 2009

substrate with ketone bodies[1-3] without altering oxygen consumption[4,5]. Classic studies of ketosis induced by fasting or starvation in humans showed that brain function was maintained which was attributed to the utilization (oxidation) of ketone bodies as alternate energy substrates to glucose by the brain[6]. Rats that have been fasted for 2-3 days showed no difference in cerebral blood flow (CBF) or $CMRO_2$[7].

One mechanism by which ketosis might be beneficial is through the metabolic step where ketones enter the TCA cycle at the level of citrate bypassing glycolysis, the step after pyruvate dehydrogenase complex where the enzyme activity is often impaired. Through feed-back regulation, ketones are known to down regulate glycolytic rates at various levels such as citrate, phosphofructokinase and/or hexokinase. In addition, particularly in brain, ketones are a carbon source for glutamate (anaplerosis) and thus help to balance glutamate/glutamine homeostasis through stabilization of energy metabolism in astrocyte following recovery from a hypoxic/ischemic event.

Based on our experiments and evidence in the literature, we have developed the hypothesis that ketones are effective against pathology associated with altered glucose metabolism, the rationale being that ketosis helps to regulate glucose metabolism. In this study, the effects of ketosis on the local cerebral metabolic rate of glucose consumption (CMRglu) were investigated in an *in vivo* rat model of ketosis using positron emission tomography (PET) with 2-[^{18}F] fluoro-2-deoxy-D-glucose (FDG).

2.   MATERIALS AND METHODS

Animal preparation and Diets

All procedures were performed in strict accordance with the National Institutes of Health "Guide for the Care and Use of Laboratory Animals" and were approved by the Institutional Animal Care and Use Committee of Case Western Reserve University. Adult male Wistar rats (final weights: 175-250 g) were allowed to acclimate in the CWRU animal facility for one week before experiments. All rats were housed two-three in a cage, maintained on a 12:12 light-dark cycle with standard/ketogenic rat chow and water available *ad libitum.*

Table 1. Diet composition

	Fat (%)	Protein (%)	Carbohydrate (%)
Ketogenic (KG)	89.5	10.4	0.1
Standard (STD)	27.5	20.0	52.6

Diet Protocols

The ketogenic (KG) diet was purchased from Research Diets (New Brunswick, NJ, U.S.A.) and standard (STD) diet, Teklab 8664, was provided by the CWRU animal facility (see table 1). The KG diet protocol was chosen for its proven effectiveness for inducing moderate and stable ketosis in young adult rats and for its palatability[8-10].

Surgical Procedures

On the day of the experiment, anesthesia was induced by gas mixture (3% isoflurane in O_2) and maintained with 1-2% isoflurane in O_2 through a nasal cone. Ketogenic and control rats underwent the same procedures. Cannulae were placed in: (i) a ventral tail artery using polyethylene tubing (PE-50, 0.023" i.d., 0.038" o.d.), plasma glucose, lactate and ketone concentrations (ii) an external jugular vein into the right atrium using a silastic catheter (0.025" i.d., 0.047" o.d.) for FDG injection.. Lidocaine (1%) was infiltrated on the site. About 2cm incision was then made on the area to be cannulated and the tail artery was isolated. An intramedic(R) polyethylene tubing, size PE 50 with an I.D. of 0.58mm and O.D. of 0.965mm was then advanced into the artery. Two sutures were used to secure the catheter in place while the other end of the catheter was connected through a tubing to a Blood Pressure Analyzer for continuous recording and monitoring of the animal's blood pressure. The neck area was also infiltrated with 1% lidocaine solution. An incision was made and the external jugular vein was isolated and catheterized with a silastic tubing (0.025" i.d., .047" o.d.) which was advanced into the right atrium. The catheter was secured and the incision wound closed with one or two sutures. The other end of the catheter was used for administration of the radioactive indicator to the animal. A small blood sample taken by tail stick from the KG group on the day of surgery was tested using a MediSense Precision Xtra ketone meter to document ketonemia (BHB > 1.0 mM).

MicroPET analysis

The animal was placed on a bed apparatus and moved into a 7-Tesla Bruker scanner for magnetic resonance imaging (MRI) to get anatomical reference for PET imaging. After MRI, the apparatus was moved into the gantry of a Concord Microsystems R4 microPET (Siemens, Knoxville,TN). A transmission scan was performed with a cobalt (^{57}Co) point source for attenuation correction. A 90-minute dynamic PET scan was started simultaneously with bolus injection of 10 MBq/100 g of FDG. Plasma glucose, lactate and BHB concentrations were directly measured. The blood radioactivity concentration was automatically sampled from the tail artery for 3 min using a blood acquisition monitor (BAM) and manual blood samples were taken at 3.5, 4, 4.5, 5, 7, 10, 15, 30, 45, 60, 75 and 90 min. The emission data were rebinned into 17 time frames (3 × 20 sec, 2 × 1 min, 1 × 2 min, 6 × 5 min, 6 × 10 min). ROIs were manually defined on the frontal cerebral cortex and the cerebellum using the high resolution MRI images and mapped onto dynamic PET images. Patlak analysis was finally applied to extract the CMRGlu in both tissue types, assuming a lumped constant (LC) of 0.705[11,12].

3. RESULTS

The diet conditions did not affect the physiological parameters body weight, plasma BHB, plasma glucose and plasma lactate (Table 2). All rats in each of the groups gained weight consistently over the experimental period. Weight gains did not show any significant differences between diet groups as well as with plasma lactate and glucose. BHB levels in blood were significantly higher in KG compared to the STD diet group.

Table 2. Physiological Parameters (mean ± standard deviation). *$p < 0.05$

	STD (n=6)	KG (n=10)
Blood BHB, mM	0.9 ± 0.6	3.78 ± 1.1*
Plasma glucose, mM	10.73 ± 3.78	10.63 ± 1.55
Plasma lactate, mM	1.123 ±0.46	0.80 ± 0.06
Weights, g	267.56 ± 59.11	252.71 ± 47.45

The CMRglu calculated with Patlak analysis was plotted as a function of the measured blood BHB concentration (**Figure 1**). These data showed that CMRglu decreases with increasing ketosis in brain of rats that were either short-term fasted or fed a KG diet for three weeks, *(the blood [BHB] "0' value represents non-ketotic rats)*. The calculated CMRglu linearly decreased with increasing blood ketone concentration in the frontal cortex (CMRGlu = -3.9* [BHB] + 42.7; R^2 = 0.484) (data not shown) and in the cerebellum (CMRGlu = -4.1*[BHB] + 61.4; R^2 = 0.30). A linear relationship between the CMRGlu in the frontal cortex and cerebellum was also observed (R^2 = 0.881). Most striking is that the data show a 10% decrease in CMRglu for every 1mM increase in blood ketone (BHB) levels. These data are consistent with our previous studies that, in combination with increased ketosis and increased transport of ketone bodies at the blood brain barrier, there are more ketones available for brain metabolism[10]

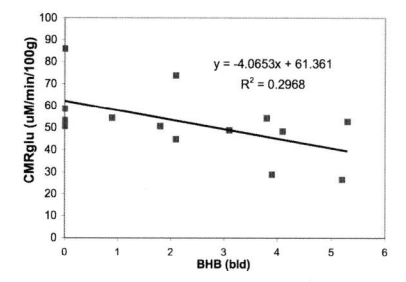

Figure 1. Cerebellum CMRglu.

Each rat was scanned on the PET scanner for 90 minutes post-injection and the list-mode data was rebinned in 17 timeframes. The PET and MRI images were registered and ROIs were drawn on the cerebellum and cortex (Figure 2). The micro PET data were used to detect upregulation of glycolysis and it provided insight into fates of glucose uptake and metabolism (CMR_{glu}) during acute and chronic ketosis in rat brain. Pet scans using [18]FDG, were used to examine the relationship between glucose uptake (CMRglu) and ketosis (ketone plasma levels). Our results showed differences in [18]FDG uptake in ketotic vs non-ketotic rat brain.

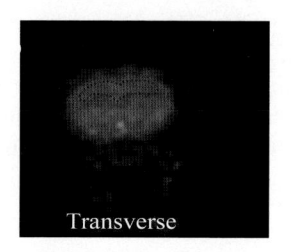 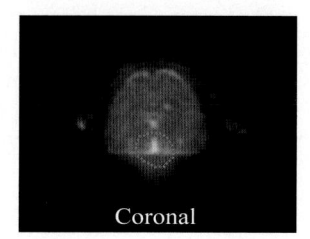 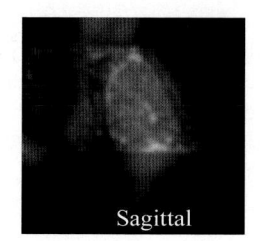

Figure 2. MicroPET images of rat brain.

4. DISCUSSION

From these data, we estimated that in the non-ketotic state with a blood concentration of 0.04 mM BHB and a cerebral blood flow of 1.0 ml/g/min, an extraction fraction of 8%[10] is expected to result in a ketone influx rate of about 3 nmol/g/min. For ketone bodies to replace at least 10% of glucose oxidation an increase in blood ketone concentration and an up regulation of monocarboxylate transporters is required. The utilization of ketones has been shown to be proportional to the uptake[3,13]. Thus, for a 1.0 mM blood BHB, a 12% extraction fraction and a cerebral blood flow of 1.0 ml/g/min a ketone influx of 120 nmol/g/min can be expected. Glucose consumption rates in the anesthetized rat cerebral cortex are approximately 500 nmol/g/min, thus there would be sufficient ketone flux to replace 10% of the glucose oxidation.

It is hypothesized that ketone bodies play a neuroprotective role through an improvement in metabolic efficiency, by sparing glucose, and the degradation of muscle-derived amino acids for substrates[15]. During hypoxia, ketone bodies have been shown to be neuroprotective[16,17] by depressing glucose uptake and CMRglu possibly due to metabolic bocks as a result of oxidative damage. Ketone bodies are thought to stabilize the lactate/pyruvate ratio and bypass the metabolic blocks associated with oxidative stress induced impairment of glucose metabolism.

5. ACKNOWLEDGEMENTS

The authors would like to thank Deb Sim, Dean Fang and John Richey for their technical support with the imaging work. This research has been supported by the National Institutes of Health, R01-NS38632 and P50 GM066309.

6. REFERENCES

1. R. A. Hawkins and J.F. Biebuyck: Ketone bodies are selectively used by individual brain regions. *Science* 205:325-327 (1979).
2. S. G. Hasselbalch, P.L. Madsen, L.P. Hageman, K.S. Olsen, N. Justesen, S. Holm, and O.B. Paulson: Changes in cerebral blood flow and carbohydrate metabolism during acute hyperketonemia. *Am J Physiol* 270:E746-E751(1996).
3. E. O. Balasse and F. Fery: Ketone body production and disposal: effects of fasting, diabetes, and exercise. *Diabetes Metab Rev.* 5:247-270 (1989).
4. D. H. Corddry, S.I. Rapoport, and E.D. London: No effect of hyperketonemia on local cerebral glucose utilization in conscious rats. *J Neurochem.* 38:1637-1641 (1982).
5. N. B. Ruderman, P.S. Ross, M. Berger, and M.N. Goodman: Regulation of glucose and ketone-body metabolism in brain of anaesthetized rats. *Biochem.J* 138:1-10 (1974).
6. O. E. Owen, A.P. Morgan, H.G. Kemp, J.M. Sullivan, M.G. Herrera, and G.F.jr. Cahill: Brain metabolism during fasting. *J.Clin.Invest.* 46:1589-1595 (1967).
7. G. Dahlquist and B. Persson: The rate of cerebral utilization of glucose, ketone bodies, and oxygen: a comparative in vivo study of infant and adult rats. *Pediatr.Res* 10:910-917 (1976).
8. A. S. al Mudallal, B.E. Levin, W.D. Lust, and S.I. Harik: Effects of unbalanced diets on cerebral glucose metabolism in the adult rat. *Neurol.* 45:2261-2265 (1995).
9. A. S. Al-Mudallal, J.C. LaManna, W.D. Lust, and S.I. Harik: Diet-induced ketosis does not cause cerebral acidosis. *Epilepsia* 37:258-261 (1996).
10. M. A. Puchowicz, K. Xu, X. Sun, A. Ivy, D. Emancipator, and J.C. LaManna: Diet-induced ketosis increases capillary density without altered blood flow in rat brain. *Am J Physiol Endocrinol.Metab* (2007).

11. V. Lebon, K.F. Petersen, G.W. Cline, J. Shen, G.F. Mason, S. Dufour, K.L. Behar, G.I. Shulman, and D.L. Rothman: Astroglial contribution to brain energy metabolism in humans revealed by 13C nuclear magnetic resonance spectroscopy: elucidation of the dominant pathway for neurotransmitter glutamate repletion and measurement of astrocytic oxidative metabolism. *J Neurosci.* 22:1523-1531 (2002).

12. I. Tkac, Z. Starcuk, I.Y. Choi, and R. Gruetter: In vivo 1H NMR spectroscopy of rat brain at 1 ms echo time. *Magn Reson.Med.* 41:649-656 (1999).

13. R. A. Hawkins, A.W. Mans, and D.W. Davis: Regional ketone body utilization by rat brain in starvation and diabetes. *Am.J.Physiol.* 250:E169-E178(1986).

14. M. A. Puchowicz, D.S. Emancipator, K. Xu, D.L. Magness, O.I. Ndubuizu, W.D. Lust, and J.C. LaManna: Adaptation to chronic hypoxia during diet-induced ketosis. *Adv.Exp.Med.Biol.* 566:51-57 (2005).

15. R. L. Veech: The therapeutic implications of ketone bodies: the effects of ketone bodies in pathological conditions: ketosis, ketogenic diet, redox states, insulin resistance, and mitochondrial metabolism. *Prostaglandins Leukot.Essent.Fatty Acids* 70:309-319 (2004).

16. A. S. Y. Chang and L.G. D'Alecy: Hypoxia and beta-hydroxybutyrate acutely reduce glucose extraction by the brain in anesthetized dogs. *Can.J.Physiol.Pharmacol.* 71:465-472 (1993).

17. R. Masuda, J.W. Monahan, and Y. Kashiwaya: D-beta-hydroxybutyrate is neuroprotective against hypoxia in serum-free hippocampal primary cultures. *J Neurosci.Res* 80:501-509 (2005).

FALSE POSITIVES IN FUNCTIONAL NEAR-INFRARED TOPOGRAPHY

Ilias Tachtsidis[1], Terence S. Leung[1], Anchal Chopra[1], Peck H. Koh[1], Caroline B. Reid[1], and Clare E. Elwell[1]

Abstract: Functional cranial near-infrared spectroscopy (NIRS) has been widely used to investigate the haemodynamic changes which occur in response to functional activation. The technique exploits the different absorption spectra of oxy- and deoxy-haemoglobin ([HbO$_2$] [HHb]) in the near-infrared region to measure the changes in oxygenation and haemodynamics in the cortical tissue. The aim of this study was to use an optical topography system to produce topographic maps of the haemodynamic response of both frontal cortex (FC) and motor cortex (MC) during anagram solving while simultaneously monitoring the systemic physiology (mean blood pressure, heart rate, scalp flux). A total of 22 young healthy adults were studied. The activation paradigm comprised of 4-, 6- and 8- letter anagrams. 12 channels of the optical topography system were positioned over the FC and 12 channels over the MC. During the task 12 subjects demonstrated a significant change in at least one systemic variable ($p \leq 0.05$). Statistical analysis of task-related changes in [HbO$_2$] and [HHb], based on a Student's t-test was insufficient to distinguish between cortical haemodynamic activation and systemic interference. This lead to false positive haemodynamic maps of activation. It is therefore necessary to use statistical testing that incorporates the systemic changes that occur during brain activation.

1. INTRODUCTION

When analysing cerebral haemodynamic activation data using functional neuroimaging the task-specific activation observed is due to the existence of a close coupling between regional changes in neuronal activation, brain tissue metabolism and regional changes in cerebral blood flow (CBF). Cranial functional near-infrared spectroscopy (NIRS) has been widely used to investigate the haemodynamic changes, which occur in response to functional activation of specific regions of the cerebral cortex. The technique exploits the different absorption spectra of oxy-haemoglobin (HbO$_2$) and

[1] Department of Medical Physics and Bioengineering, Malet Place Engineering Building, University College London, Gower Street, London WC1E 6BT, UK

P. Liss et al. (eds.), *Oxygen Transport to Tissue XXX*, DOI 10.1007/978-0-387-85998-9_46,

deoxy-haemoglobin (HHb) in the near-infrared region to measure the changes in oxygenation and haemodynamics in the brain cortical tissue. In order for this response to be monitored unambiguously it is important that the haemodynamic task-related activity is occurring on top of an unchanged global systemic and brain resting state.

We have previously reported that significant changes in mean blood pressure (MBP) and heart rate (HR) occur during anagram activation tasks and observed that NIRS haemodynamic changes were in some volunteers significantly correlated with changes in these systemic variables.[1] Most recently,[2] we reported that during a frontal lobe anagram activation task, task-related haemodynamic changes were observed both over the frontal cortex (activated region) and motor cortex (control region). The task-related changes were correlated with increases in MBP and scalp blood flow (flux) measured with laser Doppler. This implies the possibility of some systemic "global interference" in our NIRS measured data. It is possible that the anagram task elicits an emotional response, which produces changes in blood pressure that are likely to cause passive changes in the scalp blood flow. These changes can produce small task-related, but non cortical alterations in the [HbO$_2$] and [HHb] signals as measured by cranial NIRS.

Over the last decade or so, many studies have been published describing the use of the optical topography (OT) technique to map functional brain activation.[3-5] By making simultaneous NIRS measurements at multiple brain sites, one can produce spatial maps of the haemoglobin concentration changes that correspond to specific regions of the cerebral cortex. OT can therefore potentially discriminate between regional activated cortical areas and global haemodynamic changes.

The aim of this study is to investigate the functional haemodynamic changes during frontal lobe anagram activation using optical topography both over the activated and control area while continuously monitoring systemic and scalp blood flow changes.

2. MATERIAL AND METHODS

This study was approved by the UCL Research Ethics Committee. We studied 22 young healthy subjects with English as their first language (15 male, 7 female, median age 22 years, range 20-39).

NIRS measurements were conducted with the ETG-100 Optical Topography System (Hitachi Medical Co., Japan) using two 12-channel arrays. Each optode array consisted of 5 source optodes (each delivering light at 780 and 830 nm) and 4 detector optodes. The source-detector interoptode spacing was 30mm and data were acquired at 10Hz. The optodes were placed over the subject's left frontal cortex and positioned according to the international 10-20 system of electrode placement such that channels 1-12 were centred approximately over the frontopolar region (Fp) and channels 13-24 were centred approximately over the left primary motor cortex (C3). A schematic illustration of optode placement is show in Figure 1.

A Portapres® system (TNO Institute of Applied Physics) was used to continuously and non-invasively measure MBP and HR from the finger. A laser Doppler probe (FloLab, Moore Instruments) was placed over the forehead to monitor the changes in scalp blood flow (flux).

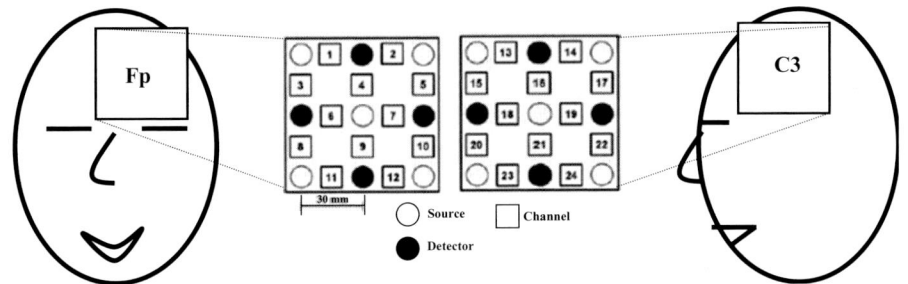

Figure 1. A schematic diagram illustrating the approximate positions of the NIRS light sources, detectors and locations of corresponding measuring positions/channels. One array was centred on the frontopolar region (Fp), the other on the left motor cortex (C3).

All the volunteers were positioned in a comfortable sitting position. Data were recorded during two minutes of the subject at rest (baseline), followed by 45 seconds of the subject solving 4-letter anagrams (9 anagrams, 5 seconds per anagram), 30 seconds rest, 45 seconds of solving 6-letter anagrams (5 anagrams, 9 seconds per anagram), 30 seconds of rest, 45 seconds of solving 8-letter anagram (5 anagrams, 9 seconds per anagram) and 30 seconds of rest. Each anagram-solving period was repeated a total of four times, with the study ending after a 2-minute rest period (total study time 19 minutes). In this study solving an anagram was defined as producing one coherent word using only the letters from another word (e.g. icon – coin). Subjects were encouraged to solve as many anagrams as possible and were instructed to say possible solutions out loud (without moving).

All optical data were subjected to an identical processing procedure using the functional Optical Signal Analysis program[6] (fOSA, University College London, UK) to convert the relative changes in light intensities to concentration changes in haemoglobin (HbO_2, HHb and their sum, HbT) using a differential pathlength factor correction of 6.26. All the signals including MBP, HR and flux, were then decimated from 10Hz to 1Hz and low pass filtered at 0.08Hz. The data were filtered using a 5th order low pass Butterworth digital filter in forward backward directions to avoid introducing a phase delay. The last pre-processing stage, prior to statistical analysis was to de-trend the time-course to remove both drift introduced by the system and any slowly changing unrelated physiological signals. A first-order linear baseline was drawn as the reference and then subsequently subtracted from the activation signal.

The response to stimulation was calculated for each subject as the difference between the average of 10 seconds worth of baseline data at the end of the rest period, and the average of 10 seconds of data commencing 15 seconds after the onset of the 4, 6 or 8 letter anagram solving periods respectively. A 'Student's t-test' was used to assess the significance of these responses (the threshold of significance was set at $p \leq 0.05$ from baseline). For the optical topographic data we then calculated the cumulative total number of channels across subjects in which we observed activation. We define activation as a statistical significant increase in $[HbO_2]$, a statistical significant decrease or no change in [HHb] and a statistical significant increase in [HbT]. Systemic interference was measured by using the Pearson correlation model to calculate correlations between the systemic variables and changes in $[HbO_2]$ and [HHb] in all of the OT channels.

3. RESULTS

A summary of the activation data for the whole group is shown in Figure 2. Each paradigm is shown separately and data are normalised to the number of valid channels. Across paradigms similar activation response was observed in both frontal cortex and motor cortex. Channels in which the highest number of subjects showed activation were channel 23 (55.56%) for the 4-letter task, channel 1 (52.94%) for the 6-letter task, and channels 6 and 21 (33.33%) for the 8-letter task. Taking into account all of the tasks, an average of 30% of the subjects showed activation (range 25-35%) on the frontal cortex and 27% (range 17-37%) on the motor cortex.

Analysis of the systemic variables show that at least 50% of the subjects demonstrated a change in at least one systemic variable. Table 1 shows the mean changes in each systemic variable for those subjects that showed a significant change.

Correlation analysis of the NIRS and systemic data shows a large variability across different OT channels and across subjects. Figure 3 shows the results of the correlation analysis between MBP and the NIRS data, across all channels for (a) subject 3 who showed generally high correlations ($r>0.5$), and (b) subject 18 who showed generally low correlations ($r<0.5$). Both subjects showed significant changes in systemic variables during the anagram tasks and both subjects had channels that showed activation. This trend was observed across subjects. The correlation between the systemic data and the NIRS data from the frontal cortex channels show no difference from the correlation between the systemic data and the NIRS data from the motor cortex channels.

Table 1. Group changes from rest to activation are presented as mean ± standard deviation for those subjects that demonstrated a significant change.

Systemic Variables	4-letter task	6-letter task	8-letter task
Δ[MBP] (mmHg)	(n=11) 6.9±2.7	(n=12) 6.3±4.9	(n=12) 6.9±1.9
Δ[HR] (beats/min)	(n=5) 4.3±4.9	(n=6) -0.4±6	(n=6) 2.4±4.3
Δ[Flux] (%)	(n=4) 14.3±31.1	(n=3) 20.3±10.3	(n=1) -17.8

4. DISCUSSION

In this study we used an optical topography system to investigate the changes in [HbO$_2$] and [HHb] during anagram solving over the frontal lobe (activated area) and motor cortex (control area) while simultaneously monitoring systemic variables. We used a Student's t-test to define significant changes in [HbO$_2$], [HHb] and [HbT] for each OT channel and for each subject during the different anagram solving tasks and used these data to define where and when activation was detected. The same analysis was performed on the systemic variables. We observed a large variability in activated OT channels across subjects. The OT results failed to define specific regional areas of activation. 50% of subjects showed a significant change in at least one systemic variable. These systemic changes appear in some subjects to correlate with the observed functional changes in [HbO$_2$] and [HHb] across the OT channels. Figure 4 shows an example of changes in [HbO$_2$] and [HHb] from an OT channel over the frontal cortex and an OT channel over the motor cortex with the simultaneously recorded changes in MBP and scalp flux.

Clearly systemic interference during the anagram task can lead to false positives in defining activated OT channels.

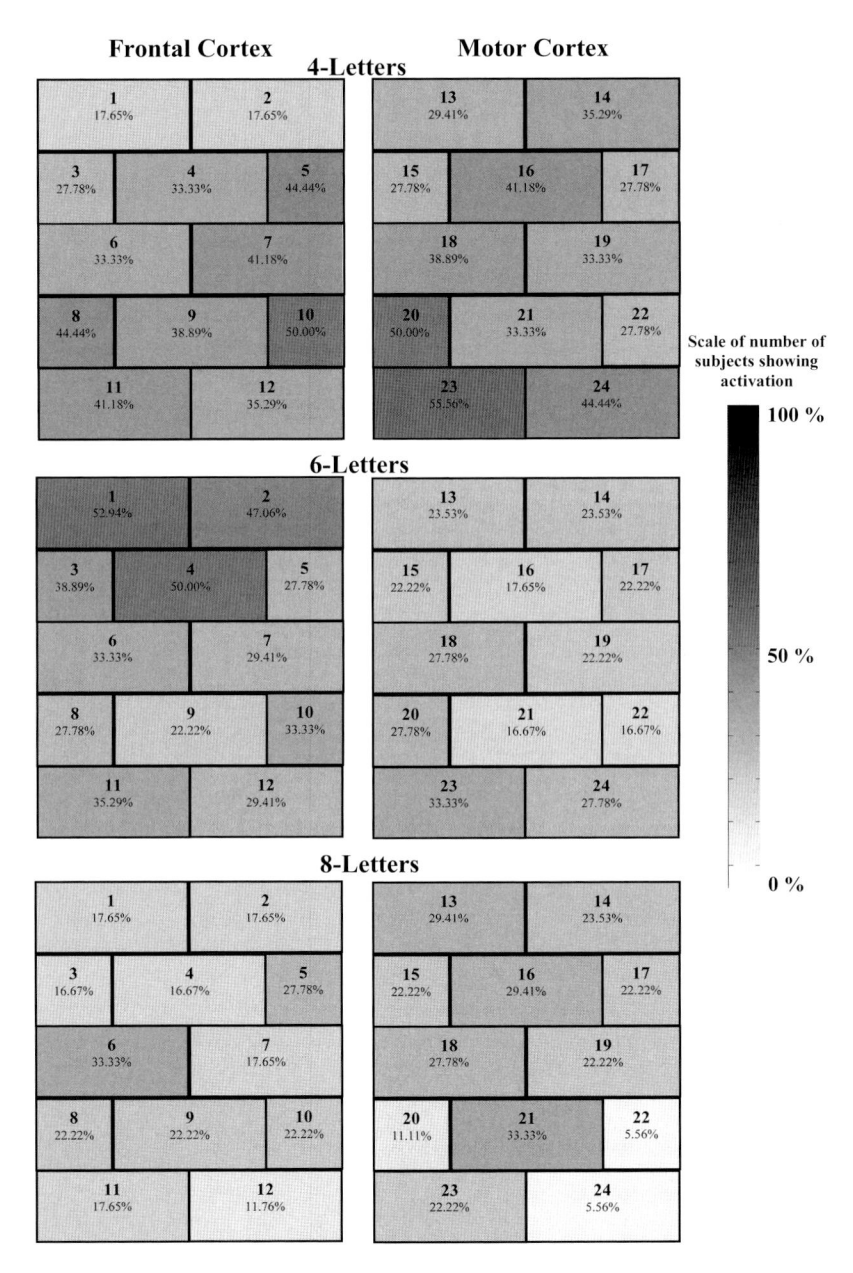

Figure 2. Group analysis shows the percentage of subjects that demonstrated activation in specific channels during the three different anagram solving paradigms.

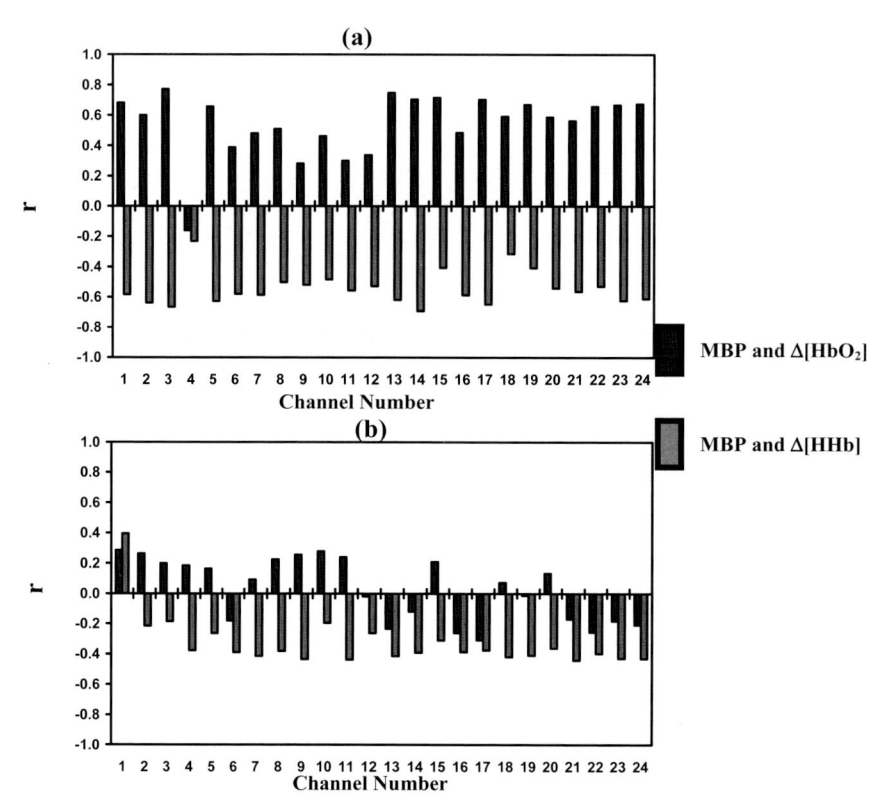

Figure 3. Individual correlation coefficients between MBP and Δ[HbO₂] and MBP and Δ[HHb] across all channels for (a) subject 3 and (b) subject 18.

In this study we used the classical approach to define significant changes in haemoglobin concentrations by employing a "Student's t-test". This approach compares two different states of the brain, i.e. "rest" versus "activation". The "rest" period is usually defined as a baseline period before the stimulus onset and the "activation" period is defined as the period 10-20 seconds after the onset of the stimulus. By keeping the rest and activation periods constant across subjects one can investigate the functional response to specific tasks. Whilst a simplistic approach of this kind helps to provide a quick assessment of the haemodynamic response to the task it does not consider any spatial coherence in the OT data. It also assumes that the measured changes in haemoglobin concentrations are due solely to the neuronal activation, and that there are no tasks-related systemic effects. We have shown that this latter assumption is not true for all subjects performing an anagram solving task. One can include a priori information regarding systemic changes and can de-correlate the physiological noise (cardiac, respiratory and vasomotion related fluctuations) from the evoked haemodynamic response, by using techniques such as Principal Component Analysis,[7] Independent Component Analysis,[8] and more recently Statistical Parametric Mapping (SPM).[6] SPM has been widely used for the analysis of functional activation data from other neuroimaging modalities such as the BOLD response in fMRI studies.[9] SPM uses mass univariate approach to modelling the spatiotemporal neuroimaging data by assigning a statistic value to every brain voxel. It enables the construction of spatial statistical processes to test hypotheses about regional specific effects in the brain. Unlike the classical approach mentioned earlier, where the two different time courses compared, SPM employs a modelling approach for each brain voxel. In our study all of the explanatory variables (HbO_2, MBP, HR and flux) were treated as regressors in the linear model. To treat the variability of haemodynamic responses arising from different events between different brain voxels, SPM allows the modelling of latency and dispersion derivatives as additional regressors to its canonical response function. The associated parameter estimates are the coefficients for each of the regressors that best model the observed response for the voxel in question (here a voxel is defined as an OT channel). To account for the spatial coherence of the functional data, SPM provides the necessary family-wise correction based on the theory of Gaussian random field to resolve the multiple comparison problem.

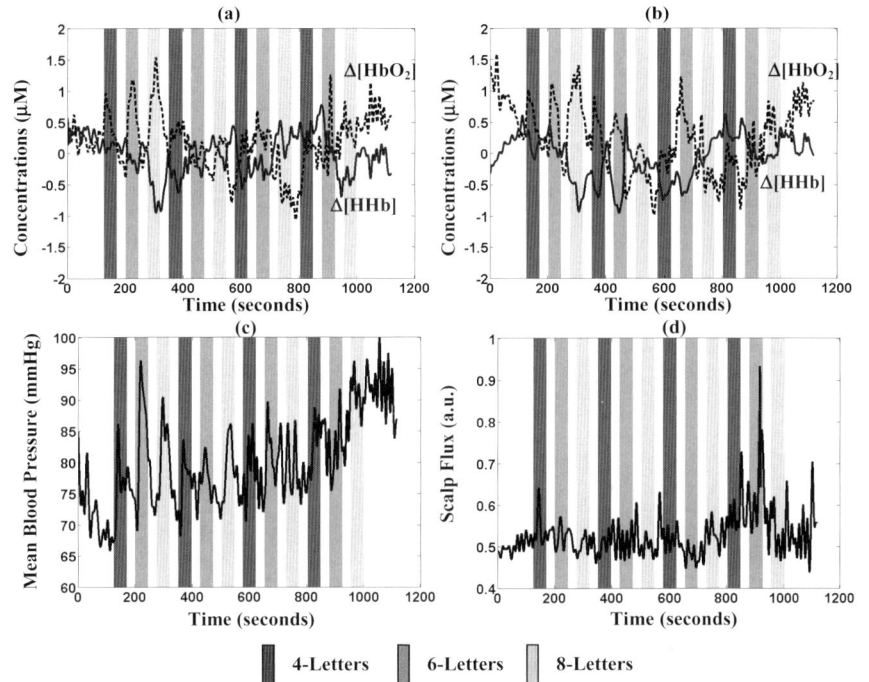

Figure 4. Results from one subject showing changes in; (a) NIRS data from channel 3 (frontal cortex) (b) NIRS data from channel 15 (motor cortex); (c) MBP and (d) scalp flux.

As an example of this method we have used fOSA-SPM software[6] to analyse NIRS and systemic data from one subject collected during the 6-letter anagram solving task. Using the "Student's t-test" analysis, this subject demonstrated activation across all OT channels. Figure 5 shows the results of the SPM analysis on the same subject's data. These are presented as an SPM t-result for the HbO_2 signal over all channels and show a spatial localisation of the haemodynamic response. Unlike the "Student's t-test" approach which compares the difference between two specific physiological states, SPM offers a more rigorous approach to analysing functional OT data by taking into account the global systemic effects by means of fitting a haemodynamic response function and performing spatial correlations across all channels.

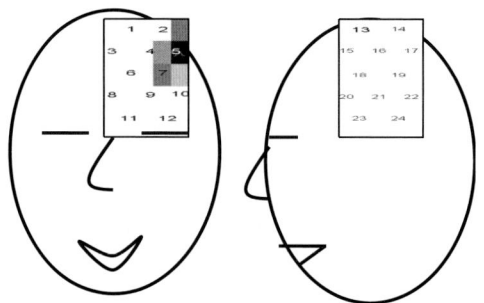

Figure 5. The SPM t-results with a threshold of significance of p≤0.05; darker pixels correspond to higher significant t-values.

In conclusion, when analysing OT data for evidence of functional activation the effect of task-related changes in systemic variables should be taken into account. SPM may be a useful tool for analysing simultaneously measured multi-channel OT NIRS data and systemic variables.

5. ACKNOWLEDGMENTS

The authors would like to acknowledge the EPSRC (Grant No EP/D060982/1).

6. REFERENCES

1. I. Tachtsidis, T.S. Leung, L. Devoto, D.T. Delpy, and C.E. Elwell, Measurement of frontal lobe functional activation and related systemic effects: a near-infrared spectroscopy investigation, *Adv. Exp. Med. Biol.* In Press (2008).
2. I. Tachtsidis, T.S. Leung, M.M. Tisdall, D. Presheena, M. Smith, D.T. Delpy, and C.E. Elwell, Investigation of frontal cortex, motor cortex and systemic haemodynamic changes during anagram solving, *Adv. Exp. Med. Biol.* In Press (2008).
3. Y. Hoshi, B. H. Tsou, V. A. Billock, M. Tanosaki, Y. Iguchi, M. Shimada, T. Shinba, Y. Yamada, and I. Oda, Spatiotemporal characteristics of hemodynamic changes in the human lateral prefrontal cortex during working memory tasks, *NeuroImage* **20**(3), 1493-1504 (2003).
4. B. Chance, S. Nioka, S. Sadi, and C. Li, Oxygenation and blood concentration changes in human subject prefrontal activation by anagram solutions, *Adv. Exp. Med. Biol.* **510**, 397-401 (2003).
5. R.P. Kennan, D. Kim, A. Maki, H. Koizumi, and R.T. Constable, Non-invasive assessment of language lateralization by transcranial near infrared optical topography and functional MRI, *Hum. Brain Mapp.* **16**(3), 183-189 (2002).
6. P.H. Koh, D.E. Glaser, G. Flandin, S. Kiebel, B. Butterworth, A. Maki, D.T. Delpy, and C.E. Elwell, Functional optical signal analysis (fOSA): a software tool for NIRS data processing incorporating statistical parametric mapping (SPM), *JBO* In Press (2007).
7. X. Zhang, V. Toronov, and A. Webb, Simultaneous integrated diffuse optical tomography and functional magnetic resonance imaging of the human brain, *Opt. Express* **13**(14), 5513-5521 (2005).
8. I. Schiessl, M. Stetter, J.E.W. Mayhew, N. McLoughlin, J.S. Lund, and K. Obermayer, Blind signal separation from optical imaging recordings with extended spatial decorrelation, *IEEE Transactions on Biomedical Engineering,* **47**(5), 573-577 (2000).
9. K.J. Friston, A.P. Holmes, J.B. Poline, P.J. Grasby, S.C. Williams, R.S. Frackowiak, and R. Turner, Analysis of fMRI time-series revisited, *NeuroImage* **2**(1), 45-53 (1995).

RELATIONSHIP BETWEEN BRAIN TISSUE HAEMODYNAMICS, OXYGENATION AND METABOLISM IN THE HEALTHY HUMAN ADULT BRAIN DURING HYPEROXIA AND HYPERCAPNEA

Ilias Tachtsidis[1], Martin M. Tisdall[2], Terence S. Leung[1], Caroline Pritchard[2], Christopher E. Cooper[3], Martin Smith[2], and Clare E. Elwell[1]

Abstract: This study investigates the relationship between changes in brain tissue haemodynamics, oxygenation and oxidised cytochrome-c-oxidase ([oxCCO]) in the adult brain during hyperoxia and hypercapnea. 10 healthy volunteers were studied. We measured the mean blood flow velocity of the right middle cerebral artery (Vmca) with transcranial Doppler (TCD) and changes in concentrations of total haemoglobin ($[HbT]=[HbO_2]+[HHb]$), haemoglobin difference ($[Hbdiff]=[HbO_2]-[HHb]$) and [oxCCO] with broadband near-infrared spectroscopy (NIRS). We also measured the absolute tissue oxygenation index (TOI) using NIR spatially resolved spectroscopy. During hyperoxia there was an increase in TOI ($2.33\pm0.29\%$), [Hbdiff] ($4.57\pm1.27\mu M$) and in the oxidation of [oxCCO] ($0.09\pm0.12\mu M$); but a reduction in Vmca ($5.85\pm4.85\%$) and HbT ($1.29\pm0.91\mu M$). During hyperoxia there was a positive correlation between [oxCCO] and TOI and [Hbdiff] (r=0.83 and r=0.95) and a negative association between [oxCCO] and Vmca and [HbT] (r=-0.74 and r=-0.87). During hypercapnea there was an increase in TOI ($2.76\pm2.16\%$), [Hbdiff] (7.36 ± 2.64), [HbT] ($2.61\pm2.7\mu M$), Vmca ($14.92\pm17.5\%$) and in the oxidation of [oxCCO] ($0.25\pm0.17\mu M$). Correlation analysis shows that there was association between [oxCCO] and TOI, [Hbdiff] and [HbT] (r=0.83, r=0.93 and r=0.82) but not with Vmca (r=0.33). We conclude that an increase in [oxCCO] was seen during both challenges and it was highly associated with brain tissue oxygenation.

[1] Department of Medical Physics and Bioengineering, Malet Place Engineering Building, University College London, Gower Street, London WC1E 6BT, UK
[2] Department of Neuroanaesthesia and Neurocritical Care, The National Hospital for Neurology and Neurosurgery, Queen Square, London WC1N 3BG, UK
[3] Department of Biological Sciences, University of Essex, Wivenhoe Park, Colchester Essex, CO4 3SQ, UK

P. Liss et al. (eds.), *Oxygen Transport to Tissue XXX*, DOI 10.1007/978-0-387-85998-9_47,
© Springer Science+Business Media, LLC 2009

1. INTRODUCTION

Measurement and monitoring of changes in haemodynamics, oxygenation and metabolism in the brain, reliably, non-invasively and at the bedside is an important aim in diagnosis and management of patients in neurocritical care. This may be achieved with a combination of two techniques, near-infrared spectroscopic (NIRS) and transcranial Doppler (TCD).

NIRS is a non-invasive technique which exploits the fact that biological tissue is relatively transparent to near infrared light allowing interrogation of the cerebral cortex by optodes placed on the scalp. Biological tissue is a highly scattering medium but if the average pathlength of light through tissue is known, the modified Beer-Lambert law,[1] which assumes constant scattering losses, allows calculation of absolute changes in chromophore concentration. NIRS has been used in animals and humans to measure the change in concentration of oxy-haemoglobin (Δ[HbO$_2$]), deoxy-haemoglobin (Δ[HHb]), and oxidized cytochrome oxidase (Δ[oxCCO]).[2-4] Cytochrome c oxidase (CCO) is the terminal electron acceptor of the mitochondrial electron transfer chain and catalyses over 95% of oxygen metabolism, thereby driving aerobic adenosine triphosphate (ATP) synthesis and playing a central role in the maintenance of mitochondrial function.[5]

Transcranial Doppler (TCD) is a non-invasive ultrasound technique which uses the Doppler shift from moving red blood cells to calculate cerebral blood flow velocity. TCD is not able to provide absolute measurements of cerebral blood flow (CBF), but if the angle of insonation and the diameter of the insonated vessel remain constant then changes in TCD measured cerebral blood flow velocity correlate with changes in CBF.[6] Several studies have shown minimal changes in the diameter of basal cerebral arteries during various physiological challenges.[7] Typically the TCD signal is acquired from the middle cerebral artery.

Hyperoxia and hypercapnea are routinely used as a treatment technique and an intracranial compliance test respectively in patients with brain injury. In the human brain the changes in brain blood flow and oxygenation during these challenges are well documented; however the changes in brain metabolism and their relationships are not.

In this study we use the combination of NIRS and TCD to investigate the relationship between changes in brain tissue haemodynamics, oxygenation and metabolism in the healthy human adult brain during hyperoxia and hypercapnea.

2. MATERIAL AND METHODS

This study was approved by the Joint Research Ethics Committee of the National Hospital for Neurology and Neurosurgery and the Institute of Neurology. We studied 10 healthy subjects (7 male, 3 female, median age 32 years, range 30-39).

The optodes from a broadband spectrometer (BBS) previously described by Tisdall et al.[4] were placed 3.5 cm apart on the right side of the forehead. NIR spectra between 650 and 980 nm were collected at 1Hz with a spectral resolution of ~5nm. Absolute Δ[oxCCO], Δ[HbO$_2$], and Δ[HHb] were calculated from changes in light attenuation using a multiple regression technique termed the UCLn algorithm.[8] Individual baseline optical pathlength was calculated using second differential analysis of the 740-nm water feature of the initial 60 seconds of spectral data.[9] Change in total haemoglobin concentration $\Delta[HbT]$ was defined as $\Delta[HbO_2]+\Delta[HHb]$ and change in haemoglobin difference concentration $\Delta[Hbdiff]$ as $\Delta[HbO_2]-\Delta[HHb]$.

The optodes from the NIRO 300 (Hamamatsu Photonics KK) were placed below the BBS optodes, with and interoptode spacing of 5cm and were used to measure absolute cerebral tissue oxygenation index (TOI) over the frontal cortex using the SRS technique[10].

Blood flow velocity in the basal right middle cerebral artery was collected at 50 Hz using a 2 MHz transcranial Doppler ultrasonography (Pioneer TC2020, Nicolet, UK) fixed in place over the right temporal region. Mean velocity of the middle cerebral artery (Vmca) was calculated from the velocity envelope using a trapezoidal integration function (MatLab, Mathworks Inc.). A modified pulse oximeter probe (Novametrix Medical Systems Inc., USA) measured SaO_2, and a Portapres finger cuff (Biomedical Instrumentation, TNO Institute of Applied Physics, Belgium) measured mean blood pressure (MBP). A modified anaesthetic machine was used to alter FiO_2 and $EtCO_2$ which were measured using an inline gas analyser (Hewlett Packard, UK) and a CO_2SMO optical sensor (Novametrix Medical Systems Inc.) respectively.

During hyperoxia FiO_2 was increased to 100% for five minutes and then returned to normoxia for five minutes. The cycle was repeated three times and the subjects adjusted their minute ventilation to maintain normocapnea throughout the study. During hypercapnea approximately 6% carbon dioxide (CO_2) was added to the inspired gases and was titrated to induce an increase in $EtCO_2$ of 1.5kPa. The elevated $EtCO_2$ was maintained for ten minutes and the inspired carbon dioxide fraction was then returned to zero for a further five minutes. The start and end of each hyperoxia and hypercapnea period was identified from the FiO_2 and $EtCO_2$ data respectively. To enable description of the group data, each individual hyperoxia and hypercapnea was divided into equal time periods, with each time point representing an eighth of the total time course of the challenge. This produced nine time points with point 1 representing the point just prior to the start of the challenge and point 9 the end point of the challenge. The same technique was applied separately to the recovery period, producing points 9 just prior to start of recovery to 17 at the end of recovery period. At each time point, the mean of the preceding 10 seconds of data was calculated. For the hyperoxia challenge data from the three experimental cycles were averaged to give a single course of hyperoxia and recovery for each subject. Group mean changes from baseline at each time point were produced. Statistical analysis was carried out using the SPSS software (version 13 for windows) and p values ≤ 0.05 were considered significant. Group changes were compared with baseline using a two-tail Student's t-test. Correlations between variables were assessed by applying Spearman rank correlation to data from the 9 time points start to the end of challenge.

3. RESULTS

Group summary data are shown in Figure 1 and correlation analysis shown in Figure 2. During hyperoxia there was a significant increase in TOI (2.33±0.29%), [Hbdiff] (4.57±1.27μM) and in the oxidation of [oxCCO] (0.092±0.117μM); but a reduction in Vmca (5.85±4.85%) and [HbT] (1.29±0.91μM). Correlation analysis shows a high positive association between [oxCCO] and TOI and [Hbdiff] (r=0.83 and r=0.95) and a high negative association between [oxCCO] and Vmca and [HbT] (r=-0.74 and r=-0.87).

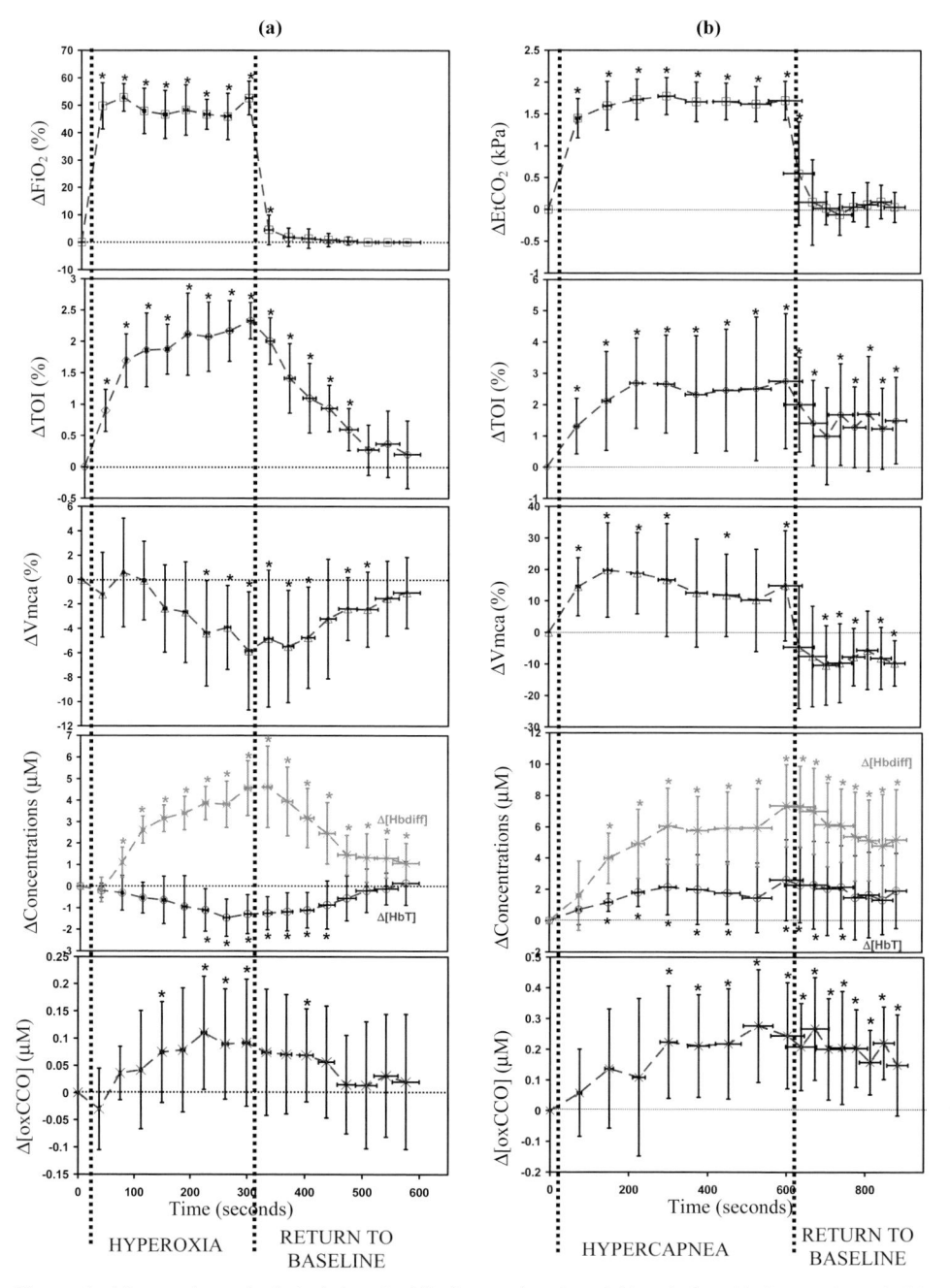

Figure 1. Mean and standard deviation (n=10) for monitored variables during (a) hyperoxia and (b) hypercapnea (*$p \leq 0.05$ for changes from baseline).

During hypercapnea there was an increase in TOI (2.76±2.16%), [Hbdiff] (7.36±2.64 μM), HbT (2.61±2.6μM), Vmca (14.91±17.49%) and in the oxidation of [oxCCO] (0.245±0.172 μM). Correlation analysis showed that there was a linear association between [oxCCO] and TOI, [Hbdiff] and HbT (r=0.83, r=0.93 and r=0.82) but not between [oxCCO] and Vmca (r=0.33).

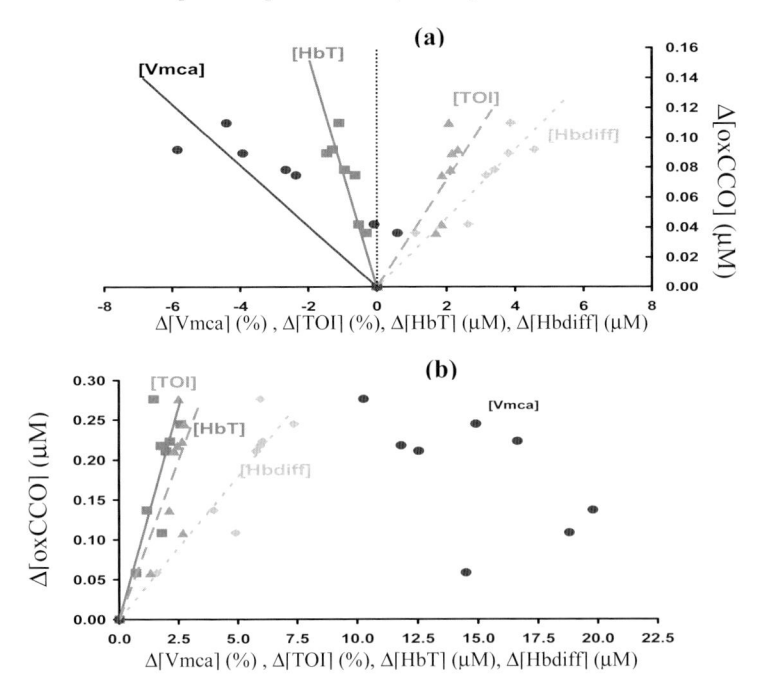

Figure 2. Scatter plot using group mean changes (n=10) from baseline to the end of (a) hyperoxia and (b) hypercapnea. Lines show the linear regression estimation between Δ[oxCCO] and each variable.

4. DISCUSSION

We investigated two paradigms that increase cerebral oxygen delivery to the healthy adult brain by different methods – increase in arterial oxygen content (hyperoxia) and increase in cerebral blood flow (hypercapnea). A significant decrease in Vmca and [HbT] during hyperoxia confirmed blood flow reduction, presumably related to the known vasoconstrictive effects of 100% oxygen; and an increase in both Vmca and [HbT] during hypercapnea confirmed the known increase in CBF secondary to rises in PaCO$_2$. In both cases, however, oxygen delivery seemed to increase as evidenced by the rise in TOI and [Hbdiff] – the rise in arterial oxygen content in hyperoxia more that compensating for the drop in flow. Both scenarios showed an increase in the oxidation of the mitochondrial cytochrome oxidase CuA centre ([oxCCO]). *In-vivo* studies in animals confirm that cerebral CCO can be oxidised by increases in oxygenation induced by hypercapnea or reactive hyperaemia.[2,11] Our data support the conclusion that, at normoxic normocapnea, cerebral CCO is not fully oxidized in the human adult and that further oxidation is possible.

In order to further investigate the above claim, correlation analysis was done between the haemodynamic, oxygenation and [oxCCO] variables. We found high correlations in the hyperoxia challenge between the changes in flow velocity, oxygenation and [oxCCO]; however during the hypercapnea challenge no association was found between flow velocity and [oxCCO]. CCO oxidation state is affected by factors other than oxygen tension, for example electron flux through the enzyme, pH changes and changes in ADP concentration.[12,13] In the case of hyperoxia it is difficult to see these other factors being as important as the rise in oxygenation and we may be seeing a direct effect of oxygen on CCO in this case. However, changes in CO_2 will have metabolic effects further to the changes in oxygenation most notably to cause a decrease in pH. It is possible that the oxidation of [oxCCO] during hypercapnea may also be related to secondary metabolic effects, whereas those of hyperoxia are more likely to be direct effects of oxygen tension.

5. ACKNOWLEDGMENTS

The authors would like to acknowledge the EPSRC (Grant No EP/D060982/1), the BBSRC (Grant No D017858/1) and the Wellcome Trust (Grant No 075608).

6. REFERENCES

1. D. T. Delpy, M. Cope, Z. P. van der, S. Arridge, S. Wray, and J. Wyatt, Estimation of optical pathlength through tissue from direct time of flight measurement, *Phys. Med. Biol.* **33**(12), 1433-1442 (1988).
2. V. Quaresima, R. Springett, M. Cope, J. T. Wyatt, D. T. Delpy, M. Ferrari, and C. E. Cooper, Oxidation and reduction of cytochrome oxidase in the neonatal brain observed by in vivo near-infrared spectroscopy, *Biochim. Biophys. Acta* **1366**(3), 291-300 (1998).
3. I. Tachtsidis, M. Tisdall, T. S. Leung, C. E. Cooper, D. T. Delpy, M. Smith, and C. E. Elwell, Investigation of in vivo measurement of cerebral cytochrome-c-oxidase redox changes using near-infrared spectroscopy in patients with orthostatic hypotension, *Physiol. Meas.* **28**(2), 199-211 (2007).
4. M. M. Tisdall, I. Tachtsidis, T. S. Leung, C. E. Elwell, and M. Smith, Near-infrared spectroscopic quantification of changes in the concentration of oxidized cytochrome c oxidase in the healthy human brain during hypoxemia, *J. Biomed. Opt.* **12**(2), 024002 (2007).
5. O. M. Richter and B. Ludwig, Cytochrome c oxidase--structure, function, and physiology of a redox-driven molecular machine, *Rev. Physiol. Biochem. Pharmacol.* **14**, 747-774 (2003).
6. J. M. Valdueza, J. O. Balzer, A. Villringer, T. J. Vogl, R. Kutter, and K. M. Einhaupl, Changes in blood flow velocity and diameter of the middle cerebral artery during hyperventilation: assessment with MR and transcranial Doppler sonography, *AJNR Am. J. Neuroradiol.* **18**(10), 1929-1934 (1997).
7. C. A. Giller, G. Bowman, H. Dyer, L. Mootz, and W. Krippner, Cerebral arterial diameters during changes in blood pressure and carbon dioxide during craniotomy, *Neurosurgery* **32**(5), 737-741 (1993).
8. S. J. Matcher, C. E. Elwell, C. E. Cooper, M. Cope, and D. T. Delpy, Performance comparison of several published tissue near-infrared spectroscopy algorithms, *Anal. Biochem.* **227**(1), 54-68 (1995).
9. S. J. Matcher, M. Cope, and D. T. Delpy, Use of the water absorption spectrum to quantify tissue chromophore concentration changes in near-infrared spectroscopy, *Phys. Med. Biol.* **39**(1), 177-196 (1994).
10. S. Suzuki, S. Takasaki, T. Ozaki, and Y. Kobayashi, A tissue oxygenation monitor using NIR spatially resolved spectroscopy, *Proc. SPIE* **3597**, 582-592 (1999).
11. R. J. Springett, M. Wylezinska, E. B. Cady, V. Hollis, M. Cope, and D. T. Delpy, The oxygen dependency of cerebral oxidative metabolism in the newborn piglet studied with 31P NMRS and NIRS, *Adv. Exp. Med. Biol.* **530**, 555-563 (2003).
12. P. E. Thornstrom, P. Brzezinski, P. O. Fredriksson, and B. G. Malmstrom, Cytochrome c oxidase as an electron-transport-driven proton pump: pH dependence of the reduction levels of the redox centers during turnover, *Biochemistry* **27**(15), 5441-5447 (1988).
13. C. E. Cooper, S. J. Matcher, J. S. Wyatt, M. Cope, G. C. Brown, E. M. Nemoto, and D. T. Delpy, Near-infrared spectroscopy of the brain: relevance to cytochrome oxidase bioenergetics, *Biochem. Soc. Trans.* **22**(4), 974-980 (1994).

USE OF A CODMAN® MICROSENSOR INTRACRANIAL PRESSURE PROBE: EFFECTS ON NEAR INFRARED SPECTROSCOPY MEASUREMENTS AND CEREBRAL HEMODYNAMICS IN RATS

Helga Blockx, Willem Flameng, Geofrey De Visscher

Abstract: Intracranial pressure (ICP) monitoring is indispensable in the assessment of neurotrauma in humans and animal models. It was shown that cerebellar ICP, leaving the cortical area intact, can replace cerebral ICP in rats. While cerebral probes may induce spreading depression, the effects of a miniature cerebellar probe on near infrared spectroscopy (NIRS) measurements and cerebral hemodynamics are not known. We therefore compared a group with an ICP probe to a control group.

Our experiments revealed decreased optical path lengths at 840 nm and 960 nm (both p=0.026) and a decreased cerebral blood flow (CBF, p=0.015). Despite these changes, the data found using NIRS agree with the blood sample analysis. An increased deoxyhemoglobin concentration (p=0.041) and a decreased sagittal sinus oxygen saturation (p=0.041), were found in the ICP probe group. Because the decreased CBF was accompanied by an increased arterio-venous oxygen difference (p=0.026) and unaltered cerebral metabolic rate of oxygen (p=0.485), this suggests an uncoupling.

These data suggest that a cerebellar miniature Codman® ICP probe induces an uncoupling of cerebral metabolism and CBF. In addition, NIRS is found to be a robust technique: even when path lengths are altered after probe insertion, physiological alterations can still be examined.

1. INTRODUCTION

ICP measurements are of utmost importance in the assessment of severe head injury in humans. These measurements are commonly used for the confirmation of traumatic brain injury in animal models. In humans, fluid-filled ventricular catheters are considered to be the most accurate method[1], but other probes have been developed and validated.

Laboratory of Cardiovascular Research, Catholic University of Leuven, Leuven, Belgium
Corresponding Author: Dr. G. De Visscher, CEHA, KULeuven, Minderbroedersstraat 17, B3000 Leuven, Belgium, Email: Geofrey.DeVisscher@med.kuleuven.be, tel: +32 16 337120, fax: +32 16 337122

Intraparenchymal microtransducers, like the Codman® microsensor probe, are a good alternative. In vitro accuracy tests of the Codman® probe, revealed a maximal error of 5 mmHg from 60 mmHg onwards[2]. To combine ICP measurements with NIRS in rats, it was shown that cerebellar ICP measured with this probe can be used as an alternative to cerebral ICP in rats[3]. This leaves the cortical area intact, presumably leading to a better assessment of other parameters in the brain.

Previously, miniature probes placed in a cerebral position were found to induce spreading depression and decrease cerebral blood flow[4]. To date it has not been investigated if a miniature cerebellar probe has effects on NIRS measurements and hemodynamics. We therefore compared a group with an ICP probe to one without.

2. MATERIALS AND METHODS

2.1. Animal Preparation and Experimental Protocol

Animal housing and treatment conditions complied with the European Union directive # 86/609 on animal welfare. Twelve male Sprague-Dawley rats (380-430 g, Charles River, Germany) were randomly assigned to one of the two groups. They were allowed free access to food and water (12/12-hour day-night cycle).

The procedure for preparing rats for NIRS measurement has previously been described[5]. In short, the animals were anaesthetised using 4% isoflurane in 30% O_2 / 70% N_2O during 5 min. After endotracheal intubation, the isoflurane concentration was switched to 2% and the animals were surgically prepared. A silicon-tipped PE50 catheter was inserted into the right external jugular vein towards the vena cava for indocyanine green (ICG) bolus injection. To monitor mean arterial blood pressure (MABP) and heart rate (HR), the left femoral artery was cannulated.

The rats were fixed in a stereotaxic apparatus (Model 900, David Kopf, USA) with thermally isolated ear bars. A thermistor inserted into the tip of an ear bar was used to measure the tympanic temperature (T-ear), an accurate measurement of brain temperature. A rectal thermistor probe was used to monitor and control body temperature (T-rectal) with a heating pad connected to a temperature control unit. End-tidal CO_2 (EtCO$_2$) and breathing rate (BrR) were monitored with an EtCO$_2$-monitor (Capnogard, Novametrix, USA). The parietal and temporal bones were exposed by removing parts of the temporal muscles and sites of possible bleeding were cauterised.

A burr hole not penetrating the skull was made at the site of the sagittal sinus. With a 23 gauge needle the remaining bone and upper wall of the sagittal sinus were punctured. After withdrawal of the needle, a PE10 catheter with a 1.5cm PE50 cuff was inserted into the sagittal sinus and sealed with a few drops of 2% (in saline) low meting point agarose (SeaPlaque, FMC BioProducts, USA). This cannula is used for cerebral venous blood sampling required for accurate measurement of the cerebral oxygen consumption.

For cerebellar ICP assessment, ICP group only, the trapezial muscle was detached from the skull and cauterised. A burr hole was drilled into the right part of the occipital bone, 2mm caudal of the cranial edge and 2mm lateral from the midline. The microsensor ICP probe (ICP Neuromicrosensor, $\varnothing$ 1.2mm, length 4mm, Codman & Shurtleff Inc., USA) was inserted through this burr hole. The tip was carefully positioned in the cerebeller parenchyma in the sagittal plane at a 35° angle relative to the horizontal plane.

Emitting and receiving optical fibres ($\varnothing$ 3mm) were placed onto the right and left temporal bones, respectively (frontal pole of the fibres at the level of the bregma, angle of 10° relative to both the coronal and sagittal plane). To prevent blood entering the space between the skull and the optical fibres, optical gel (Optical Gel code 0608, R.P. Cargille Laboratories Inc., USA) was used to fill in this space. The skull was covered with black ink and black clay (Modelling clay No. 8401, Eberhard Faber GmbH, Germany).

The isoflurane concentration was switched to and maintained at 1.2% in 30% O_2 / 70% N_2O until the end of the experiment. The study comprised 2 groups: GROUP 1 receiving no ICP probe (control) and GROUP 2 receiving an ICP probe (n=6 in each group). After the 10min stabilisation period, NIRS measurement was started. Ten minutes later an arterial blood sample was taken followed by an ICG-bolus injection at 12.5min. Subsequently a cerebral venous blood sample was taken. At the end of the experiment, the animals were killed by terminal anoxia.

2.2. NIRS Equipment

The NIRS system and algorithms were developed at University College London. The system allows assessment of absolute values and absolute changes in the concentration of oxyhaemoglobin ([HbO_2]), deoxyhaemoglobin ([Hb]) and oxidised Cu_A.

ICG can be measured by this system, when its absorption spectrum is included in the algorithm. A sterile 1ml/mg ICG (IR-125, laser grade, Acros, Belgium) solution was used. 5% bovine serum albumin (BSA fraction V, Sigma, Belgium) was added to this solution to bind ICG. NIR spectra were collected contiguously with a period of 100ms and 100 spectra were averaged to obtain a time resolution of 10 s (0.1Hz). For ICG bolus transit detection the sampling frequency was switched from 0.1Hz to 10Hz.

2.3. Data and Statistical Analysis

A Blood Flow Index (BFI) was calculated as previously described[6] and corrected for individual differences in blood volume (BFIw) by multiplying BFI with the body weight expressed in grams[5]. The arterio-venous O_2 content ($avDO_2$) difference was calculated from the arterial (SaO_2) and venous (sagittal sinus) oxygen (SvO_2) saturations and the average hemoglobin concentration ([Hb]blood) obtained form the same blood samples according to the following formula: $avDO_2 = 1.34 \times [Hb]blood \times (SaO_2-SvO_2)$

The index of $CmrO_2$ ($ICmrO_2$) was calculated from the $avDO_2$ and BFIw according to the following formula based on the Fick equation: $ICmrO_2 = avDO_2 \times BFIw$

All results are expressed as median and 95% confidence interval (CI). A two-sided Wilcoxon-Mann-Whitney rank-sum test was used for between group analysis. Probability values of less than 0.05 were regarded as statistically significant.

3. RESULTS

3.1. Physiological data

The physiological parameters (Table 1) were obtained from either continuous monitoring or blood sample analysis. No difference was found in weight, T-rectal, T-ear, HR, MABP, BrR, EtCO$_2$, arterial CO$_2$ pressure (PaCO$_2$), arterial oxygen pressure (PaO$_2$) and arterial oxygen saturation (SaO$_2$), and these data were within physiological range.

3.2. Path Lengths

NIRS assessed path lengths were significantly decreased at 840 nm and 960 nm in the ICP-probe group compared to controls, while path length at wavelength 740 nm may not have been altered significantly due to one outlier (Figure 1).

Table 1: Physiological data obtained from control animals and animals with an ICP probe.

	Control (n = 6)	ICP probe (n = 6)	p-value
Weight (g)	416 [407, 430]	425 [404, 429]	0.900
T-rectal (°C)	37.7 [37.6, 37.9]	37.7 [37.5, 38.2]	0.937
T-ear (°C)	34.6 [34.4, 35.1]	35.2 [33.4, 36.2]	0.485
Heart Rate (BPM)	386 [371, 415]	387 [326, 416]	1.000
MABP (mmHg)	85.5 [65.2, 95.4]	79.3 [67.6, 98.5]	0.699
BrR (BPM)	74.0 [68.9, 97.3]	83.4 [67.0, 94.4]	0.699
EtCO$_2$ (mmHg)	54.1 [45.1, 57.1]	49.1 [43.0, 51.3]	0.240
PaCO$_2$ (mmHg)	46.2 [38.7, 47.5]	42.6 [36.7, 47.0]	0.310
PaO$_2$ (mmHg)	133 [114, 144]	128 [117, 158]	0.937
SaO$_2$ (%)	96.7 [94.7, 97.5]	96.6 [95.5, 98.7]	0.909
ICP (mmHg)	---	8.18 [4.08, 11.71]	---

Values are expressed as median [95% CI]. ICP: intracranial pressure probe; T-rectal: rectal temperature; T-ear: ear temperature; BPM: beats per minute; MABP: mean arterial blood pressure; BrR: breathing rate; EtCO$_2$: end tidal CO$_2$; PaCO$_2$: arterial CO$_2$ pressure; PaO$_2$: arterial O$_2$ pressure; SaO$_2$: arterial oxygen saturation; ICP: intracranial pressure. P-values < 0.05 were considered statistically significant.

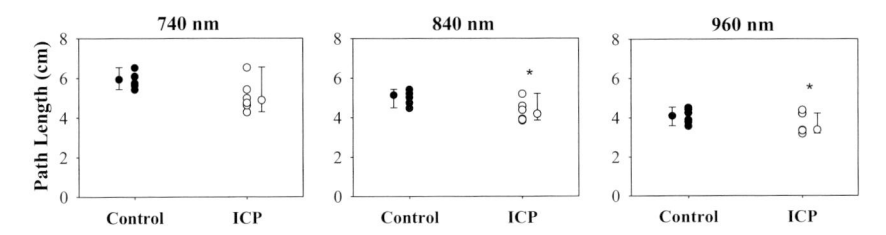

Figure 1. Path lengths obtained from controls (black dots) and animals with an ICP probe (open dots). Medians with 95% confidence intervals are given. * indicate significantly different from control group (p < 0.05).

3.3 Cerebral Data

All cerebral data (Table 2) are derived from NIRS measurements or blood sample analyses. [HbO₂] and total haemoglobin concentration ([HbT]) did not differ between the two groups, while [Hb] and SvO₂ significantly increased and decreased after ICP-probe insertion, respectively.

BFIw significantly decreased and avDO₂ significantly increased after ICP-probe insertion, but no difference in cerebral rate of oxygen metabolism index (ICMRO₂) was observed.

Table 2: Brain oxygenation data obtained from control animals and animals with an ICP probe.

	Control (n = 6)	ICP probe (n = 6)	p-value
[HbO₂] (μM)	76.0 [65.2, 88.3]	72.6 [47.8, 146.1]	0.818
[Hb] (μM)	18.0 [16.5, 23.9]	26.9 [17.7, 38.9]	**0.041**
[HbT] (μM)	93.3 [81.9, 106.2]	96.0 [65.4, 185.0]	0.937
Cu_A (μM)	0.520 [-0.432, 0.669]	1.011 [-0.192, 1.26]	**0.041**
svO₂ (%)	81.8 [70.0, 84.1]	73.8 [68.5, 80.2]	**0.041**
BFIw	16.9 [13.4, 19.0]	10.5 [7.6, 16.6]	**0.015**
avDO₂ (ml/dl)	2.91 [2.61, 4.92]	4.39 [3.22, 5.14]	**0.026**
ICMRO₂	47.4 [41.4, 86.4]	44.9 [36.8, 68.9]	0.485

Values are expressed as median [95% CI]. ICP probe: intracranial pressure probe; [HbO₂]: oxyhaemoglobin concentration; [Hb]: deoxyhaemoglobin concentration; [HbT]: total haemoglobin concentration; Cu_A: changes in oxidised cytochrome oxidase versus baseline; SvO₂: sinus sagittalis venous oxygen saturation; BFIw: weight corrected blood flow index; avDO₂: arterio-venous O₂ content difference; ICMRO₂: metabolic rate of oxygen index. P-values < 0.05 were considered statistically significant.

4. DISCUSSION

In this study, we investigated whether a miniature cerebellar probe influences NIRS measurements and cerebral hemodynamics in general. The Codman® ICP probe did not have an effect on the systemic physiologic parameters, which were within normal range and comparable to controls. Furthermore the ICP values of the ICP probe group were in accordance with previous studies using the same ICP probe in rats[7].

Insertion of a Codman® ICP probe had 2 major effects: (1) a marked decrease in optical path length and (2) a decreased CBF. Possible explanations are spreading depression (SD), vasospasm or an uncoupling of cerebral blood flow and cerebral metabolism. SD, a wave of depolarization associated with a lowered neuronal bioelectrical activity[8], is known to reduce CBF[8] and to occur after cerebral insertion of larger probes[4]. In addition, SD can induce oedema[8], by causing extreme shifts in ion balance, leading to cell swelling[8]. Because the only assumption of NIRS is a water content of 80%, cerebral oedema might influence path length, especially the path length at 960 nm. However, SD is known to increase scattering and therefore increase, instead of decrease, path lengths[9]. Also, SD causes an increase in glucose and oxygen consumption[8]. In our experiment, however, $ICMRO_2$ remained unaltered. In accordance with the fact that convoluted cortex is less likely to produce SD compared to smooth cortex (primates vs. rats), insertion of a probe in cerebellar position is less likely to induce SD in the cerebrum because of discontinuity[9]. Although SD might be present in the cerebellum at the time of insertion of the probe, it seems unlikely to be determinant for observed changes in cerebral parameters. Verhaegen and colleagues[4] found a similar decrease in CBF after insertion of a probe as small as 300 μm, unassociated with SD. Since our Codman® probe is 1.2 mm thick, a decrease in BFIw in our experiment was inevitable.

An alternative explanation is vasospasm induced by subarachnoid hemorrhage. However, no bleeding was found at the site of probe insertion after the experiment, but microscopic volumes might be present. It is known however that the amount of blood in the subarachnoid space is well correlated with vasospasm development[10]. In addition, vasospasm only occurs after a few days[10], excluding this hypothesis.

Because the BFIw decrease is not due to an altered $ICMRO_2$, we believe an uncoupling occurred, a hypothesis that is affirmed by a decreased SvO_2, an increased $avDO_2$ and [Hb]. Hayashi and colleagues[11] found that neuronal NOS inhibition can be responsible for this uncoupling. Interestingly, nNOS positive neurons are not homogenously distributed in the brain and are especially present in cerebellum[11]. However, the limitations of this study do not allow a final conclusion as to what the underlying mechanism is that caused these changes.

In conclusion, despite the altered path lengths, potentially influencing calculation of NIRS variables, [HbO2] and [Hb] are in complete accordance with the data of the blood sample analysis. This leads to the conclusion that NIRS is a robust system, allowing assessment of real physiologic alterations despite these changes. Additionally, our data suggest that a cerebellar miniature Codman® ICP probe induces an uncoupling of cerebral metabolism and cerebral blood flow.

5. REFERENCES

1. Czosnyka,M. & Pickard,J.D. Monitoring and interpretation of intracranial pressure. J. Neurol. Neurosurg. Psychiatry 75, 813-821 (2004).
2. Morgalla,M.H., Mettenleiter,H., Bitzer,M., Fretschner,R. & Grote,E.H. ICP measurement control: laboratory test of 7 types of intracranial pressure transducers. J. Med. Eng Technol. 23, 144-151 (1999).
3. Rooker,S. et al. Comparison of intracranial pressure measured in the cerebral cortex and the cerebellum of the rat. J. Neurosci. Methods 119, 83-88 (2002).
4. Verhaegen,M.J., Todd,M.M., Warner,D.S., James,B. & Weeks,J.B. The role of electrode size on the incidence of spreading depression and on cortical cerebral blood flow as measured by H2 clearance. J. Cereb. Blood Flow Metab 12, 230-237 (1992).
5. De Visscher,G. et al. Cerebral blood flow assessment with indocyanine green bolus transit detection by near-infrared spectroscopy in the rat. Comp Biochem. Physiol A Mol. Integr. Physiol 132, 87-95 (2002).
6. Kuebler,W.M. et al. Noninvasive measurement of regional cerebral blood flow by near-infrared spectroscopy and indocyanine green. J. Cereb. Blood Flow Metab 18, 445-456 (1998).
7. Rooker,S., Jorens,P.G., Van Reempts,J., Borgers,M. & Verlooy,J. Continuous measurement of intracranial pressure in awake rats after experimental closed head injury. J. Neurosci. Methods 131, 75-81 (2003).
8. Gorji,A. Spreading depression: a review of the clinical relevance. Brain Res. Brain Res. Rev. 38, 33-60 (2001).
9. Somjen,G.G. Mechanisms of spreading depression and hypoxic spreading depression-like depolarization. Physiol Rev. 81, 1065-1096 (2001).
10. Pluta,R.M. Delayed cerebral vasospasm and nitric oxide: review, new hypothesis, and proposed treatment. Pharmacol. Ther. 105, 23-56 (2005).
11. Hayashi,T. et al. Neuronal nitric oxide has a role as a perfusion regulator and a synaptic modulator in cerebellum but not in neocortex during somatosensory stimulation--an animal PET study. Neurosci. Res. 44, 155-165 (2002).

EFFECT OF SEVERE HYPOXIA ON PREFRONTAL CORTEX AND MUSCLE OXYGENATION RESPONSES AT REST AND DURING EXHAUSTIVE EXERCISE

Thomas Rupp[*] and Stéphane Perrey[*]

Abstract: Near infrared spectroscopy (NIRS) may provide valuable insight into the determinants of exercise performance. We examined the effects of severe hypoxia on cerebral (prefrontal lobe) and muscle (gastrocnemius) oxygenation at rest and during a fatiguing task. After a 15-min rest, 15 healthy subjects (age 25.3 ± 0.9 yr) performed a sustained contraction of the ankle extensors at 40% of maximal voluntary force until exhaustion. The contraction was performed at two different fractions of inspired O_2 fraction ($F_{IO2} = 0.21/0.11$) in randomized and single-blind fashion. Cerebral and muscle oxy-(HbO_2) deoxy-(HHb) total-hemoglobin (HbTot) and tissue oxygenation index (TOI) were monitored continuously by NIRS. Arterial O_2 saturation (SpO_2) was estimated by pulse oximetry throughout the protocol. Muscle TOI did not differ between normoxia and hypoxia after the 15-min rest, whereas SpO_2 and cerebral TOI significantly dropped (-6.5 $\pm$ 0.9% and -3.9 $\pm$ 1.0%, respectively, P<0.05) in hypoxia. The muscle NIRS changes during exercise were similar in normoxia and hypoxia, whereas the increased cerebral HbTot and HbO_2 near exhaustion were markedly reduced in hypoxia. In conclusion, although F_{IO2} had no significant effect on endurance time, NIRS patterns near exhaustion in hypoxia differed from normoxia.

1. INTRODUCTION

Since its first application 30 years ago,[1] near infrared spectroscopy (NIRS) has been shown to be an effective tool for monitoring non-invasively central and peripheral changes in oxygenation. It thus may provide valuable insight into the determinants of performance during exercise.[2-7]

[*] Faculty of Sport Sciences, Motor Efficiency & Deficiency Laboratory, Avenue du Pic Saint Loup, 34090 Montpellier, France. thomas.rupp@univ-montp1.fr

P. Liss et al. (eds.), *Oxygen Transport to Tissue XXX*, DOI 10.1007/978-0-387-85998-9_49,
© Springer Science+Business Media, LLC 2009

To date, only a small number of papers have simultaneously measured cerebral (Cox) and muscle (Mox) oxygenation during fatiguing exercise under normoxia and/or hypoxia.[2, 5-9] Previous studies[9, 10] demonstrated that cerebral, but not muscle, tissue shows greater deoxygenation during acute hypoxia at rest and some suggested that Cox was more likely than Mox to limit the maximal whole-body exercise capacity.[5]

The purpose of this study was to explore the effects of severe hypoxia on cerebral (prefrontal lobe) and muscle (gastrocnemius) tissue oxygenation at rest and during a local sustained fatiguing task. We hypothesized that (i) cerebral and muscle deoxygenation would be prevalent during fatiguing exercise under hypoxia and (ii) Cox would drop before the failure in motor performance.

2. METHODS

2.1. Subjects

Fifteen healthy right-footed males (mean ± SE: age 25.3 ± 0.9 yr, height 178.9 ± 1.6 cm, body mass 70.7 ± 1.9 kg) volunteered and gave informed written consent to participate in this study. Subjects were physically active (training 4.9 ± 0.9 h·wk^{-1}); had no history of cardiovascular, respiratory, musculoskeletal or neurological disorders; and were free of medication. Subjects were requested to refrain from training on the two days prior to testing and to avoid caffeine and alcohol ingestion for the 12 h preceding the tests. The study procedures complied with the Declaration of Helsinki for human experimentation and were approved by the local ethics committee.

2.2. Experimental Protocol

Each subject visited the laboratory on three occasions (familiarization session and testing sessions) separated by 24 to 48 h. In the preliminary session, a full explanation of the experimental protocol and recommendations were given. Subjects were familiarized with the experimental procedures and individually appropriate adjustments to the ankle-extension device were made. In the testing sessions, they performed a sustained isometric contraction of the ankle extensors at 40% of their maximal voluntary isometric contraction (MVIC) up to exhaustion. This procedure was completed while subjects were exposed in a randomized and single-blind fashion to either room air (normoxia, F_{IO2} = 0.21) or an hypoxic gas mixture (hypoxia, F_{IO2} = 0.11) corresponding to an altitude of ~4500 m. Normobaric hypoxic condition (F_{IO2} = 0.11) was simulated by diluting ambient air with nitrogen via a mixing chamber, with the dilution constantly controlled by a PO_2 probe (Alti-Trainer200, Sport and Medical Technology, Geneva, Switzerland).

The subjects were seated on a padded bench connected to a homemade ankle-ergometer. The foot of the strongest leg (right leg in all instances) was fixed by several semi-rigid Velcro bands on a pedal fitted out with a strain gauge (DEC 60, Captels, St Mathieu de Treviers, France). The knee angle was set at 80° compared with complete extension (0°), and the ankle formed a 90° angle with the pedal. The ergometer was connected to a computer for data acquisition at 1000 Hz and subsequent analysis (MP30, Biopac Systems, Santa Barbara, CA, USA). A monitor provided subjects with visual feedback on the ankle force. Baseline values were first collected for 2 minutes while subjects rested in the exercising position on the ergometer. Then, to reach a steady state, subjects breathed the appropriate gas mixture at rest for 15 min before continuing the experiments. After a standardized warm-up of the calf muscle, they performed MVICs. The force level target (40% of MVIC) was then determined and represented by a horizontal line on a monitor during the exercise. Subjects were instructed to quickly match their force level to the target line as precisely and as long as possible. Strong verbal encouragement was given to each subject during MVICs and exercise. Exhaustion was defined as a drop in produced force under 30% of MVIC for more than 3 consecutive seconds. Exercise was immediately followed by a post-exercise MVIC to assess force loss. Arterial O_2 saturation (SpO_2) was estimated by pulse oximetry (Pulsox 300i, Konica Minolta Sensing Inc., Osaka, Japan) throughout the protocol.

2.3. Prefrontal Cortex and Muscle Oxygenation

NIRS techniques have been described elsewhere.[4, 11] Cox and Mox were measured continuously and simultaneously with two channels of a four-wavelength (775, 810, 850, 905 nm) high temporal resolution NIRS device (NIRO-300, Hamamatsu Photonics, Hamamatsu City, Japan) throughout the experiment. The probes consisted of one emitter and one detector housed in a black holder. The holders were stuck on the skin with double-sided adhesive tape to ensure no change in their relative positions and to minimize the intrusion of extraneous light and the loss of transmitted NIR light from the field of interrogation. A differential optical pathlength factor (DPF) that takes into account light scattering in tissue is often inserted in the modified Lambert-Beer law used to describe optical attenuation. We decided not to use DPF values as they may vary from one wavelength to another, across subjects, and even over time for a given subject and tissue.[12]

To assess Cox, the detection probes were positioned over the left prefrontal cortical area between Fp1 and F3, according to the modified international EEG 10-20 system. The inter-optode distance was 5 cm and the probe holder was covered and maintained with a homemade black Velcro headband. The prefrontal cortex is known to project to pre-motor areas and to be responsible for movement planning and pacing strategies, as well as decision-making.[13-14] It was also recently shown that decreased frontal cortical oxygenation was associated with reduced muscle force-generating capacity.[15]

To determine local Mox profiles, the probes were attached to the skin on the belly of the right gastrocnemius medialis and in parallel with the long axis of the muscle. The distance between the transmitting and receiving optodes was fixed at 4 cm and the probe holder was covered with a black sweatband maintained with an elastic muff net.

Both cerebral and muscle NIRS data were collected with a sampling frequency of 2 Hz. Relative concentration changes ($\Delta\mu$mol·cm) were measured from resting baseline and pre-exercise level of oxy-($\Delta[HbO_2]$), deoxy-($\Delta[HHb]$) and total-hemoglobin ($\Delta[HbTot]$ = $[HbO_2]$ + $[HHb]$). [HbTot] reflects the changes in tissue blood volume within the illuminated area, [HHb] is known to be a reliable estimator of changes in tissue de-oxygenation status, while $[HbO_2]$ seems to be the most sensitive indicator of regional cerebral blood flow (CBF) modifications.[16] Finally, we used a multidistance spatially resolved tissue oximeter (NIRO-300) that is able to quantify tissue oxy-hemoglobin saturation directly as a tissue oxygenation index (TOI), which is a surrogate measure for cerebrovenous saturation when applied to the head.[17]

2.4. Statistical Analysis

NIRS values were averaged at different time epochs of the protocol: the last 30 s of a 2-min rest period in normoxia, the last 30 s of a 15-min rest period while breathing the appropriate gas mixture, and the last 30 s before the exercise onset. Total exercise time was considered as 100% and data were averaged every 20% of this total time to obtain mean values at 20, 40, 60, 80 and 100% of the performance duration. F_{IO2} (normoxia-hypoxia) and time effects (pre-post 15-min rest period or 0-20-40-60-80-100% of total exercise time) on each of the dependent variables were analyzed using a two-way ANOVA for repeated measures and the Fisher LSD post-hoc procedure when appropriate. Endurance times in normoxia vs. hypoxia were compared with Student's paired t-tests. All parameters are expressed as mean ± SE, and the P-value for significance was established at 0.05.

3. RESULTS

3.1. Rest Data

After the 15-min rest period while breathing the hypoxic gas mixture, SpO_2 dropped significantly (from 97.2 ± 0.3 to 90.7 ± 0.8%), whereas no modification was noticed in normoxic condition (from 97.3 ± 0.3 to 97.5 ± 0.2%).

Cerebral [HbTot] was not affected by rest period in either normoxia or hypoxia. Cerebral [HHb] was significantly increased and [HbO₂] was significantly decreased after the 15-min rest in hypoxia but not in normoxia. Cerebral TOI did not statistically differ between normoxia and hypoxia before the 15-min rest period (74.0 ± 1.7 and 73.9 ± 1.2%, respectively) but diverged after it (74.5 ± 1.7 and 69.5 ± 1.0%, respectively), due to a marked cerebral TOI decrease in hypoxia.

Muscle [HbTot] and [HbO₂] were not affected by the rest period in normoxia or hypoxia. Muscle [HHb] was significantly increased after the 15-min rest in hypoxia but not in normoxia. Although muscle TOI did not statistically differ between normoxia and hypoxia before (64.6 ± 1.2 and 65.6 ± 1.4%, respectively) or after (64.3 ± 1.0 and 63.1 ± 1.3%, respectively) the 15-min rest, muscle TOI was slightly decreased in hypoxia but not in normoxia.

3.2. Exercise Data

F_{IO2} had no significant effect on endurance time (458.4 ± 11.0 and 449.4 ± 12.6 s in normoxia and hypoxia, respectively). MVIC dropped similarly after exercise whatever the F_{IO2} condition (-18.0% in normoxia vs. -16.2% in hypoxia). SpO_2 was significantly lower in hypoxia compared with normoxia both pre- (89.7 ± 1.3 and 97.5 ± 0.3%, respectively) and post-exercise (92.6 ± 1.0 and 97.7 ± 0.6%, respectively) but was not significantly affected by exercise.

As presented in Fig. 1, F_{IO2} had a significant effect on cerebral NIRS changes during exercise. Cerebral Δ[HbTot] and Δ[HbO₂] increased from pre-exercise level to exhaustion in normoxia, whereas they stabilized in the last part of exercise (from 80% of total exercise time) in hypoxia. Concomitantly, cerebral Δ[HHb] decreased from pre-exercise level to the last part of exercise in both normoxia and hypoxia and then stabilized (from 60 and 80% of total exercise time, respectively). Last, cerebral TOI was significantly higher in normoxia compared with hypoxia throughout exercise. Cerebral TOI increased from onset to the last part of exercise in hypoxia and then tended to drop, whereas this parameter was not significantly affected by exercise in normoxia.

Muscle NIRS changes during exercise were not different in normoxia compared with hypoxia (Fig. 1). Muscle $\Delta[HbTot]$ and $\Delta[HHb]$ increased throughout exercise while $\Delta[HbO_2]$ dropped significantly. After an increase at exercise onset compared with pre-exercise level, muscle TOI dropped significantly until exhaustion.

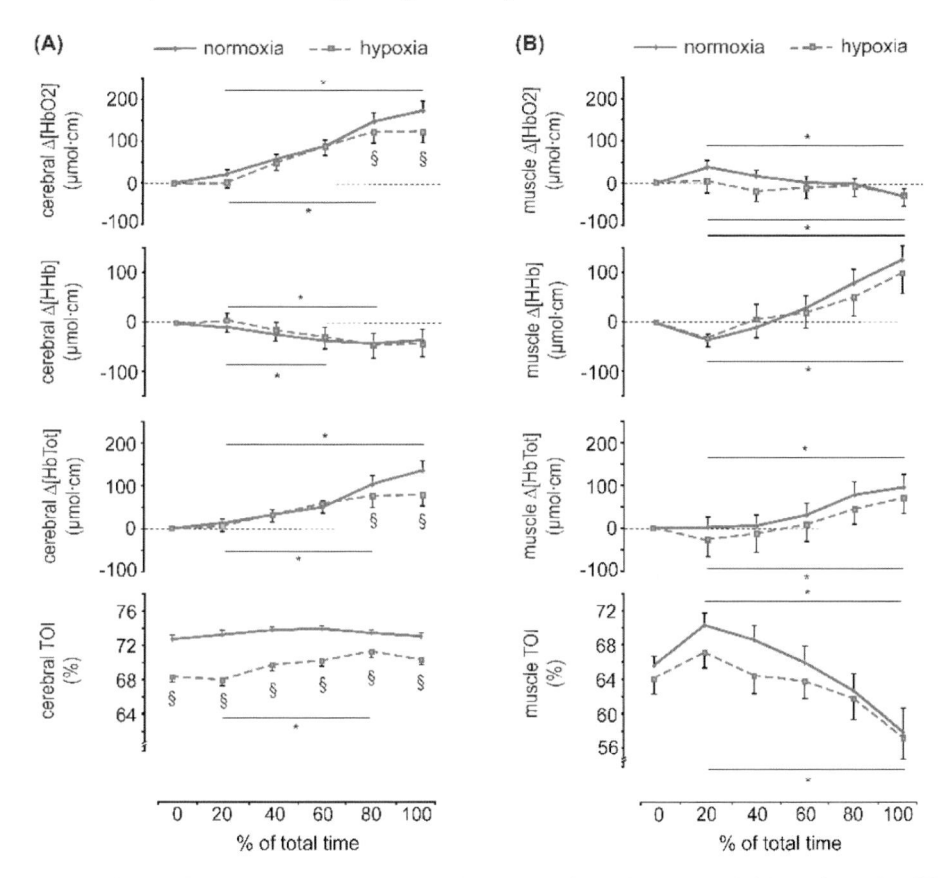

Figure 1. Changes from pre-exercise values in cerebral (A) and muscle (B) oxy-($[HbO_2]$), deoxy-($[HHb]$), total-hemoglobin ($[HbTot]$), and TOI during exercise in normoxia and hypoxia. Values are means $\pm$ SE. § different compared with normoxia ($P \leq 0.05$); * different from 20% of total exercise time value ($P \leq 0.05$).

4. DISCUSSION

Local Cox and hemodynamics measured by NIRS reflect cortical activation for various motor tasks[3, 18, 19] while Mox assessed by NIRS is known to reflect the metabolic changes that occur at the muscle site.[4] Our results confirmed recent data[9, 10] and showed that hypoxia during rest and exercise markedly decreased Cox, whereas Mox was not different between normoxia and hypoxia. Although F_{IO2} had no significant effect on endurance time, NIRS patterns during exercise in hypoxia differed from those in normoxia, as a significant lower cerebral HbTot and HbO_2 was observed near exhaustion, suggesting possible changes in CBF[16]. Finally, Cox patterns might reflect reduced cerebral cortex activity at volitional fatigue in response to dynamic global exercise,[3, 6] but this does not seem to be the case at the end of a local sustained fatiguing exercise.

5. REFERENCES

1. F. F. Jöbsis, Noninvasive, infrared monitoring of cerebral and myocardial oxygen sufficiency and circulatory parameters, *Science* **198**(4323), 1264-1267 (1977).
2. R. A. De Blasi, S. Fantini, M. A. Franceschini, M. Ferrari and E. Gratton, Cerebral and muscle oxygen saturation measurement by frequency-domain near-infra-red spectrometer, *Med. Biol. Eng. Comput.* **33**(2), 228-230 (1995).
3. K. Shibuya and M. Tachi, Oxygenation in the motor cortex during exhaustive pinching exercise, *Respir. Physiol. Neurobiol.* **153**(3), 261-266 (2006).
4. M. Ferrari, L. Mottola and V. Quaresima, Principles, techniques, and limitations of near infrared spectroscopy, *Can. J. Appl. Physiol.* **29**(4), 463-487 (2004).
5. H. B. Nielsen, R. Boushel, P. Madsen and N. H. Secher, Cerebral desaturation during exercise reversed by O2 supplementation, *Am. J. Physiol. Heart Circ. Physiol.* **277**(3), H1045-1052 (1999).
6. A. W. Subudhi, A. C. Dimmen and R. C. Roach, Effects of acute hypoxia on cerebral and muscle oxygenation during incremental exercise, *J. Appl. Physiol.* **103**(1), 177-183 (2007).
7. M. Amann, L. M. Romer, A. W. Subudhi, D. F. Pegelow and J. A. Dempsey, Severity of arterial hypoxemia affects the relative contributions of peripheral muscle fatigue to exercise performance, *J. Physiol.* **581**(Pt1), 389-403 (2007).
8. Y. Bhambhani, R. Malik and S. Mookerjee, Cerebral oxygenation declines at exercise intensities above the respiratory compensation threshold, *Respir. Physiol. Neurobiol.* **156**(2), 196-202 (2007).
9. P. N. Ainslie, A. Barach, C. Murrell, M. Hamlin, J. Hellemans and S. Ogoh, Alterations in cerebral autoregulation and cerebral blood flow velocity during acute hypoxia: rest and exercise, Am J Physiol Heart Circ Physiol **292**(2), H976-983 (2007).
10. J. E. Peltonen, J. M. Kowalchuk, D. H. Paterson, D. S. Delorey, G. R. Dumanoir, R. J. Petrella and J. K. Shoemaker, Cerebral and muscle tissue oxygenation in acute hypoxic ventilatory response test, *Respir. Physiol. Neurobiol.* **155**(1), 71-81 (2007).
11. C. E. Elwell, M. Cope, A. D. Edwards, J. S. Wyatt, D. T. Delpy and E. O. Reynolds, Quantification of adult cerebral hemodynamics by near-infrared spectroscopy, *J. Appl. Physiol.* **77**(6), 2753-2760 (1994).
12. A. Duncan, J. Meek, M. Clemence, C. Elwell, P. Fallon, L. Tyszczuk, M. Cope and D. Delpy, Measurement of cranial optical path length as a function of age using phase resolved near infrared spectroscopy, *Pediatr. Res.* **39**(5), 889-894 (1996).
13. E. K. Miller and J. D. Cohen, An integrative theory of prefrontal cortex function, *Annu. Rev. Neurosci.* **24**,167-202 (2001).
14. M. Suzuki, I. Miyai, T. Ono, I. Oda, I. Konishi, T. Kochiyama and K. Kubota, Prefrontal and premotor cortices are involved in adapting walking and running speed on the treadmill: an optical imaging study, *Neuroimage* **23**(3), 1020-1026 (2004).
15. P. Rasmussen, E. A. Dawson, L. Nybo, J. J. van Lieshout, N. H. Secher and A. Gjedde, Capillary-oxygenation-level-dependent near-infrared spectrometry in frontal lobe of humans, *J. Cereb. Blood Flow Metab.* **27**(5), 1082-1093 (2007).
16. Y. Hoshi, N. Kobayashi and M. Tamura, Interpretation of near-infrared spectroscopy signals: a study with a newly developed perfused rat brain model, *J. Appl. Physiol.* **90**(5), 1657-62 (2001).
17. G. Greisen, Is near-infrared spectroscopy living up to its promises?, *Semin. Fetal. Neonatal. Med.* **11**(6), 498-502 (2006).
18. H. Obrig, C. Hirth, J. G. Junge-Hulsing, C. Doge, T. Wolf, U. Dirnagl and A. Villringer, Cerebral oxygenation changes in response to motor stimulation, *J. Appl. Physiol.* **81**(3), 1174-1183 (1996).
19. L. Mottola, S. Crisostomi, M. Ferrari and V. Quaresima, Relationship between handgrip sustained submaximal exercise and prefrontal cortex oxygenation, *Adv. Exp. Med. Biol.* **578**, 305-309 (2006).

AUTHOR INDEX

SUBJECT INDEX

Printed in the United States of America